CHARLES DARWIN

Evolutionary Writings

———

Edited by
JAMES A. SECORD

OXFORD
UNIVERSITY PRESS

OXFORD

UNIVERSITY PRESS

Great Clarendon Street, Oxford OX2 6DP

Oxford University Press is a department of the University of Oxford.
It furthers the University's objective of excellence in research, scholarship,
and education by publishing worldwide in

Oxford New York

Auckland Cape Town Dar es Salaam Hong Kong Karachi
Kuala Lumpur Madrid Melbourne Mexico City Nairobi
New Delhi Shanghai Taipei Toronto

With offices in

Argentina Austria Brazil Chile Czech Republic France Greece
Guatemala Hungary Italy Japan Poland Portugal Singapore
South Korea Switzerland Thailand Turkey Ukraine Vietnam

Oxford is a registered trade mark of Oxford University Press
in the UK and in certain other countries

Published in the United States
by Oxford University Press Inc., New York

Editorial material and selection © James A. Secord 2008

British Library Cataloguing in Publication Data

Data available

Library of Congress Cataloging-in-Publication Data

Data available

Typeset by Cepha Imaging Private Ltd., Bangalore, India
Printed in Great Britain
on acid-free paper by
Clays Ltd., St Ives plc

ISBN 978-0-19-920863-0

CONTENTS

ACKNOWLEDGEMENTS

THIS anthology stems from my involvement with the Darwin Correspondence Project, which is editing all the letters to and from Charles Darwin. The Project receives essential and very generous support from the Andrew W. Mellon Foundation, British Ecological Society, Isaac Newton Trust, John Templeton Foundation, National Endowment for the Humanities, and National Science Foundation. I wish to thank William Huxley Darwin for permission to publish the *Recollections* and 1838 'Life', and the Syndics of Cambridge University Library, the Geological Society of London, Special Collections and Archives of Knox College Library, Galesburg, Illinois, Trustees of the National Library of Scotland, and Trustees of the Natural History Museum, London, for permission to publish manuscripts in their possession. Permission to publish material from the *Correspondence of Charles Darwin*, ed. F. Burkhardt et al. (1985–) has been granted by the Syndics of Cambridge University Press. Heather Brink-Roby provided invaluable assistance throughout, and the biographical index is largely her work. Many individuals mentioned in the early pages of the *Recollections* were identified by Donald F. Harris. Paul White offered excellent suggestions and references for the Introduction, which was also much improved after discussion by the Past versus Present project of the Cambridge Victorian Studies Group, funded by the Leverhulme Trust. John van Wyhe has been unfailingly generous in sharing information, particularly through his remarkable Darwin website. I especially wish to thank Marwa Elshakry, who with the assistance of Ahmed Ragab has provided fresh translations from Arabic; Adriana Novoa and Alex Levine, for access to their forthcoming collection of Argentine responses; and Shelley Innes, for help with Russian and German translations. The capable support of Alison Pearn made it possible to finish this in time. I am also grateful to Janet Browne, Rosy Clarkson, Diana Donald, Samantha Evans, Nicola Gauld, Melanie Keene, Sam Kuper, Randal Keynes, David Kohn, David Livingstone, Peter Mandler, Clare Pettitt, Kees Rookmaaker, Liz Smith, and many other friends. Judith Luna has been an exemplary editor: patient, accommodating, and enthusiastic. I am most indebted to Anne Secord, whose encouragement and constructive criticism have been vital at every stage.

INTRODUCTION

IF the intellectual landscape of the twentieth century was dominated by Marx, Freud and Darwin, it is clear that the reputation of only one of that heroic triumvirate has survived intact into the twenty-first. With so-called 'Darwinian' views on economic and social competition in the ascendant, the international pre-eminence of Darwin is more marked than ever, his books more widely read and discussed than at any time since they first appeared. Darwin's views on human origins, the beginnings of life, and the nature of the fossil record play key roles in controversies about religion and science, particularly in relation to the teaching of evolution in schools. His account of the human mind has proved central in the development of psychology. His subtle analyses of the interconnectedness of life and environment are reference points in debates about species extinction and climate change. He is hailed as a visionary in fields as diverse as linguistics and global geology. His theory of evolution by natural selection is the coping stone of the modern life sciences.

Darwin's fame grew out of the reception of his books, and although he wrote thousands of letters and hundreds of scientific articles and occasional pieces, reading Darwin means reading books. Of these, three were instrumental in establishing his reputation during his lifetime: the revised edition of the *Journal of Researches* (1845), an account of his voyage around the world on HMS *Beagle*, which touched implicitly on evolutionary themes; *On the Origin of Species* (1859), which outlined his novel theory of evolution by natural selection; and *The Descent of Man* (1871), which applied his ideas to the study of humans. In terms of his personal reputation, the central text is *Recollections of the Development of My Mind and Character*, published in 1887 as the opening chapter in a memoir edited by one of his sons.

Darwinism is a global phenomenon. *Origin* has been translated into over thirty languages, more than any scientific work other than Euclid, while the *Recollections* and *Descent of Man* are each available in twenty. The power of Darwin's writings derives from their ability to challenge, surprise, and inspire readers in the widest possible range of circumstances. It is because these books have been read in so many ways that it is vital to confront the texts in the originals and not just as pithy quotations or through piecemeal searches on the internet. Yet with the collected works occupying twenty-nine volumes, this is not an easy task. Existing selections often leave out Darwin's most controversial and innovative ideas, and have been edited with an eye towards current evolutionary biology. Reading a single work, although an obvious starting point, is only a partial solution. Even the 500-plus pages of *Origin* say almost nothing about humans and give a tactically skewed view of what its author is trying to do.

So much has been at stake in reading Darwin's deceptively simple prose that understanding his books has proved elusive. Darwin is often presented as believing in natural selection as the sole mechanism for evolution, but this was never the case, not even in the *Origin*'s first edition. His views on heredity are still typically seen as a blank waiting to be filled in by the discoveries of Gregor Mendel, the rise of laboratory-based genetics, and the discovery of the DNA structure; yet nothing could be further from the truth. Darwin was intensely interested in variation, reproduction, and inheritance, as is clear from the early chapters of *Origin*. To take another example, his views on the status of women and the extinction of races are often minimized or misunderstood. Darwin did not, as was long thought to be the case, turn to these issues only towards the end of his life; they were there from the first. And finally, it is only through reading a range of his works, and reactions to them, that we gain an idea of his complex and ambiguous attitudes towards religion. He was certainly not an atheist. Darwin may say in his *Recollections* that Christianity is a 'damnable doctrine' (p. 392),[1] but his ideals of moral virtue in *Descent* are carefully grounded in the golden rule preached by Christ in the Sermon on the Mount (p. 255).

Darwin's writings still hold the power to shock. His pages include scenes of surpassing beauty in nature, described in passages of glowing prose; but we are to understand these as outcomes of war, conquest, invasion, and extermination. A raw sense of the violence of nature is combined with an appreciation for its interconnectedness and fragility. The energy of life is possible only through the hovering presence of death. The coral reefs of the Indian Ocean grow on

---

[1] Page numbers in the text refer to this edition.

miles-high cemeteries of dead ancestors. Parasitic grubs eat the bodies of their hosts from the inside out. Patrician landowners dwell in luxury by means of a hidden economy of slave labour. The living world of animals and plants, for all its apparent order and design, is the outcome of a multitude of individual acts of casual violence. The face of nature is bloodied by a force like that of a hundred thousand wedges. These are not rhetorical set pieces or concluding flourishes; rather, they appear in the context of a cumulative weight of examples expressed in plain and simple prose. The occasional awkwardness in construction and the tendency of the later works to mirror Darwin's self-conception as 'a kind of machine for grinding general laws out of large collections of facts' gives a sense of an author concerned about substance (p. 422). Readers are invited to observe closely, even obsessively: to share a fascination for the instinctual habits of ants, the finer points of pigeon fancying, and the sexual antics of barnacles. Although the books are only intermittently autobiographical, the reader gains a strong sense of their author, whose self-deprecating enthusiasm is infectious. In detailing the courtship rituals of the Australian bower bird, the intelligence of earthworms, or the fertilization mechanisms of orchids, Darwin's appreciation of nature in all its aspects is evident.

A remarkable number of Darwin's books have remained continuously in print, and they occupy a unique status in the canon. Not only are they almost the only scientific books from the past three centuries which attract a non-specialist readership decades after publication, but the number of readers across the world is growing. How have Darwin's writings achieved this? On his death in 1882 Darwin was already heralded as a great man, and his significance was acknowledged even as his theories were widely challenged at the end of the nineteenth century. But it is only in the fifty years since the centenary of the publication of *Origin* in 1959 that Darwin has emerged as the epitome of the scientific celebrity. His bearded image is everywhere from television documentaries to postage stamps, banknotes, magazine covers, religious tracts, advertisements, caricatures, and cereal packages. Creative genius, racist, opponent of slavery, murderer of God, patient observer, engineer of western imperialism, apologist for capitalism, gentle prophet of evolution: there is potential validity, to a larger or lesser degree, in each of these readings. The multiplicity of images point to Darwin's books as fault-lines for controversy, opening new questions and fresh lines of enquiry. Their power derives from an ability to express simultaneously the expansive confidence of the century in which he lived, and the ambiguity, uncertainty, and fragility of the place of humans in nature.

An Imperial Voyage

When the writer Elizabeth Gaskell needed a model for a young traveller-naturalist in her last great novel *Wives and Daughters* (1864–6) there was one figure familiar to all: Darwin, whose narrative of his travels around the world her readers could be expected to know. His *Journal of Researches* closely followed the conventions of the expedition narratives produced by naturalists writing in the wake of the explorer James Cook and the Prussian naturalist Alexander von Humboldt, combining lively adventure with scientific observations, accounts of encounters with indigenous peoples, and evocations of exotic scenery. Publication of the *Journal* secured Darwin's reputation as a noted man of science and literary lion in the London salons. It was thus in the context of the burgeoning programme of European imperial exploration that Darwin first came to public attention.

Darwin's status on the *Beagle* had been highly unusual, for he was not the ship's naturalist, but a gentleman companion to the captain, Robert FitzRoy. Darwin had special privileges, including first choice of natural history specimens and a place at the captain's table. Not surprisingly, the surgeon officially appointed as naturalist left in disgust almost as soon as the *Beagle* docked in South America. Darwin claimed to be a novice in natural history at the start of the voyage, but this is misleading. Born in February 1809, he had come from gentry stock at the heart of the English scientific enlightenment. His mother, who died when Charles was only 8, was a daughter of Josiah Wedgwood, the celebrated founder of the pottery manufacturing dynasty. His father, Robert Darwin, was a successful physician and son of the poet Erasmus Darwin, a renowned author of evolutionary speculations and erotic botanical verse. Young Charles, like his father and grandfather, studied medicine. Disinclined to practise after training in Edinburgh (he couldn't stand the sight of blood), he proceeded to Cambridge to prepare for a career as an Anglican priest. Throughout his education, however, his real passion was for natural history, and he became acquainted with leading men of science.

Darwin may have dismissed the Edinburgh naturalist Robert Jameson as 'that old brown, dry stick', but attended dozens of his lectures and spent hours in his fabulous museum.[2] By the summer after graduating from Cambridge in 1831, Darwin was at 22 probably the best-educated naturalist of his age in Britain, particularly skilled in invertebrate zoology and with some knowledge of geological surveying and natural history collecting. Inspired by Humboldt's magnificent travel writings, he was plotting an expedition to the mid-Atlantic island of Tenerife.

But larger prospects loomed. The sciences were tied to global trade, with European governments supporting large-scale expeditions and surveys. The *Beagle*'s aim was to provide better charts of the South American coasts, including economically significant harbours and the treacherous straits around Tierra del Fuego. The continent was just opening up to English trade after centuries of domination by Spain and Portugal. FitzRoy, who came from one of the oldest aristocratic families in England, was a keen supporter of this enterprise, and had asked the Cambridge scientific men to suggest someone to travel with the *Beagle*, who could contribute to its scientific aims, pay for his passage, and provide much-needed genteel company. The trip, as Darwin acknowledged, was the most important event of his life. He learned to hunt wild rheas (a kind of South American ostrich) on horseback, witnessed the mass killing of aboriginal peoples, and saw slaves beaten by their masters. He collected the gigantic bones of extinct sloths and armadillos, walked through tropical jungles, climbed high peaks in the Andes, and observed the effects of an earthquake on a great city. He kept meticulous notebooks and collected thousands of specimens, each carefully numbered and tagged in preparation for description by specialist naturalists back home. As his father noted on Darwin's return, the years of concentrated scientific observation had extended even the ridge of his eyebrows.[3]

Of all the extraordinary things Darwin gathered, the most striking were visual impressions of human diversity. Wherever he went, Darwin categorized racial types: the tall Tehuelches of Patagonia; the

---

[2] C. Darwin to J. D. Hooker, 29 [May 1854], in *The Correspondence of Charles Darwin*, ed. F. Burkhardt et al., v (1989), 195. Hereafter referred to as *Correspondence*.

[3] L. A. Nash, 'Some Memories of Charles Darwin', *Overland Monthly and Out West Magazine*, 16 (Oct. 1890), 404–8, at 405.

meat-eating Gauchos of the Pampas; Spanish ladies with their great combs and dark hair; Australian aboriginals; and Malay peoples of the Indian Ocean. From his first landing on the mid-Atlantic island of St Jago, he had focused on the aesthetics of race, noting how 'black skins and snow-white linen' were set off by 'coloured turbans and large shawls' (p. 6). The *Beagle* itself displayed in microcosm a range of human variation among the crew, and most clearly in three natives of Tierra del Fuego (one Alakaluf man, and a Yamana man and young woman) who had been captured during the previous voyage. After several years in England, the adaptation of the three to their new circumstances was reflected in their European clothes and ability to speak some English. In the early weeks of the voyage, o'run-del'lico (named Jemmy Button by the crew) consoled Darwin on his seasickness, who considered him sympathetic, intelligent, almost civilized. It was encountering 'Fuegians' in the very different circumstances of their homeland in the southernmost part of South America that gave Darwin a shock that would last to the end of his life. Their culturally rich, although hard life of hunting and fishing appeared to him as little better than wretched misery; their complex languages seemed an undifferentiated primitive babble; their customs and body painting were signs of demonic bestiality.[4] An ardent believer in the unity of the human race and a passionate opponent of racial oppression, Darwin was barely able to define these 'savages' as members of his own species.

In terms of self-perception, Darwin's greatest discovery on the voyage was his vocation for science. Among the new disciplines which came into being in the early nineteenth century, the most exciting was geology, which had been freshly confected from mineral surveying, biblical chronology, and the study of fossils. Upon his first landing, Darwin began to write of himself as a geologist, and his notebooks reveal that he was already thinking of global subsidence and uplift on a global scale. He was especially indebted to Charles Lyell's *Principles of Geology* (1830–3), the first volume of which was a gift from FitzRoy just before the *Beagle* set sail. Darwin's most successful and audacious speculation answered one of the

---

[4] N. Hazelwood, *Savage: The Life and Times of Jemmy Button* (New York, 2001), offers an accessible overview; important parts of the main anthropological work are translated in J. Wilbert (ed.), *Folk Literature of the Yamana Indians: Martin Gusinde's Collection of Yamana Narratives* (Berkeley and Los Angeles, 1977).

government's chief desiderata for the voyage—a convincing explanation of coral reefs. Darwin suggested, before he ever saw a reef, that they grew up over thousands of years, millimetre by millimetre, as the ocean floor submerged. Although this theory overturned Lyell's belief that coral reefs grew upwards from the rims of underwater volcanoes, in another way it illustrated just how deeply Darwin had come to see the world with Lyellian eyes, for it relied on Lyell's belief in the cumulative effects of tiny processes over long periods of time. It was also the outcome of his interest, present since his student days in Edinburgh, in invertebrate zoology and the microscopic study of living matter.

For Darwin, the voyage offered the opportunity to secure a place among the leading men of science in London, and his main work after returning was to produce three volumes on the geology of the places visited during the voyage, as well as writing several important scientific papers. With the help of a generous Treasury grant, he also edited a sumptuous set of folios illustrated with colour plates on the voyage's zoological findings. These were reference works targeted at specialists, which secured his credibility as a man of science. The unexpected publishing triumph was Darwin's *Journal of Researches*. Based on his informal voyage diary, the book combined engaging incidents of travel with scientific observation and speculation. This genre, epitomized by Humboldt's celebrated *Personal Narrative*, had attracted Darwin to scientific voyaging in the first place. The title under which it commonly appears today, *The Voyage of the Beagle*, is a twentieth-century invention and rightly belongs to FitzRoy's book about the voyage. Initially Darwin's *Journal* had a limited circulation, as the third volume of an expensive set of publications dominated by FitzRoy's *Narrative*. Early in the voyage the captain had praised the quality and interest of Darwin's diary, and he generously offered to include it in the final reports. Only a wealthy readership could be expected to consult this kind of work, although the publisher Henry Colburn also issued it as a stand-alone volume, and it was widely reviewed. Its long-term future was secured when the adventurous London publisher John Murray decided to include it in his Colonial and Home Library series in 1845. Darwin had revised heavily, streamlining his text and adding expert opinions from the naturalists who had contributed to the *Beagle* zoology. The chapter on the indigenous peoples of Tierra del Fuego was expanded, and his coral reef theory was given in more detail.

The most significant change, however, was in format and readership, for Murray advertised his series as 'cheap literature for all classes', and in three inexpensive paper-covered parts at half the original price, it could now be afforded by professional families, merchants, and working-class libraries. It became, for example, one of the books that the fictional tailor in Charles Kingsley's *Alton Locke* (1850) could reasonably have been expected to have read. With its small type and narrow margins, this version of the *Journal of Researches* made Darwin's reputation as a writer and introduced his ideas to a wider public in the English-speaking world. Crucial to this process was the transformation of printing, publishing, and bookselling in the mid-nineteenth century, which made it possible to produce and distribute books far more cheaply than before. The book was widely acclaimed, and there were cheap pirated editions in the United States, beginning with a two-volume version from Harper in New York. From the 1870s it began to appear in other languages; the only early one, an 1844 German translation of the original edition, had been instigated by Humboldt. Widely read in the late nineteenth century, Darwin's *Journal* had a second life as the archetype of a 'boy's own' imperial adventure, all the more attractive as the romantic setting for a great discovery.

The Mystery of Mysteries

Although Darwin was convinced of the truth of evolution when the second edition of his *Journal* appeared in 1845, he only hinted at his beliefs in public. Yet this revised version, with its expanded discussions of geographical distribution and the succession of animals in the fossil record, was effectively his first evolutionary book. It asked the right questions and posed the appropriate puzzles. For the first time, the Galapagos were recast into a laboratory for the study of how one species might change into another. 'Hence, both in space and time,' he wrote, 'we seem to be brought somewhat near to that great fact—that mystery of mysteries—the first appearance of new beings on this earth' (p. 44). Darwin here recalled a famous discussion of the problem of species by the imperial astronomer John Herschel, with whom he had dined at the Cape of Good Hope on the final leg of the *Beagle*'s return home.

We will probably never know if Herschel and Darwin discussed the mystery of species at this crucial juncture, but there can be no doubt that both men recognized that the issue was best canvassed in private conversation rather than in publication. The marginal status of theorizing about the evolution of new species — or 'transmutation', as it was generally called — at the time Darwin wrote up his *Beagle* results can only be understood in relation to the extraordinary transformation of the sciences in the late eighteenth and early nineteenth centuries. This involved the greatest change in the organization and practice of knowledge since the early Middle Ages, when the cathedral schools and universities had introduced the scientific learning of Islam and the Greeks (particularly Aristotle) into Christian education. At that time an understanding of nature was divided into natural philosophy, which investigated the causes of things, and natural history, which offered description and inventory. The ultimate aim of studying nature, as Thomas Aquinas had shown in his great theological synthesis, was to understand God. The culmination of this approach appeared in the seventeenth and early eighteenth centuries, in the physical theologies of René Descartes, Isaac Newton, and John Ray.

This synthesis achieved a late expression in the poetry of Erasmus Darwin, Charles's grandfather, whose works from the end of the eighteenth century began from the sexuality of plants to show that nature was ascending an evolutionary ladder through laws sustained by a benevolent deity. To 'darwinize' in Regency England was to write verses like these:

> First forms minute, unseen by spheric glass,
> Move on the mud, or pierce the watery mass;
> These, as successive generations bloom,
> New powers acquire, and larger limbs assume;
> Whence countless groups of vegetation spring,
> And breathing realms of fin, and feet and wing.[5]

In Protestant Europe and America, a belief that God's attributes and existence could be inferred from the natural world retained exceptional potency, as indicated by sales of the Revd William Paley's *Natural Theology* for decades following its publication in 1802. Paley argued that the natural world, like the watch in the book's celebrated opening, displayed intelligent design in every aspect of its construction. The anatomy of the eye, for example, demonstrated complete mastery

---

[5] E. Darwin, *The Temple of Nature* (London, 1803), ll. 295–302.

of the same optical laws that an instrument maker used for designing telescopes. Lenses of different shapes and sizes had been provided for creatures with different ways of life, while focusing was managed by exquisitely delicate muscles.[6] Paley's God was the ideal utilitarian engineer, whose creations were so perfect that transmutation was unnecessary. Even poverty, illness, and other trials of human existence were part of the divine plan; if we were too happy, we might not look to Christ for salvation. Evidence of adaptation was everywhere.

Around the time that Paley and Erasmus Darwin wrote, the centuries-old system of organizing knowledge about the natural world was being transformed. The division between natural philosophy and natural history began to break down, replaced by a focus on new fields such as geology, physiology, and physics that aimed to analyse the workings of nature. The new disciplines created a cadre of experts, important for the state and economy. The jars in cabinets of curiosities were broken open, with specimens dissected to show how they functioned to produce the totality of a living organism. The classification of the Swedish naturalist Carl Linnaeus, which employed a self-confessedly 'artificial' method based on sexual characters, was gradually replaced by 'natural' systems using a combination of features. New theories stressed the location of organisms in space and time, showing how their internal workings meshed with the external environment. Above all, enquiry into nature began to be distinguished from the overall goal of understanding God. Beginning in the eighteenth century, it began to be possible to pursue the study of the natural world without reference to an understanding of the divine. Theology, once the 'queen of the sciences', could be seen as separate and distinct.

In the wake of the French Revolution, the politics of knowledge became acutely sensitive, especially in Britain where an Evangelical revival of Christianity was in progress. The terror of the guillotine and the destruction of the Church were attributed to the misuse of rational learning, and the notion that organisms somehow evolved into other forms was condemned not only as dangerous but incredible. Erasmus Darwin's versifying was ridiculed and the late-Enlightenment Parisian transmutationist Jean-Baptiste Lamarck was dismissed as a wild speculator; as Lyell damningly said, Lamarck's

[6] W. Paley, *Natural Theology* (1802; Oxford 2006), esp. pp. 7–26.

Zoological Philosophy (1809) read more like a fashionable novel than serious science. Evolution was the province of godless radicals, dissolute Frenchmen, and lecturers at the cut-price London medical schools. In Edinburgh, Darwin had heard the invertebrate specialist Robert Edmond Grant 'burst forth in high admiration of Lamarck & his views on evolution' (p. 371). But this was not a path Darwin wanted to follow; the practitioners he admired pushed questions about origins to the margins.

In this constrained situation, one of the pleasures for readers of Darwin's post-*Beagle* publications was the way in which speculation could be managed within the context of genres usually limited to description and narration. What was not apparent, even between the lines, was just how very much further their author had gone. The journals that Darwin had used during the voyage to jot down ideas on geology and zoology, began to record one of the most extraordinary intellectual inquiries ever undertaken. 'Origin of man now proved', he wrote triumphantly in one of his secret notebooks, '— Metaphysic must flourish. — He who understands baboon would do more towards metaphysics than Locke'. Darwin revelled in his audacity: 'love of the deity effect of organization, oh you Materialist! . . . Why is thought, being a secretion of brain, more wonderful than gravity a property of matter? It is our arrogance, it our admiration of our-selves. — ' What Darwin called his 'mental rioting'[7] had led him to question not only the stability of species, but an entire structure of belief.

Species transmutation had been a slowly dawning conviction, expressed in vague doubts as Darwin sorted his collections on the final boring months of the *Beagle*'s return to England. But it was after his arrival home, when London specialists had a chance to work out the complex affinities of his specimens, that Darwin began to believe that species evolved. He became obsessed by the geological and geographical distribution of certain species in South America. Organisms living at the present time shared many characteristics with those buried beneath them as fossils; animals and plants in one province had affinities with their neighbours. A pattern of relationships

⁷ C. Darwin to J. D. Hooker, 6 May 1847, in *Correspondence*, iv (1988), 40. The citations are from Notebook M: 84 and C: 166 in *Charles Darwin's Notebooks, 1836–1844: Geology, Transmutation of Species, Metaphysical Enquiries*, ed. P. H. Barrett, P. J. Gautrey, S. Herbert, D. Kohn, and S. Smith (Ithaca and London, 1987), 539, 291.

could thus be traced both in time and space. A good example of geographical affinity was provided by the inhabitants of the islands off
the coast of South America, notably in the bizarre fauna and flora of
the Galapagos. Although Darwin recognized relationships with species he had seen on the continent, he had no flash of inspiration while
he was there, certainly nothing that would lead him to become an
evolutionist. In the case of the finches, which later became a textbook
case of how varieties could become species through geographical
isolation, he even failed to notice that the birds varied from island to
island. The labels on his specimens, still preserved in the Natural
History Museum in London, make almost no mention of the specific
islands on which they were collected.[8] As the 1845 *Journal of
Researches* admitted (p. 59), Darwin realized his mistake only after
specialists such as the expert 'bird man' John Gould examined his
collections.

At its heart, Darwin's earliest evolutionary theorizing combined
a microscopic understanding of living matter with a vision of large
global processes. Early in the summer of 1837 he opened his first
notebook on species with the bold heading 'Zoonomia'—the laws
of life. There had been plenty of works advocating the development of one species into another, not least by his grandfather under
that very title. What made Darwin's attempt different is that he
approached evolutionary speculations with the full empirical fervour
of the new sciences. He believed that generation, reproduction, and
inheritance—subjects that became the foundation for what we today
would call genetics—would provide the key to the way in which new
species came into being. Arguments from Paley and other scientific
works in natural theology underlined the significance of adaptation,
the way that organisms were fitted to their circumstances and to each
other. Paley's book also brought home lessons about literary structure and the nature of explanation.

As an independently wealthy gentleman, Darwin enjoyed the freedom to speculate, imagining himself as a 'devil's chaplain'—a term
that had been used to condemn a notorious atheist radical. But he
also knew he was expected to act responsibly in an era when memories of the French Revolution remained vivid. As he self-mockingly

[8] F. Sulloway, 'Darwin and his Finches: The Evolution of a Legend', *Journal of the
History of Biology*, 15 (1982), 1–53.

confided in 1844 to his friend, the botanist Joseph Hooker, to admit to being an evolutionist was a bit like confessing a murder.[9] Convinced that evolution was a fact, his first explanation had involved organic germs, or monads, which developed through a process of generation into higher forms. He pored over manuals of stock breeding and gooseberry growing, elaborating his theory to explain the full range of reproductive phenomena. He also read widely in contemporary debates about philosophy, aesthetics, and political economy. In the latter field—a new science focused on the creation, distribution, and consumption of wealth—controversy centred on Britain's adaptation to the early stages of a transformation in manufacturing that would replace human labour with machines. Each new invention, each encroachment of the factory system, had the potential to throw thousands out of work. The ideals of a social order based on hierarchy and deference was being replaced by a belief in individual competition. God worked through the iron laws of scientific economics, which ensured the best possible outcome for society. It was in reading the founding work of this tradition of political economy, Thomas Malthus's *Essay on the Principle of Population* (6th edn., 1826), that Darwin came to see the potential of his reproductive theory in a new way. Malthus had argued that populations, unchecked, reproduced geometrically. But resources, especially food, were scarce, and increased only arithmetically. Targeting utopian philosophers who believed in human perfectibility, Malthus argued there were limits to growth, that war, famine, plague were necessary to keep the population in check.

The battle between different species of animals and plants, what the poet Tennyson would call 'nature red in tooth and claw', was already familiar both from Lyell's *Principles* and the writings of the French naturalist Augustin-Pyramus de Candolle. In reading Malthus, Darwin identified the crucial issue as competition for resources between *individuals* of the same species. The first expression of the new idea was scribbled in his notebook on 28 September 1838:

We ought to be far from wondering of changes in number of species, from small changes in nature of locality. Even the energetic language of Decandoelle does not convey the warring of the species as inference from Malthus.—in Nature

[9] C. Darwin to J. D. Hooker [11 Jan. 1844], *Correspondence*, iii (1987), 1–2.

production does not increase, whilst no checks prevail, but the positive check of famine and consequently death . . population in increase at geometrical ratio in FAR SHORTER time than 25 years—yet until the one sentence of Malthus no one clearly perceived the great check amongst men.—take Europe on an average, every species must have same number killed, year with year, by hawks, by, cold &c—. . even one species of hawk decreasing in number must effect instantaneously all the rest.—One may say there is a force like a hundred thousand wedges trying force every kind of adapted structure into the gaps in the œconomy of Nature, or rather forming gaps by thrusting out weaker ones.[10]

It was an image of terrible, mechanical violence, embodying all the waste and destruction of individual competition. Reproduction, which in Darwin's pre-Malthus theory was the mechanism for species change, now became a source for the tiny variations that led one group of individuals towards reproductive success and others to die. Only a few wedges, those sharpest, hardest, best adapted—would thrust their way to survival. The others would perish. As Darwin would later recall, 'being well prepared to appreciate the struggle for existence . . . it at once struck me that under these circumstances favourable variations would tend to be preserved & unfavourable ones to be destroyed. The result of this would be the formation of new species.' 'I had at last', he wrote, 'got a theory by which to work' (p. 411).

Evolution in Public

Almost two decades later, virtually the same idea occurred to the naturalist and collector Alfred Russel Wallace. Like Darwin, Wallace had travelled extensively and was engrossed in questions of the distribution of animals and plants in time and space. And like Darwin, his thoughts about a mechanism for the origin of species crystallized around Malthus, whose ideas he recalled on a tropical island during a bout of fever. As soon as he was able to get out of bed, Wallace drafted his theory and sent it to Darwin, who was known to be sympathetic to theoretical views and had good contacts in the scientific world.

Darwin, who was writing up his own theory, was appalled by the prospect of an ugly priority squabble. A 'delicate arrangement' was

[10] Notebook D: 134–5 in *Charles Darwin's Notebooks*, 374.

soon agreed among his friends, whereby both Wallace's and Darwin's manuscripts were read at the Linnean Society in July 1858 and published in its proceedings. Although there were significant differences between the two, the resemblance was indeed striking, with Wallace's key phrases almost matching the headings in a sketch Darwin had composed in 1842. As he told Lyell, 'Your words have come true with a vengeance that I sh^d. be forestalled.'[11] Busy with his *Beagle* publications and in poor health, fearful about being condemned as a dangerous materialist, Darwin had kept his speculations private. In 1854 he had started writing on species in earnest, but in the intervening period had told only a handful of friends and family about his continuing evolutionary speculations. He was clearly anxious about the delay, especially after a little paper on barnacles grew uncontrollably into a project that took eight years and four published volumes to complete. Even Lyell, among his closest scientific associates, did not know the specifics of Darwin's Malthusian theory until the 1850s.

More than anyone else, Wallace recognized that the real work was not coming up with a theory of species change, but making it convincing. As he explained a decade later to the novelist Charles Kingsley:

As to C Darwin, I know exactly our relative positions, & my great inferiority to him. I compare myself to a Guerilla chief, very well for a skirmish or for a flank movement, & even able to sketch out the plan of a campaign, but reckless of communications & careless about Commissariat;—while Darwin is the great General, who can manœuvre the largest army, & by attending to his lines of communication with an impregnable base of operations, & forgetting no detail of discipline, arms or supplies, leads on his forces to victory.

'I feel truly thankful', Wallace concluded, 'that Darwin had been studying the subject so many years before me, & that I was not left to attempt & to fail, in the great work he has so admirably performed.'[12] It should be no surprise that their joint papers at the Linnean Society had little impact, for the question could not be solved by a few pages in a journal. Although the Linnean Society's president has been mocked for announcing that 1858 had been one of those years in which nothing particularly significant had happened

---

[11] C. Darwin to C. Lyell, 18 [June 1858], in *Correspondence*, vii (1991), 107.
[12] A. Wallace to C. Kingsley, 7 May 1869, Knox College Library, Galesburg.

in natural history, in fact he was entirely correct. Tactics, as in the American Civil War to which Wallace's letter implicitly referred, were everything.

The tactical problem at the end of the 1850s was very specific. Species transmutation had been widely debated in public since *Vestiges of the Natural History of Creation*, an anonymous work that had created a huge sensation on both sides of the Atlantic on its publication in 1844. This was the one evolutionary book that all English-speaking readers could be expected to know. In striking and readable prose, *Vestiges* narrated a cosmic epic beginning with the formation of the solar system from a condensing fire-mist and ending with the emergence of the human mind. The underlying principle was a law of progressive development, with all nature giving birth to the next higher form through a 'universal gestation of nature'. In its narrative connections, use of particulars, and assumption of progress in history, it was deeply indebted to the historical novels of Sir Walter Scott. At the same time, the general argument of *Vestiges* was also based on the latest scientific findings, and had some affinity with Darwin's pre-Malthus reproductive model for transmutation. Most men of science, however, guessed that there were too many mistakes for it to have been written by one of their own (it was, in fact, from the pen of the Scottish journalist and author Robert Chambers). Yet *Vestiges* was most definitely not a work of popular science, for that would imply that there was a secure body of evolutionary knowledge to simplify and elucidate. Rather, it embodied a democratic vision of science directed against the canons of elite practitioners in the laboratory and field.

If the advocates of specialist science were to fight back, some form of credible naturalistic explanation for species origins was vital. Europe's leading comparative anatomist, Richard Owen, advocated reopening the question, and others clearly agreed. For his part, by June 1858 Darwin had already completed substantial portions of a multi-volume evolutionary treatise on the model of Lyell's *Principles*, a philosophical natural history bringing together all the facts he had gathered from correspondence and reading since the *Beagle* voyage. Wallace's bombshell forced a rethinking of his publication strategy. His aim was unchanged, but at this juncture he attempted a series of articles that would get the main aspects of his views into print as quickly as possible. Realizing that this would be open to criticism for lack of supporting facts, he produced an 'abstract of an essay', a shorter version of his unfinished manuscript.

On the Origin of Species by Means of Natural Selection, or the Preservation of Favoured Races in the Struggle for Life was published by John Murray in November 1859. It is a very unusual book. Unlike most other works of its kind, *Origin* unfolds no story of geological progress or evolutionary development. Moreover, the book, as the novelist George Eliot lamented, was 'ill-written' (p. 213). Its attractions were as unlike those of a novel as could be, being centred on extended argument and analytical exposition rather than narrative continuity of the kind found in *Vestiges*. Thus the origins of the universe, the standard opening of the evolutionary epic, are reduced in *Origin* to a tantalizing reference to planetary gravitation in the final sentence (p. 211). Developmental embryology, the traditional starting place for discussions of transmutation, was likewise placed in a late chapter. Instead Darwin started— most unexpectedly—with an analogy between the actions of domestic breeders in selecting stock, and the way in which certain individuals were preserved in nature. The violent metaphor of competitive wedging was replaced by gentle references to the farmyard, garden, and aviary, a rhetorical transformation of his theory that had taken place in the months after reading Malthus. Nature, Darwin could now explain, was like a pigeon fancier, aware of the tiniest differences that led one bird towards success and another towards failure.

Darwin's story of his experience in breeding pigeons is only one of dozens of captivating bits of autobiography scattered throughout his books. He describes watching ants and aphids, his success in proving the crossing of hermaphrodite barnacles, and his dissections of the rudimentary eyes of a blind burrowing rodent from South America. There are stories about nature, too, although these had to be handled carefully lest they be read as fantasy. For example, *Origin* explains how an aquatic bear, catching insects with its open mouth, could be transformed into a creature 'as monstrous as a whale' (p. 181). Such stories, although scarcely disrupting the flow of argument, offered ways of hinting at larger narratives of evolutionary transformation. These were so familiar to readers from *Vestiges* and other evolutionary epics that they scarcely needed elaborating. In this particular case, however, Darwin had gone too far and failed to avoid being classed as a cosmic speculator.

Mind, Morals, and Man

Just as Darwin had delayed publishing until his theory had the possibility of a good reception, so did he wait to comment on human
origins. The evolutionary manuscript interrupted by Wallace had
featured a chapter on man, but this was left out of *Origin* as
too speculative. But by the late 1860s the debate had become much
more open. Rather than providing further instalments of his fully
documented work (as in his *Variation of Animals and Plants under
Domestication*, published in 1868), Darwin felt confident enough to
publish on human evolution. As he admitted in the opening pages of
Descent, human evolution was old news by 1871: books by naturalists
such as John Lubbock, Ernst Haeckel, Carl Vogt, and T. H. Huxley
had already demonstrated the continuity between apes and humans.
What mattered was that Darwin was saying it too, and that he
focused on religion, music, language, the moral sense, and other
characteristics traditionally identified as uniquely human.

Darwin's naturalized system of morals went back to his theoretical
work of the 1830s, which tackled the problem of evil in the world that
had been at the heart of post-Malthusian political economy and
theological discussion in England. This tradition, which has been
identified by historians as 'Christian political economy', had targeted
revolutionary optimism on the one hand and the imposition of moral
codes by the state on the other. As a succession of Anglican clerics
had asked in the wake of Malthus's *Principle of Population*, what
was the purpose of shortages, starvation, and overpopulation among
humans?[13] For Darwin, the answer was that scarcity explained adaptation. As Darwin wrote in his notebook immediately after reading
Malthus, 'the final cause of all this wedgings, must be to sort out
proper structure & adapt it to change'.[14] In Darwin's reformulation
of Christian political economy, which drew especially on Paley's
stress on adaptation, all of nature became subject to the scarcity that
had previously been seen to govern human affairs.

Darwin, however, went further in recasting the tradition. Malthus
and his followers had argued that the only way that humans could
avoid the disaster of overpopulation (other than immigration) was
through moral restraint. In Darwin's view, however, morals were no

[13] On this tradition, see A. M. C. Waterman, *Revolution, Economics and Religion:
Christian Political Economy, 1798–1833* (Cambridge, 1991).

[14] D: 135, *Charles Darwin's Notebooks*, 375.

more a special characteristic of humans than was the tendency towards overpopulation. Morals were not unique, he argued in the early chapters of *Descent*, but existed in rudimentary form through-out the animal kingdom as instincts. ('Our descent, then, is the origin of our evil passions!! — ', he had scribbled in his notebook. 'The Devil under form of Baboon is our grandfather!')[15] Even the noblest aspects of humanity derived from the social instincts found in the higher apes. The theological basis of Christian political economy had been turned on its head. If God was manifest in the creation, it was not through caring design, but through the dearth and death that led the fittest to survive. Morals and instincts alike had to work in conjunction with Malthusian principles, not against them. In effect, Darwin advocated Christian political economy without the Christianity.

The greatest novelty of *Descent* was in explaining the relation between sex and race. Here Darwin enlarged upon the mechanism of sexual selection that had been broached in *Origin*. Why did men have beards and women not? Why were women (as most male contempor-aries agreed) less intelligent and creative but more patient, selfless, and intuitive than men? To explain these differences — in fact differ-ences between the sexes of any species — Darwin postulated a strug-gle among males to obtain the females. Those successful in this competition would leave progeny sharing their characteristics, gradu-ally leading to more and more strongly marked characters differ-entiating the sexes. Much of *Descent* is a vast catalogue of these distinctions as they can be found throughout the animal kingdom, especially in birds. By analogy, Darwin argued, sexual selection was largely responsible for human racial distinctions, through the long-continued exercise of particular aesthetic preferences. There was one important difference: in most animals, female choice was paramount, but in humans males did the selecting.

Even in the changed climate of England in the 1870s, discussions of sex were potentially open to accusations of impropriety, particu-larly in a book that based religion and morals on instinct rather than revealed truth. Darwin's publisher, for example, advised against using the adjective 'sexual' in the title, guidance that resulted in the anodyne *The Descent of Man, and Selection in Relation to Sex*.[16]

[15] D: 123, *Charles Darwin's Notebooks*, 550.
[16] G. Dawson, *Darwin, Literature and Victorian Respectability* (Cambridge, 2007), 35.

Contemporary caricatures show how readily Darwinism could be equated with racial mixing, miscegenation, and promiscuity. Yet it is striking how muted such criticisms were. Darwin's lively anecdotes and coy talk of 'choosing' encouraged readers to associate his writing not so much with the murderous revolutions and moral dissolution feared by a reviewer in *The Times* (pp. 335–6), but with popular romantic fiction and contemporary travel literature.

An Evolutionary Self-Analysis

After *Descent* and its sequel, *The Expression of the Emotions in Man and Animals* (1872), Darwin avoided publishing on controversial topics, focusing instead on less contentious subjects related to his theory, such as the movement of plants and the intellectual abilities of earthworms. But in private he renewed the ambitious programme of self-observation and reflection on the human mind first pursued in his notebooks of the 1830s. The first of the autobiographical texts included here (pp. 351–4) was part of this original programme. In August 1838, at the height of his evolutionary theorizing, Darwin jotted down fragmentary notes on the emergence of his memories. The earliest was being cut accidentally by a knife when a cow ran past the dining room window, one of several recollections inspired by fear. He explored the origins of his fascination with natural history collecting; other traits, such as a love of pictures and music, were (or so Darwin claimed) not really 'natural' to him. Not only were particular mental qualities inherited, but so too were tendencies towards different kinds of memories. His younger sister, for example, recalled scenes 'where others were chief actors', while Darwin remembered events chiefly relating to himself and almost nothing of his mother's early death. These differences, Darwin believed, were innate. As he wrote with satisfaction, 'I was born a naturalist' (p. 353).

Concerns about evolution and heredity were very much to the fore as Darwin sat down in 1876 to recall the events of his life. Three years earlier, his cousin Francis Galton, researching the mental attributes of prominent men of science and their fathers, had sent a questionnaire which Darwin filled out fully and enthusiastically. He would shortly publish 'A Biographical Sketch of an Infant'—his observations on the early behaviour of his first child—in the recently founded journal *Mind*. The *Recollections of the Development of My*

Mind and Character are thus as much a contribution to evolutionary psychology as they are a traditional biographical narrative. For Darwin, the crucial category was the 'development of mind and character'. (Tellingly, the 'my' was left out of the title in the first draft, and inserted only later.) To understand cognition, Darwin had long examined those minds closest around him: his sisters, his children, his father, and himself. The engaging sense of personality that comes across so strongly in Darwin's *Recollections* resulted, at least in part, from his fascination in the origins of human qualities. Nothing was outside the explanatory agenda of evolution.

Darwin concluded that his special qualities of mind were not added throughout life, but were present in potential from birth. He denied any substantive intellectual role for his father, brother, sisters, and most of his teachers. Learning classics at Shrewsbury school, he claimed, had been a complete waste of time. Yet, although an exposure to Latin and Greek may not have given him much love for Caesar and Cicero, it proved vital to his sensitivity to language and his skills as a writer. Similarly, Darwin downplayed advantages gained from medical school in Edinburgh and his degree at Cambridge. Instead, everything turned on the *Beagle* voyage, not because of help from captain and crew, but because solitude in wild nature had allowed the discovery of his true self. What Darwin did claim to acquire through experience were qualities of industry, patience, and attention, habits that stood him in good stead after the ship's return. His stress on progress and improvement throughout all his writing on his life is very much of a piece with his views on human evolution more generally. As he noted in the *Recollections*, 'The primeval instincts of the barbarian slowly yielded to the acquired tastes of the civilised man' (p. 388). His individual story recapitulated the development of the race.

At once superficial and revealing, compelling yet discursive and anecdotal, Darwin titled his work *Recollections* because they were precisely that, rather than a formal *Autobiography* (although that title is commonly and incorrectly used). Behind the undoubted charm and informality of the memoir are the same intellectual concerns that motivated all Darwin's later work. That these ambitions for the memoir are rarely acknowledged has to do with the fact that the *Recollections*, alone among Darwin's later writings, were not intended for publication during his lifetime. Commenced soon after Darwin learned of the impending birth of his first grandchild, the text was written in the first instance for his family, providing a way of speaking to future generations. During the following months he added incidents and memories as they occurred to him, and in 1881 he brought the story up to date. The longest addition was fourteen pages about his father, Robert Darwin, which completed the (male) intellectual genealogy he had undertaken two years earlier in publishing a short life of his famous grandfather Erasmus.

Would Darwin have been upset to see his autobiographical reflections in print? Writing to a German colleague, he had declared that he would never publish his own life.[17] But posthumous publication was a very different thing, and given his fame there is every likelihood that he knew that the manuscript might appear in print in some form. Darwin chose to hand over responsibility to his survivors: he wrote for his descendants, and it was up to them to judge the suitability of his family memoir as a public document. In particular, he knew that his condemnation of Christianity would be deeply distressing to his wife Emma, for whom it would mark the decisive proof that they would never meet in heaven.[18] By leaving a clear manuscript, Darwin could both have his last word to the family, but allow them to determine how much it was appropriate to make public. This had so many advantages that to do otherwise would have been odd. It allowed Darwin's story to go before the public, thereby enhancing his reputation and furthering the evolutionary cause, but simultaneously maintained his character for modesty, avoiding any appearance of self-aggrandizement or self-importance. It sidestepped the kind of unseemly scandals that had been precipitated by other contemporary memoirs, most notoriously that of Thomas Carlyle. Almost no one of Darwin's status published an autobiography during his or her lifetime, while many (such as the philosopher John Stuart Mill, the geologist Roderick Murchison, and the mathematician Mary Somerville) left manuscript recollections as the basis for a public memoir.

When Darwin's *Recollections* were published in 1887 as a chapter in *The Life and Letters* compiled by his son Francis, they crowned his

---

[17] C. Darwin to J. V. Carus, 17 July 1879, cited in J. Browne, *Charles Darwin: The Power of Place* (London, 2002), 427.

[18] The delicate problem of dealing with the religious passages in the manuscript led to a difficult family quarrel: see J. Moore, 'Of Love and Death: Why Darwin "gave up Christianity" ', in id. (ed.), *History, Humanity and Evolution* (Cambridge, 1989), 195–229, esp. 199–209.

lifelong attempt to create a new scientific persona for the evolutionist. Understandably, the family exercised its prerogative to cut out passages—especially about religion—that would appear too raw or revealing. (These passages were not restored until the 1950s, first in a Russian language edition and then in an English version edited by Darwin's granddaughter.) Darwin had himself already put considerable work into establishing a respectable image. He did everything possible to make it clear he was no closet speculator, no 'Mr. *Vestiges*', and not even the intellectual heir of his grandfather. This was by no means a simple matter, especially as evolutionary ideas had traditionally been associated with questionable morals, atheism, and sexual promiscuity. The work Darwin had done to establish his credentials as a gentleman of science paid off. By the time of *Origin*'s publication he could speak as the author of three major works on geology and four on zoological and palaeontological classification. The title page announces him as a 'Fellow of the Royal, Geological, Linnæan, etc. societies', and reminds readers of the ever-popular *Journal of Researches*, which is also recalled in the first sentence of the introduction. Both *Origin* and *Descent* testify to their author's cautious empiricism in research ('patiently accumulating and reflecting on all sorts of facts which could possibly have any bearing') and in coming to conclusions. The strategic silences on evolution in the 1845 *Journal of Researches*, and *Origin*'s cautiously worded references to man, both spoke eloquently of the author's unwillingness to offend. The same caution appears in the *Recollections*, the document that has served more than any other to establish Darwin's public image.

Darwin among the Machines

The strategies employed in Darwin's texts were, by any standard, extraordinarily successful. Thomas Henry Huxley hailed *Origin* as 'a veritable Whitworth gun in the armoury of liberalism'.[19] It was an appropriate analogy, for Joseph Whitworth's invention, a replacement for the much-criticized Enfield rifles used during the Crimean War of the 1850s, was proving so accurate that Queen Victoria had hit a bull's eye at a range of 400 yards.

[19] [T. H. Huxley,] 'The Origin of Species', *Westminster Review*, NS 17 (1860), 541–70, at [541].

Darwin may not have been quite so precise a marksman, but there can be little doubt that his works transformed the terms of the international evolutionary debate, which was far less violent and ill-tempered than it is often depicted. Within Britain, *Origin* did not so much initiate a crisis as conclude a major piece of unfinished business from the 1830s. With significant exceptions, as Darwin acknowledged, reviewers treated his argument patiently and in good faith. Out of a first edition of 1,250 copies, no less than 500 were bought by Britain's leading circulating library,[20] and the sixth edition of 1872 was an attractively priced 'people's edition', printed with stereotyped plates for inexpensive reissues. In Darwin's view, men of science — who had for the most part failed to be convinced by previous evolutionary works — were the key to the public debate. By the second half of the nineteenth century this meant reaching potential converts from across much of Europe and eastern North America, as well as colonial and trading outposts throughout the rest of the world. Towards this end, Darwin maintained a monumental correspondence, debating theology with Harvard professors, encouraging young collectors in Africa and India, cajoling leading naturalists throughout the world to change their views. He engineered translations of his books into the main European languages, pleased when they were satisfactory, working for better ones when they were not. The first German translator, the palaeontologist Heinrich Bronn, burdened *Origin* with an afterword combating its main thesis; in France, the philosophical writer Clémence Royer gave the book a new subtitle and a fiery preface advocating secular progress.

As Darwin had hoped, convincing scientific men of the validity of evolutionary questions proved crucial in winning a fair hearing for his theories. Few agreed, however, that his novel mechanism was likely to be as important as *Origin* had claimed. Some, like Huxley, argued that natural selection needed experimental proof to be a valid explanation, but still kicked themselves for failing to think of it. The source and size of the variations needed for a process based on selection was much debated. Many physicists, following the lead of William Thomson (Lord Kelvin), pointed out that Darwin's theory required a timescale far longer than that indicated by contemporary theories of the sun's heat. Still others applied natural selection theory to exciting new problems, such as the evolution of horses and birds,

[20] Browne, *Charles Darwin: The Power of Place*, 88–9.

and the way in which certain organisms mimicked the appearance of others to obtain an evolutionary advantage. Within a few months Darwin was compiling lists of supporters and opponents, keeping score to show opinion tipping in his favour. By the end of the 1860s, virtually all practising naturalists thought *Origin* had demonstrated that the problem of species origins would be resolved. Natural selection opened the door to a possible mechanism for evolutionary change, but was not seen as the final answer; what really mattered was that species—and even human morals and religion—might be brought within the realm of scientific explanation.

The reception of *Origin* and *Descent* benefited from the growing status of science, which opened up a wide range of paid professional careers in laboratories, museums, botanical gardens, and field stations around the world. Managing a vast network of correspondents from his home in the village of Downe in Kent, Darwin cultivated expert practitioners in these new institutions supported by governments and industry. That his works—unlike most science at this time—could be understood by a broader audience as well opened up the possibility of a shared discussion in general quarterly journals, literary weeklies, and newspapers. In Britain, the most important notice of *Origin* was in *The Times*, by this time approaching its later role as the imperial newspaper of record. A journalist had been asked to provide a review, but after a few sentences he demurred and sent the job on to Huxley, who penned a positive notice which set the tone for much of the rest of the debate. The quarterly *Edinburgh Review* featured an anonymous essay by the leading comparative anatomist Richard Owen, who was irritated by Darwin's belittling of other mechanisms for evolutionary change, but simultaneously condemned 'sacerdotal revilers' who would repress scientific debate on theological grounds. On this latter point the *Edinburgh* expressed a view that had become increasingly prevalent from the 1850s. With a few exceptions, notably Darwin's old geology teacher Adam Sedgwick, Anglican divines followed Charles Kingsley in believing that a naturalistic explanation of species origins could readily be accommodated with a belief in God. For most—as indeed for Wallace, Lyell, and many others—the only exception was the origin of man's moral and spiritual qualities. Yet even here, there was a willingness to discuss which was encouraged by the absence of strong statements on the issue in the *Origin*. There were criticisms of the more forthright positions taken in *Descent*, notably scathing accusations in *The Times* of gross moral culpability; but generally reviewers were serious and respectful, reflecting the tone of debate set by Darwin himself.

The Darwinian dispute offered the Established Church in England, scarred by decades of theological conflict, an opportunity to demonstrate liberal virtues of accommodation and compromise. In the United States, the presence of a host of competing sects and the comparatively decentralized scientific community meant that evolutionary discussion was both more diverse and more vocal. Although some Protestants agreed with the Presbyterian Charles Hodge that 'atheism' was the answer to the title of his book, *What is Darwinism?*, most agreed with the Harvard botanist Asa Gray that Christianity could be accommodated with Darwinism. Advocates for causes from the rights of women to reform of Judaism saw in Darwin's books a natural justification of change. Even campaigners for racial equality such as Frederick Douglass did not oppose Darwinism in anything like the way that might be expected.[21] Progress was equated with evolution, and evolution was equated with Darwinism.

As part of an expanding international network of communications, Darwinism became a symbol of the virtues of open discussion and modern ways of life. This was the case whether natural selection was thought to license cooperation and the 'peace biology' of the Russian anarchists, the rapacious competition of American monopoly capitalism, or the concordance of evolution with Japanese Shinto or Hindu reincarnation. Liberal values, progress among the hierarchy of nations, and a tendency to seeing a divine hand in invariant laws rather than miracles: this summed up the dominant meanings of Darwinism.

The evolutionary epic, leading the reader from cosmic chaos to technological civilization, emerged as one of the global publishing phenomena of the modern age. This genre had a long history, going back to *Vestiges*, the universal histories of the Enlightenment, and even to Lucretius' *On the Nature of Things*; but in the late nineteenth century it spawned a host of global best-sellers, ranging from Haeckel's *Natural History of Creation* (1868) and Herbert Spencer's

---

[21] On these issues, see the essays in R. L. Numbers and J. Stenhouse (eds.), *Disseminating Darwinism: The Role of Place, Race, Religion and Gender* (Cambridge, 1999).

Synthetic Philosophy (1862–96), to Arabella Buckley's *Winners in Life's Race* (1883) and Wilhelm Bölsche's *Love-Life in Nature* (1898–1902).[22] Books of this kind, with their monad-to-man stories of progress and stress on alternative mechanisms for species change (often involving embryological development), depended relatively little on the specific contents of Darwin's writings but heavily upon their reputation. Such works enjoyed extraordinary sales in the newly reunited Germany, where rapid industrialization was built upon an impressive corps of university-trained scientists and engineers. In many countries newly emerging onto the world scene, the classic evolutionary works were taken up with enthusiasm. In Japan under the Meiji regime, evolutionary discussion was an integral part of rapid scientific modernization, with Herbert Spencer's philosophy being especially popular.

In traditional societies such as China, evolutionary works were advocated by oppositional political movements. Writing in an ultra-reforming journal printed in Japan and smuggled into China through the treaty ports, Liang Qichao attacked three thousand years of despotic monarchy. The Manchu dynasty was only the latest regime to evolve the human political equivalent of the blind cave fish discussed in *Origin*, which had lost its sight through lack of use. This put the Chinese people, all four hundred million of them, at risk of extinction in the global competition for power. 'Alas,' he wrote, 'to pit such people against the races of Europe in this world of struggle and survival of the fittest—What hope is there?'[23] Such views, which quickly became known as 'Social Darwinism', were widely canvassed throughout the industrialized world. Capitalist entrepreneurs, especially in the United States, argued that their success reflected the laws of individual competition, and preached a 'gospel of wealth' based on evolutionary science. It was not difficult to find support for such positions in Darwin's own writings. As one of his books put it, 'Hard cash, paid down over and over again, is an excellent test of inherited superiority.'[24] Darwin encouraged research into eugenics, opposed the passing on of wealth solely to eldest sons, and assumed the inevitable extinction of supposedly 'inferior' races.

[22] B. Lightman, *Victorian Popularizers of Science: Designing Nature for New Audiences* (Chicago, 2007), 219–94, discusses this tradition in Britain.

[23] Cited in J. R. Pusey, *China and Charles Darwin* (Cambridge, Mass., 1983), 185.

[24] C. Darwin, *The Variation of Animals and Plants under Domestication*, 2 vols. (London, 1868), ii. 3.

In the global marketplace for knowledge, evolutionary science was news. To open a discussion about Darwinism was to argue that new technologies of communication ought to go hand in hand with new ways of thinking. The growth of literacy and middle-class readerships, combined with mechanized printing and fast distribution, encouraged the emergence of cheap periodicals as a worldwide publishing phenomenon. Weekly science magazines and newspaper columns, sprouting up everywhere from Stockholm to Calcutta, offered lively forums for debate about the wider meanings of Darwinism. In these periodicals, local contributions could appear alongside extracts from the writings of leading men of science, who could (and did) contribute their views directly. The same passage read very differently in an Argentine newspaper or a French popular science monthly, a colonial literary weekly in South Africa or a quarterly journal from New England. Controversy was often heated. Notably, many religious writers welcomed industrial progress, and believed that cheap periodicals, steam-powered printing, machine-made paper, and other science-based innovations were only the latest and greatest contributions of a creator god towards the advance of civilization. From this perspective, disseminating Darwinism was no way to give thanks for providential gifts.

Certainly there was (and is) no inherent conflict between faith and reason, but evolution—particularly as applied to humans—offered the potential for real difficulty in a number of theological traditions, particularly those that stressed the unique role of a creator and the special status of humans. Most Catholics, particularly in the conservative theological atmosphere engendered by Pope Pius IX, condemned naturalistic evolution in the strongest terms, although it was never formally subject to papal sanction. In England, successive Catholic archbishops pointed out contradictions with the doctrines of miracles and the human soul. At the other end of the theological spectrum, some Evangelical Protestants believed that Darwinism was allied to religious doubt, materialism, even atheism. In this they were encouraged by the tiny but vocal corps of freethinkers, who welcomed *Origin* and *Descent* as scientifically credible replacements for earlier evolutionary classics, but deplored their ambiguities and compromises. The first edition of *Origin*, for example, implied the possibility for a divine origin for life, speaking of life being 'breathed' into a single or several forms (p. 211); the second edition strengthened this by specifying 'breathed by the Creator'. Freethinkers around the world accused their hero of masking his real character as an atheist. Darwin privately regretted that he had 'truckled' and removed the phrase from one place in the book, but he left it in the final sentence, where it remained in all subsequent editions as a lifeline to readers hoping to reconcile evolutionary naturalism with a belief in God.[25] The possibility of such an accommodation was crucial to Darwin's success.

These discussions, first played out in north-west Europe and the eastern United States, were repeated in many other countries. What was the relation between industrial modernization and religious revival? How were migrants to the large urban centres to retain their traditions of faith? In many regions, the appearance of *Origin* and *Descent* injected scientific life into long-standing debates about materialism. The founding act of these exchanges was often the publication of books by Darwin or his followers in a local language. The appearance of *Origin* in Russian in 1864 offered the radical intelligentsia the basis for a philosophical materialism that would renew the nation in the wake of defeat in the Crimean War. Facing an apathetic peasantry on the one hand, and an increasingly repressive aristocracy on the other, the new generation read Darwin's books as the epitome of progressive change and a way of attacking the power of the Orthodox Church. The palaeontologist Vladimir Kovalevsky fought on several fronts, diffusing scientific knowledge to the public as well as researching the evolution of the horse. Among similarly minded reformers in Spain, the translation of *Origin* in 1877 was welcomed as 'a most happy symbol of our progress',[26] though in fact most readers continued to consult both it and Darwin's other works in French. Here, as throughout the Spanish-speaking world, the controversy was polarized, with conservative politicians and clerics confronting radical positivists about the moral issues raised in *Descent*.

The problem of introducing the economic benefits of modern technology and science, while avoiding the secularizing tendencies that could follow in their wake, was most acutely faced in the Arabic-speaking world. The Lebanese physician Shiblī Shumayyil argued

---

[25] C. Darwin to J. D. Hooker, 29 Mar. 1863, *Correspondence*, xi (1999); C. Darwin, *The Origin of Species: A Variorum Text*, ed. M. Peckham (Philadelphia, 1959), 753, 759.

[26] *Revista Contemporánea*, cited in T. F. Glick, 'Spain', in Glick (ed.), *The Comparative Reception of Darwinism* (Austin, Tex., 1974), 307–45, at 311.

for the wholesale adoption of evolutionary ideas, viewing Darwin through the lens of German freethinkers such as Ludwig Büchner. In contrast, the Persian modernizer Jamal al-Din al-Afgani advocated the need to industrialize, but warned against the temptations of evolutionary philosophy. His *Refutation of the Materialists* (1880–1), widely read in Arabic translation, ridiculed Darwin as a 'wretch' who claimed that a mosquito could become an elephant in a few centuries.[27] In later writings, however, Afgani was willing to praise Darwin's writings as offering more than mere evolutionary materialism. A leading Islamic theologian, Ḥusayn al-Jisr, accepted that key passages in the Qur'an might need to be re-examined if evolution could be proved unequivocally true. The strongest backlash against Darwin was among Christian missionaries who had embraced the sciences as part of their evangelizing. During the early 1880s, the professor of geology and chemistry at Beirut's Protestant College, the Harvard graduate Edwin Lewis, delivered a speech in Arabic praising Darwin as the exemplar of a man of science patiently conducting empirical research. Although censured by the college's American Evangelical administrators, who dismissed Lewis forthwith, the lecture was published in the weekly *al-Muqtaṭaf*, where it led to a vigorous and fruitful debate. The journal became a leading forum for the discussion of new ideas.[28]

In the Arabic-speaking world, as elsewhere across the globe, the commencement of evolutionary debates was taken to signal the arrival of the modern age, involving the recasting of social relations, an industrializing economy, intellectual innovation, and a rethinking of ancient systems of patronage and belief. Any discussion of evolution was necessarily entangled with questions of free trade, imperial expansion, and mass communication. With the introduction of electric lighting, cheap illustrated newspapers, and an international telegraph network, technologies based on science began to shape expectations for future progress among millions of people. Through the application of electricity and steam, the intellectual distance from London to Calcutta, Budapest to San Francisco, Beirut to Philadelphia, could effectively disappear. In discussing evolution, people contemplated what these

---

[27] N. R. Keddie, *An Islamic Response to Imperialism: Political and Religious Writings of Sayyid Jamāl ad-Dīn 'al-Afghāni'* (Berkeley and Los Angeles, 1983), 136.

[28] M. Elshakry, 'The Gospel of Science and American Evangelism in Late Ottoman Beirut', *Past and Present*, 196 (2007), 173–214, esp. 207–14.

unprecedented powers for understanding, destroying, and controlling nature were going to mean. Darwinism came to symbolize the way in which all aspects of human existence, for better or worse, would be brought within the realm of scientific explanation.

Whatever side of the debate was aired, discussing Darwinism was part of the global rush towards the future so widely canvassed in the later nineteenth century. This is the period that witnessed the birth of prophetic fiction by Samuel Butler, Camille Flammarion, Jagadananda Roy, and H. G. Wells, in novels that employed evolutionary science to predict 'things to come'. In an anonymous letter to a colonial New Zealand newspaper in 1863 entitled 'Darwin among the Machines', Butler argued that the instruments of industrial civilization were evolving faster then their human masters and would eventually inherit the earth. 'Day by day. . .', he wrote, 'the machines are gaining ground upon us; day by day we are becoming more subservient to them . . . Our opinion is that war to the death should be instantly proclaimed against them' (p. 221). Ironically, it was only through those machines that news of Darwin's work could spread so quickly across the globe, to reach the pages of a Christchurch daily. It is entirely appropriate that one of the originating points of modern science fiction is a response to Darwin's writing, published just about as far from Darwin's home in Kent as it was possible to go.

NOTE ON THE TEXTS

This book has three aims: to encourage further exploration of the writings of Charles Darwin; to offer the complete text of his most engaging work, the autobiographical *Recollections*, in a reliable text as free as possible from editorial intervention; and finally, to encourage an understanding of Darwin's works informed by their reception, especially in a global context.

As the focus is on Darwin as a public figure, in dealing with writings published during his lifetime I have chosen those editions in which his work first reached a wide audience. Wherever possible I have included complete chapters, unabridged except for the omission of Darwin's notes. All passages omitted from the text have been briefly summarized in passages marked off with square brackets.

The chapters from the *Journal of Researches* are taken from the first issue of the second edition of his *Beagle* book, entitled *Journal of Researches into the Natural History and Geology of the Countries Visited during the voyage of H.M.S. Beagle Round the World, under the Command of Capt. Fitz Roy, R.N.* (London: John Murray, 1845). This included important new material that reflected Darwin's still largely secret views on the transmutation of species. (Darwin's preface to the 1845 edition, which thanked his shipmates and listed the voyage publications, is here omitted.) The selections from *On the Origin of Species by Means of Natural Selection, or the Preservation of Favoured Races in the Struggle for Life* (London: John Murray, 1859) and *The Descent of Man, and Selection in Relation to Sex*, 2 vols. (London: John Murray, 1871) are from the first issue first editions. The review extracts and letters are taken from the sources as indicated; where possible these are from the first printings (e.g. in newspapers and periodicals) rather than later reissues. Ellipses (. . .) in these ancillary texts are the editor's; those in the texts by Darwin are his own.

The text of the *Recollections of the Development of my Mind and Character* is complete, unabridged, and edited for the first time to the standards of modern scholarship. For the past fifty years, the standard text has remained Nora Barlow's pioneering edition of 1958, which regularized spelling, expanded abbreviations, italicizied titles, and made other changes that tend to reduce the informality and immediacy of a document which Darwin wrote without thought of style. The aim here is to provide an accurate and complete transcription of Darwin's final text. For the first time in any edition of the *Recollections*, all original spellings, abbreviations, and underlinings have been preserved, following the conventions of *The Correspondence of Charles Darwin*. Unlike the *Correspondence*, however, no attempt has been made to record systematically in textual notes evidence of the process of composition, although the more substantial of Darwin's later additions have been noted. The text has been established by Anne Secord from the original manuscript (DAR 26) in Cambridge University Library (henceforth CUL).

The *Recollections* were first published in an expurgated form in Francis Darwin (ed.), *The Life and Letters of Charles Darwin, including an Autobiographical Chapter*, 3 vols. (London: John Murray, 1887), i. 11–20, 21–2, 26–107, 307–13. Significant passages, especially on religion, were left out and first became available in Nora Barlow's edition. As the cut-down version was crucial in establishing Darwin's reputation for over seventy years, the major omissions in the *Life and Letters* abridgement have been here signalled in endnotes.

The autobiographical fragment of 1838 was first published in *More Letters of Charles Darwin*, ed. Francis Darwin and Alfred C. Seward, 2 vols. (London: John Murray, 1903), i. 1–5. The version here is reproduced by permission, and is from *The Correspondence of Charles Darwin*, ed. Frederick Burkhardt et al., ii (Cambridge: Cambridge University Press, 1986), 438–42, based on a fresh transcription from the manuscript (DAR 91: 56–62). The extensive records of Darwin's alterations and annotations have not been reproduced here.

SELECT BIBLIOGRAPHY

Editions of Works by Darwin

Charles Darwin's Beagle *Diary*, ed. R. D. Keynes (Cambridge, 1988).
Charles Darwin's Notebooks, 1836–1844: Geology, Transmutation of Species, Metaphysical Enquiries, ed. P. H. Barrett, S. Herbert, D. Kohn, and S. Smith (Ithaca and London, 1987).
Charles Darwin's Shorter Publications, ed. J. van Wyhe (Cambridge, 2009).
The Descent of Man, and Selection in Relation to Sex (1871; rpt. Princeton, 1981).
The Descent of Man, and Selection in Relation to Sex, ed. J. Moore and A. Desmond (2nd edn., 1879; Penguin, 2004).
Journal of Researches into the Natural History and Geology of the Countries Visited During the Voyage of H.M.S. Beagle, ed. R. D. Keynes (1860; London, 2003).
On Evolution: The Development of the Theory of Natural Selection, ed. T. F. Glick and D. Kohn (Indianapolis, 1996).
On the Origin of Species (1859; rpt. Cambridge, Mass., 1964).
On the Origin of Species, ed. J. Endersby (1859; Cambridge, 2009).
The Works of Charles Darwin, ed. P. H. Barrett and R. B. Freeman, 29 vols. (London, 1986).

Darwin's Correspondence

The *Correspondence of Charles Darwin*, ed. F. Burkhardt et al., 16 vols. and continuing (Cambridge, 1985–); hereafter abbreviated *Correspondence*.
Charles Darwin: The Beagle Letters, ed. F. Burkhardt (Cambridge, 2008).
Origins: Charles Darwin's Selected Letters, 1821–1859, ed. F. Burkhardt (Cambridge, 2008).
Evolution: Charles Darwin's Selected Letters, 1860–1870, ed. F. Burkhardt, S. Evans, and A. Pearn (Cambridge, 2008).

Recommended Websites

http://www.darwinproject.ac.uk/: authoritative annotated transcriptions and summaries of thousands of letters to and from Darwin, with extensive supporting materials.
http://darwin-online.org.uk/: Darwin's complete publications from reliable editions, with thousands of manuscripts and a comprehensive bibliography.

http://darwinlibrary.amnh.org/: authoritative editions of Darwin's manuscripts and many other materials.

Biography and Bibliography

Browne, J., *Charles Darwin*, 2 vols. (London, 1995–2002).
—— *Darwin's* Origin of Species*: A Biography* (London, 2006).
Desmond, A., and Moore, J., *Darwin* (London, 1991).
Freeman, R. B., *The Works of Charles Darwin: An Annotated Bibliographical Handlist* (Folkestone, Kent, 1977).
Gruber, H., *Darwin on Man: A Psychological Study of Scientific Creativity* (2nd edn., Chicago, 1981).
Herbert, S., *Charles Darwin, Geologist* (Ithaca, 2005).
Keynes, R., *Annie's Box: Charles Darwin, his Daughter and Human Evolution* (London, 2001).
Keynes, R. D., *Fossils, Finches and Fuegians: Charles Darwin's Adventures and Discoveries on the* Beagle, *1832–1836* (London, 2002).
Quammen, D., *The Kiwi's Egg: Charles Darwin and Natural Selection* (London, 2007).

Critical Studies

Beer, G., *Darwin's Plots: Evolutionary Narrative in Darwin, George Eliot and Nineteenth-Century Fiction* (London, 1983).
Dawson, G., *Darwin, Literature and Victorian Respectability* (Cambridge, 2007).
Desmond, A., and Moore, J., *Darwin's Sacred Cause: Race, Slavery and the Quest for Human Origins* (London, 2009).
Donald, D. (ed.), *Endless Forms: Charles Darwin, Natural Science and the Visual Arts* (New Haven, 2009).
Gagnier, R., *Subjectivities: A History of Self-Representation in Britain, 1832–1920* (Oxford, 1991).
Hodge, J., and Radick, G. (eds.), *The Cambridge Companion to Darwin* (Cambridge, 2003).
Kohn, D. (ed.), *The Darwinian Heritage* (Princeton, 1985).
Landau, M., *Narratives of Human Evolution* (New Haven, 1991).
Neve, M., 'Introduction', in C. Darwin, *Autobiographies*, ed. M. Neve and S. Messenger (London, 2002), pp. ix–xxiii.
Ospovat, D., *The Development of Darwin's Theory: Natural History, Natural Theology, and Natural Selection* (Cambridge, 1981).
Richards, E., 'Darwin and the Descent of Woman', in D. Oldroyd and I. Langham (eds.), *The Wider Domain of Evolutionary Thought* (Dordrecht, 1983), 57–111.
Richards, R. J., and Ruse, M. (eds.), *The Cambridge Companion to Darwin's* Origin of Species (Cambridge, 2009).

Richardson, A., *Love and Eugenics in the late Nineteenth Century: Rational Reproduction and the New Woman* (Oxford, 2003).

Smith, J., *Charles Darwin and Victorian Visual Culture* (Cambridge, 2006).

Sulloway, F., 'Darwin and his Finches: The Evolution of a Legend', *Journal of the History of Biology*, 15 (1982), 1–53.

—— 'Darwin's Conversion: The *Beagle* Voyage and its Aftermath', *Journal of the History of Biology*, 15 (1982), 325–96.

—— 'Darwin and the Galapagos', *Biological Journal of the Linnaean Society*, 21 (1984), 29–59.

Young, R. M., *Nature's Metaphor: Darwin's Place in Victorian Culture* (Cambridge, 1985).

Evolution

Bowler, P. J., *Evolution: The History of an Idea*, 3rd edn. (Berkeley and Los Angeles, 2003).

Burrow, J., *Evolution and Society: A Study in Victorian Social Theory* (Cambridge, 1966).

Corsi, P., *The Age of Lamarck: Evolutionary Theories in France, 1790–1830* (Berkeley and Los Angeles, 1988).

Desmond, A., *The Politics of Evolution: Morphology, Medicine, and Reform in Radical London* (Chicago, 1989).

Larson, E. J., *Evolution: The Remarkable History of a Scientific Theory* (New York, 2004).

Radick, G., *The Simian Tongue: The Long Debate about Animal Language* (Chicago, 2007).

Richards, R. J., *Darwin and the Emergence of Evolutionary Theories of Mind and Behavior* (Chicago, 1987).

Ruse, M., *The Darwinian Revolution* (Chicago, 1979).

Secord, J., *Victorian Sensation: The Extraordinary Publication, Reception, and Secret Authorship of* Vestiges of the Natural History of Creation (Chicago, 2000).

Contexts

Bayly, C. A., *The Birth of the Modern World, 1780–1914* (Oxford, 2004).

Bowler, P. J., and Pickstone, J. V. (eds.), *The Cambridge History of Science*, vi. *The Modern Biological and Earth Sciences* (Cambridge, 2009).

Brooke, J. H., *Science and Religion: Some Historical Perspectives* (Cambridge, 1991).

Dixon, T., *Science and Religion: A Very Short Introduction* (Oxford, 2008).

Hazelwood, N., *Savage: The Life and Times of Jemmy Button* (New York, 2001).

Hilton, B., *A Mad, Bad, and Dangerous People? England 1783–1846* (Oxford, 2006).

Hoppen, K. T., *The Mid-Victorian Generation, 1846–1886* (Oxford, 1998).

Jardine, N., Spary, E. C., and Secord, J. A. (eds.), *Cultures of Natural History* (Cambridge, 1997).

Lightman, B., *Victorian Science in Context* (Chicago, 1997).

—— *Victorian Popularizers of Science: Designing Nature for New Audiences* (Chicago, 2007).

McKitterick, D. (ed.), *The Cambridge History of the Book in Britain*, vi. *1830–1914* (Cambridge, 2009).

Rudwick, M. J. S., *Worlds before Adam: The Reconstruction of Geohistory in the Age of Reform* (Chicago, 2008).

Russett, C. E., *Sexual Science: The Victorian Construction of Womanhood* (Cambridge, Mass., 1989).

Stocking, G. W., *Victorian Anthropology* (New York, 1987).

Topham, J. R., 'Scientific Publishing and the Reading of Science in Nineteenth-Century Britain: A Historiographical Survey and Guide to Sources', *Studies in History and Philosophy of Science*, 31 (2000), 559–612.

Waterman, A. M. C., *Revolution, Economics and Religion: Christian Political Economy, 1798–1833* (Cambridge, 1991).

White, P., *Thomas Huxley: Making the 'Man of Science'* (Cambridge, 2003).

Reception

Appleman, P. (ed.), *Darwin: A Norton Critical Edition*, 3rd edn. (New York, 2001).

Bowler, P., *The Non-Darwinian Revolution: Reinterpreting a Historical Myth* (Baltimore, 1988).

Cantor, G., and Swetlitz, M. (eds.), *Jewish Tradition and the Challenge of Darwinism* (Chicago, 2006).

Crook, P., *Darwinism, War and History: The Debate over the Biology of War from the 'Origin of Species' to the First World War* (Cambridge, 1994).

Ellegård, A., *Darwin and the General Reader: The Reception of Darwin's Theory of Evolution in the British Periodical Press, 1859–1872* (1958).

Elshakry, M., 'The Gospel of Science and American Evangelism in Late Ottoman Beirut', *Past and Present*, 196 (2007), 173–214.

Engels, E. M., and Glick, T. F. (eds.), *The Reception of Charles Darwin in Europe*, 2 vols. (*The Reception of British and Irish Authors in Europe*, series ed. E. Shaffer) (London, 2009).

Finney, C., *Paradise Revealed: Natural History in Nineteenth-Century Australia* (Melbourne, 1993).

Glick, T. F. (ed.), *The Comparative Reception of Darwinism* (Austin, Tex., 1974).

—— Puig-Samper, M. A., and Ruiz, R. (eds.), *The Reception of Darwinism in the Iberian World: Spain, Spanish America and Brazil* (Dordrecht, 2001).

Hawkins, M., *Social Darwinism in European and American Thought, 1860–1945: Nature as Model and Nature as Threat* (Cambridge, 1997).

Hull, D. L., *Darwin and his Critics: The Reception of Darwin's Theory of Evolution by the Scientific Community* (Chicago, 1973).

Kelly, A., *The Descent of Darwin: The Popularization of Darwinism in Germany, 1860–1914* (Chapel Hill, NC, 1981).

Kevles, D. J., *In the Name of Eugenics: Genetics and the Uses of Human Heredity* (Berkeley and Los Angeles, 1985).

Moore, J. R., *The Post-Darwinian Controversies: A Study of the Struggle to Come to Terms with Darwin in Great Britain and America, 1870–1900* (Cambridge, 1979).

Novoa, A., and Levine A., *¡Darwinistas! Evolution, Race, and Science in Nineteenth Century Argentina* (forthcoming).

Numbers, R. L., and Stenhouse J. (eds.), *Disseminating Darwinism: The Role of Place, Race, Religion and Gender* (Cambridge, 1999).

Pancaldi, G., *Darwin in Italy: Science across Cultural Frontiers* (Bloomington and Indianapolis, 1983).

Pusey, J. R., *China and Charles Darwin* (Cambridge, Mass., 1983).

Vucinich, A., *Darwin in Russian Thought* (Berkeley and Los Angeles, 1988).

Ziadat, A. A., *Western Science in the Arab World: The Impact of Darwinism, 1860–1930* (Basingstoke, Hampshire, 1986).

Implications

Dawkins, R., *The Selfish Gene* (Oxford, 1976).

Dembski, W., and Ruse M. (eds.), *Debating Design: From Darwin to DNA* (Cambridge, 2004).

Dennett, D., *Darwin's Dangerous Idea: Evolution and the Meanings of Life* (London, 1995).

Gayon, J., *Darwinism's Struggle for Survival* (Cambridge, 1998).

Kitcher, P., *Living with Darwin: Evolution, Design, and the Future of Faith* (Oxford, 2007).

Levine, G., *Darwin Loves You: Natural Selection and the Re-Enchantment of the World* (Princeton, 2006).

Lewens, T., *Darwin* (London, 2007).

Ruse, M., *Can a Darwinian Be a Christian? The Relationship between Science and Religion* (Cambridge, 2000).

Further Reading in Oxford World's Classics

Darwin, Charles, *On the Origin of Species*, ed. Gillian Beer.

Gaskell, Elizabeth, *Wives and Daughters*, ed. Angus Easson.

Gosse, Edmund, *Father and Son*, ed. Michael Newton.

Literature and Science in the Nineteenth Century, ed. Laura Otis.

Malthus, Thomas, *An Essay on the Principle of Population*, ed. Geoffrey Gilbert.

Mill, John Stuart, *On Liberty and Other Essays*, ed. John Gray.

Paley, William, *Natural Theology*, ed. Matthew D. Eddy and David Knight.

Travel Writing, 1700–1830: An Anthology, ed. Elizabeth A. Bohls and Ian Duncan.

A CHRONOLOGY OF CHARLES DARWIN

1809   Charles Robert Darwin born (12 Feb.), second son and fifth of six children, to Robert Waring Darwin, physician, and Susannah Wedgwood, daughter of the pottery manufacturer Josiah Wedgwood.

1815   Threat of French invasion ends after British victory at Waterloo.

1817   Attends Revd George Case's day school at Shrewsbury; death of mother.

1818–25   Boards at Shrewsbury School.

1825–7   Enters University of Edinburgh to study medicine; follows lectures on natural history, chemistry, and geology.

1828–31   Studies at Christ's College, Cambridge, for ministry; attends lectures on botany and geology.

1830–3   Charles Lyell's *Principles of Geology*.

1831   Graduates BA from Cambridge (Apr.).

1831–6   Companion to Captain Robert Fitzroy on HMS *Beagle*; visits Tierra del Fuego (1832–4); Galapagos (1835); and coral reefs of the Indian Ocean (1836).

1832   Reform Act enhances power of middle classes in Britain.

1833–40   Bridgewater Treatises bring natural theology up to date.

1836–7   In Cambridge working on *Beagle* collections.

1837–42   Lives in London and keeps private notebooks on transmutation of species and issues relating to man, mind, and materialism.

1838–43   Edits *The Zoology of H.M.S.* Beagle, with contributions by Richard Owen, John Gould, Thomas Bell, and others.

1838   Reads Thomas Malthus, *An Essay on the Principle of Population*, and elaborates principle of natural selection in following months.

1839   Marries Emma Wedgwood, his first cousin; 'Observations on the Parallel Roads of Glen Roy' published in *Philosophical Transactions of the Royal Society*; *Journal of Researches into the Geology and Natural History of the Various Countries Visited by H.M.S.* Beagle; after birth of his first son, takes notes on his development and expressions.

1842   *The Structure and Distribution of Coral Reefs*; writes manuscript sketch of species theory; moves to Down House in Kent, south of London.

1844   Completes manuscript of essay on species theory; anonymous evolutionary cosmology *Vestiges of the Natural History of Creation* widely discussed in Europe and America; *Geological Observations on the Volcanic Islands Visited during the Voyage of H.M.S. Beagle.*

1845   Revised edition of *Journal of Researches* published in Murray's Colonial and Home Library.

1846   *Geological Observations on South America.*

1848   Revolutions throughout continental Europe; death of father.

1851   Great Exhibition of the Works of All Nations in London.

1851–4 Barnacle monographs published.

1856   Begins long version of work on 'Natural Selection'.

1858   Papers by Darwin and Alfred Russel Wallace on natural selection read at Linnean Society (July).

1859   *On the Origin of Species by Means of Natural Selection* (Nov.).

1860   British Association for the Advancement of Science meets at Oxford.

1861–3 Richard Owen and Thomas Henry Huxley debate relation between ape and human.

1861–5 American Civil War.

1862   *On the Various Contrivances by which British and Foreign Orchids are Fertilised by Insects.*

1863   T. H. Huxley's *Man's Place in Nature and Other Anthropological Lectures.*

1865   *On the Movements and Habits of Climbing Plants.*

1867   Second Reform Act increases male suffrage in Britain.

1868   *The Variation of Animals and Plants under Domestication* (2 vols.); Ernst Haeckel's *Natürliche Schöpfungsgeschichte* ('Natural History of Creation').

1869   Francis Galton's *Hereditary Genius.*

1871   Revolutionary workers' uprising and commune in Paris; *The Descent of Man, and Selection in Relation to Sex* (2 vols.).

1872   Sixth edition of *Origin*, with final revisions; *The Expression of Emotions in Man and Animals.*

1875   *Insectivorous Plants.*

1876 Begins writing 'Recollections'; material added later during 1881; *The Effects of Cross and Self Fertilisation in the Vegetable Kingdom.*

1877 'A Biographical Sketch of an Infant' in *Mind*, based on early notes.

1880 *The Power of Movement in Plants.*

1881 *The Formation of Vegetable Mould, through the Action of Worms.*

1882 Darwin dies at Down (19 Apr.); buried in Westminster Abbey (26 Apr.).

1887 *The Life and Letters of Charles Darwin* (3 vols.) edited by Francis Darwin, including expurgated edition of the 'Recollections'.

1903 *More Letters of Charles Darwin* in two volumes, edited by Francis Darwin and A. C. Seward, with autobiographical fragment of 1838.

1909 International celebrations of the Darwin centennial.

JOURNAL OF RESEARCHES

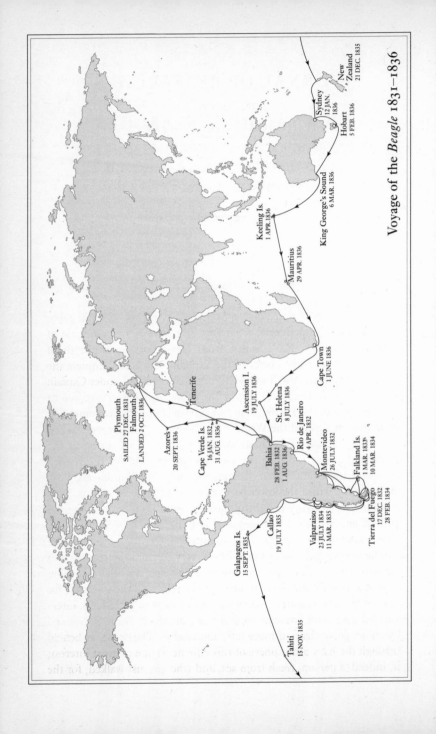

Voyage of the *Beagle* 1831–1836

Plymouth
SAILED 27 DEC. 1831
Falmouth
LANDED 2 OCT. 1836

Azores
20 SEPT. 1836

Tenerife

Cape Verde Is.
16 JAN. 1832
31 AUG. 1836

New Zealand
21 DEC. 1835

Sydney
12 JAN. 1836

Hobart
5 FEB. 1836

King George's Sound
6 MAR. 1836

Keeling Is.
1 APR. 1836

Mauritius
29 APR. 1836

Cape Town
1 JUNE 1836

Ascension I.
19 JULY 1836

St. Helena
8 JULY 1836

Bahia
28 FEB. 1832
1 AUG. 1836

Rio de Janeiro
4 APR. 1832

Montevideo
26 JULY 1832

Falkland Is.
1 MAR. 1833
10 MAR. 1834

Tierra del Fuego
17 DEC. 1832
28 FEB. 1834

Callao
19 JULY 1835

Valparaiso
23 JULY 1834
11 MAR. 1835

Galapagos Is.
15 SEPT. 1835

Tahiti
15 NOV. 1835

JOURNAL OF RESEARCHES

INTO THE
NATURAL HISTORY AND GEOLOGY OF THE
COUNTRIES VISITED DURING THE VOYAGE OF
H.M.S. BEAGLE ROUND THE WORLD,
UNDER THE COMMAND OF CAPT. FITZ ROY, R.N.

====

CHAPTER I

St. Jago — Cape de Verd Islands

AFTER having been twice driven back by heavy south-western gales, Her Majesty's ship Beagle, a ten-gun brig, under the command of Captain Fitz Roy, R.N., sailed from Devonport on the 27th of December, 1831. The object of the expedition was to complete the survey of Patagonia and Tierra del Fuego, commenced under Captain King in 1826 to 1830 — to survey the shores of Chile, Peru, and of some islands in the Pacific — and to carry a chain of chronometrical measurements round the World.* On the 6th of January we reached Teneriffe, but were prevented landing, by fears of our bringing the cholera: the next morning we saw the sun rise behind the rugged outline of the Grand Canary island, and suddenly illumine the Peak of Teneriffe, whilst the lower parts were veiled in fleecy clouds. This was the first of many delightful days never to be forgotten. On the 16th of January, 1832, we anchored at Porto Praya, in St. Jago, the chief island of the Cape de Verd archipelago.

The neighbourhood of Porto Praya, viewed from the sea, wears a desolate aspect. The volcanic fires of a past age, and the scorching heat of a tropical sun, have in most places rendered the soil unfit for vegetation. The country rises in successive steps of table-land, interspersed with some truncate conical hills, and the horizon is bounded by an irregular chain of more lofty mountains. The scene, as beheld through the hazy atmosphere of this climate, is one of great interest; if, indeed, a person, fresh from sea, and who has just walked, for the first time, in a grove of cocoa-nut trees, can be a judge of anything but his own happiness. The island would generally be considered as very uninteresting; but to any one accustomed only to an English landscape, the novel aspect of an utterly sterile land possesses a grandeur which more vegetation might spoil. A single green leaf can scarcely be discovered over wide tracts of the lava plains; yet flocks of goats, together with a few cows, contrive to exist. It rains very seldom, but during a short portion of the year heavy torrents fall, and immediately afterwards a light vegetation springs out of every crevice. This soon withers; and upon such naturally formed hay the animals live. It had not now rained for an entire year. When the island was discovered, the immediate neighbourhood of Porto Praya was clothed with trees, the reckless destruction of which has caused here, as at St. Helena, and at some of the Canary islands, almost entire sterility. The broad, flat-bottomed valleys, many of which serve during a few days only in the season as water courses, are clothed with thickets of leafless bushes. Few living creatures inhabit these valleys. The commonest bird is a kingfisher (Dacelo Iagoensis), which tamely sits on the branches of the castor-oil plant, and thence darts on grasshoppers and lizards. It is brightly coloured, but not so beautiful as the European species: in its flight, manners, and place of habitation, which is generally in the driest valley, there is also a wide difference.

One day, two of the officers and myself rode to Ribeira Grande, a village a few miles eastward of Porto Praya. Until we reached the valley of St. Martin, the country presented its usual dull brown appearance; but here, a very small rill of water produces a most refreshing margin of luxuriant vegetation. In the course of an hour we arrived at Ribeira Grande, and were surprised at the sight of a large ruined fort and cathedral. This little town, before its harbour was filled up, was the principal place in the island: it now presents a melancholy, but very picturesque appearance. Having procured a black Padre for a guide, and a Spaniard who had served in the Peninsular war* as an interpreter, we visited a collection of buildings, of which an ancient church formed the principal part. It is here the governors and captain-generals of the islands have been buried. Some of the tombstones recorded dates of the sixteenth century. The heraldic ornaments were the only things in this retired place that reminded us of Europe. The church or chapel formed one side of a quadrangle, in the middle of which a large clump of bananas were growing. On another side was a hospital, containing about a dozen miserable-looking inmates.

We returned to the Vênda* to eat our dinners. A considerable number of men, women, and children, all as black as jet, collected to watch us. Our companions were extremely merry; and everything we said or did was followed by their hearty laughter. Before leaving the town we visited the cathedral. It does not appear so rich as the smaller church, but boasts of a little organ, which sent forth singularly inharmonious cries. We presented the black priest with a few shillings, and the Spaniard, patting him on the head, said, with much candour, he thought his colour made no great difference. We then returned, as fast as the ponies would go, to Porto Praya.

Another day we rode to the village of St. Domingo, situated near the centre of the island. On a small plain which we crossed, a few stunted acacias were growing; their tops had been bent by the steady trade-wind, in a singular manner—some of them even at right angles to their trunks. The direction of the branches was exactly N.E. by N., and S.W. by S., and these natural vanes must indicate the prevailing direction of the force of the trade-wind. The travelling had made so little impression on the barren soil, that we here missed our track, and took that to Fuentes. This we did not find out till we arrived there; and we were afterwards glad of our mistake. Fuentes is a pretty village, with a small stream; and everything appeared to prosper well, excepting, indeed, that which ought to do so most—its inhabitants. The black children, completely naked, and looking very wretched, were carrying bundles of firewood half as big as their own bodies.

Near Fuentes we saw a large flock of guinea-fowl—probably fifty or sixty in number. They were extremely wary, and could not be approached. They avoided us, like partridges on a rainy day in September, running with their heads cocked up; and if pursued, they readily took to the wing.

The scenery of St. Domingo possesses a beauty totally unexpected, from the prevalent gloomy character of the rest of the island. The village is situated at the bottom of a valley, bounded by lofty and jagged walls of stratified lava. The black rocks afford a most striking contrast with the bright green vegetation, which follows the banks of a little stream of clear water. It happened to be a grand feast-day, and the village was full of people. On our return we overtook a party of about twenty young black girls, dressed in excellent taste; their black skins and snow-white linen being set off by coloured turbans and large shawls. As soon as we approached near, they suddenly all turned round, and covering the path with their shawls, sung with great energy a wild song, beating time with their hands upon their legs. We threw them some vintéms,* which were received with screams of laughter, and we left them redoubling the noise of their song.

One morning the view was singularly clear; the distant mountains being projected with the sharpest outline, on a heavy bank of dark blue clouds. Judging from the appearance, and from similar cases in England, I supposed that the air was saturated with moisture. The fact, however, turned out quite the contrary. The hygrometer* gave a difference of 29.6 degrees, between the temperature of the air, and the point at which dew was precipitated. This difference was nearly double that which I had observed on the previous mornings. This unusual degree of atmospheric dryness was accompanied by continual flashes of lightning. Is it not an uncommon case, thus to find a remarkable degree of aerial transparency with such a state of weather?

Generally the atmosphere is hazy; and this is caused by the falling of impalpably fine dust, which was found to have slightly injured the astronomical instruments. The morning before we anchored at Porto Praya, I collected a little packet of this brown-coloured fine dust, which appeared to have been filtered from the wind by the gauze of the vane at the mast-head. Mr. Lyell has also given me four packets of dust which fell on a vessel a few hundred miles northward of these islands. Professor Ehrenberg finds that this dust consists in great part of infusoria with siliceous shields,* and of the siliceous tissue of plants. In five little packets which I sent him, he has ascertained no less than sixty-seven different organic forms! The infusoria, with the exception of two marine species, are all inhabitants of fresh-water. I have found no less than fifteen different accounts of dust having fallen on vessels when far out in the Atlantic. From the direction of the wind whenever it has fallen, and from its having always fallen during those months when the harmattan* is known to raise clouds of dust high into the atmosphere, we may feel sure that it all comes from Africa. It is, however, a very singular fact, that, although Professor Ehrenberg knows many species of infusoria peculiar to Africa, he finds none of these in the dust which I sent him: on the other hand, he finds in it two species which hitherto he knows as living only in South America. The dust falls in such quantities as to dirty everything on board, and to hurt people's eyes; vessels even have run on shore owing to the obscurity of the atmosphere. It has often fallen on ships when several hundred, and even more than a thousand miles from the coast of Africa, and at points sixteen hundred miles distant in a north and south direction. In some dust which was collected on a vessel three hundred miles from the land, I was much surprised to find particles of stone above the thousandth of an inch square, mixed with finer matter. After this fact one need not be surprised at the diffusion of the far lighter and smaller sporules of cryptogamic plants.*

The geology of this island is the most interesting part of its natural history. On entering the harbour, a perfectly horizontal white band in the face of the sea cliff, may be seen running for some miles along the coast, and at the height of about forty-five feet above the water. Upon examination, this white stratum is found to consist of calcareous matter, with numerous shells embedded, most or all of which now exist on the neighbouring coast. It rests on ancient volcanic rocks, and has been covered by a stream of basalt, which must have entered the sea when the white shelly bed was lying at the bottom. It is interesting to trace the changes, produced by the heat of the overlying lava, on the friable mass, which in parts has been converted into a crystalline limestone, and in other parts into a compact spotted stone. Where the lime has been caught up by the scoriaceous* fragments of the lower surface of the stream, it is converted into groups of beautifully radiated fibres resembling arragonite.* The beds of lava rise in successive gently-sloping plains, towards the interior, whence the deluges of melted stone have originally proceeded. Within historical times, no signs of volcanic activity have, I believe, been manifested in any part of St. Jago. Even the form of a crater can but rarely be discovered on the summits of the many red cindery hills; yet the more recent streams can be distinguished on the coast, forming lines of cliffs of less height, but stretching out in advance of those belonging to an older series: the height of the cliffs thus affording a rude measure of the age of the streams.

During our stay, I observed the habits of some marine animals. A large Aplysia is very common. This sea-slug is about five inches long; and is of a dirty yellowish colour, veined with purple. On each side of the lower surface, or foot, there is a broad membrane, which appears sometimes to act as a ventilator, in causing a current of water to flow over the dorsal branchiæ or lungs. It feeds on the delicate sea-weeds which grow among the stones in muddy and shallow water; and I found in its stomach several small pebbles, as in the gizzard of a bird. This slug, when disturbed, emits a very fine purplish-red fluid, which stains the water for the space of a foot around. Besides this means of defence, an acrid secretion, which is spread over its body, causes a sharp, stinging sensation, similar to that produced by the Physalia, or Portuguese man-of-war.

I was much interested, on several occasions, by watching the habits of an Octopus, or cuttle-fish. Although common in the pools of water left by the retiring tide, these animals were not easily caught. By means of their long arms and suckers, they could drag their bodies into very narrow crevices; and when thus fixed, it required great force to remove them. At other times they darted tail first, with the rapidity of an arrow, from one side of the pool to the other, at the same instant discolouring the water with a dark chestnut-brown ink. These animals also escape detection by a very extraordinary, chameleon-like power of changing their colour. They appear to vary their tints according to the nature of the ground over which they pass: when in deep water, their general shade was brownish purple, but when placed on the land, or in shallow water, this dark tint changed into one of a yellowish green. The colour, examined more carefully, was a French grey, with numerous minute spots of bright yellow: the former of these varied in intensity; the latter entirely disappeared and appeared again by turns. These changes were effected in such a manner, that clouds, varying in tint between a hyacinth red and a chestnut brown, were continually passing over the body. Any part, being subjected to a slight shock of galvanism,* became almost black: a similar effect, but in a less degree, was produced by scratching the skin with a needle. These clouds, or blushes as they may be called, are said to be produced by the alternate expansion and contraction of minute vesicles containing variously coloured fluids.

This cuttle-fish displayed its chameleon-like power both during the act of swimming and whilst remaining stationary at the bottom. I was much amused by the various arts to escape detection used by one individual, which seemed fully aware that I was watching it. Remaining for a time motionless, it would then stealthily advance an inch or two, like a cat after a mouse; sometimes changing its colour: it thus proceeded, till having gained a deeper part, it darted away, leaving a dusky train of ink to hide the hole into which it had crawled.

While looking for marine animals, with my head about two feet above the rocky shore, I was more than once saluted by a jet of water, accompanied by a slight grating noise. At first I could not think what it was, but afterwards I found out that it was this cuttle-fish, which, though concealed in a hole, thus often led me to its discovery. That it possesses the power of ejecting water there is no doubt, and it appeared to me that it could certainly take good aim by directing the tube or siphon on the under side of its body. From the difficulty which these animals have in carrying their heads, they cannot crawl with ease when placed on the ground. I observed that one which I kept in the cabin was slightly phosphorescent in the dark.

[*After stopping at the geologically unusual St Paul's Rocks in the mid-Atlantic, described in the second half of Chapter I, the* Beagle *lands at Rio de Janeiro and spends two years charting the far southern and eastern coasts of South America. On inland excursions narrated in Chapters II–VIII, Darwin meets the gauchos of Argentina, learns to hunt the flightless birds called rheas, and laments the war of extermination against the native population. He arrives in Buenos Aires during a violent revolution.*]

CHAPTER VIII

Banda Oriental and Patagonia

[*Having been kept for two weeks as a virtual prisoner, Darwin escapes to Monte Video, which becomes his base for exploring the region of Banda Oriental, or Uruguay. He collects fossils of giant armadillos, ground sloths, and other extinct mammals, and in the chapter's final pages ponders their relation to species living today in the same places.*]

THE geology of Patagonia is interesting. Differently from Europe, where the tertiary formations* appear to have accumulated in bays, here along hundreds of miles of coast we have one great deposit, including many tertiary shells, all apparently extinct. The most common shell is a massive gigantic oyster, sometimes even a foot in diameter. These beds are covered by others of a peculiar soft white stone, including much gypsum, and resembling chalk, but really of a pumiceous* nature. It is highly remarkable, from being composed, to at least one-tenth part of its bulk, of Infusoria: Professor Ehrenberg has already ascertained in it thirty oceanic forms. This bed extends for 500 miles along the coast, and probably for a considerably greater distance. At Port St. Julian its thickness is more than 800 feet! These white beds are everywhere capped by a mass of gravel, forming probably one of the largest beds of shingle in the world: it certainly extends from near the Rio Colorado to between 600 and 700 nautical miles southward; at Santa Cruz (a river a little south of St. Julian), it reaches to the foot of the Cordillera; half way up the river, its thickness is more than 200 feet; it probably everywhere extends to this great chain, whence the well-rounded pebbles of porphyry* have been derived: we may consider its average breadth as 200 miles, and its average thickness as about 50 feet. If this great bed of pebbles, without including the mud necessarily derived from their attrition, was piled into a mound, it would form a great mountain chain! When we consider that all these pebbles, countless as the grains of sand in the desert, have been derived from the slow falling of masses of rock on the old coast-lines and banks of rivers; and that these fragments have been dashed into smaller pieces, and that each of them has since been slowly rolled, rounded, and far transported, the mind is stupified in thinking over the long, absolutely necessary, lapse of years. Yet all this gravel has been transported, and probably rounded, subsequently to the deposition of the white beds, and long subsequently to the underlying beds with the tertiary shells.

Everything in this southern continent has been effected on a grand scale: the land, from the Rio Plata to Tierra del Fuego, a distance of 1200 miles, has been raised in mass (and in Patagonia to a height of between 300 and 400 feet), within the period of the now existing sea-shells. The old and weathered shells left on the surface of the upraised plain still partially retain their colours. The uprising movement has been interrupted by at least eight long periods of rest, during which the sea ate deeply back into the land, forming at successive levels the long lines of cliffs or escarpments, which separate the different plains as they rise like steps one behind the other. The elevatory movement, and the eating-back power of the sea during the periods of rest, have been equable over long lines of coast; for I was astonished to find that the step-like plains stand at nearly corresponding heights at far distant points. The lowest plain is 90 feet high; and the highest, which I ascended near the coast, is 950 feet; and of this, only relics are left in the form of flat gravel-capped hills. The upper plain of S. Cruz slopes up to a height of 3000 feet at the foot of the Cordillera. I have said that within the period of existing sea-shells Patagonia has been upraised 300 to 400 feet: I may add, that within the period when icebergs transported boulders over the upper plain of Santa Cruz, the elevation has been at least 1500 feet. Nor has Patagonia been affected only by upward movements: the extinct tertiary shells from Port St. Julian and Santa Cruz cannot have lived, according to Professor E. Forbes, in a greater depth of water than from 40 to 250 feet; but they are now covered with sea-deposited strata from 800 to 1000 feet in thickness: hence the bed of the sea, on which these shells once lived, must have sunk downwards several hundred feet, to allow of the accumulation of the superincumbent strata. What a history of geological changes does the simply-constructed coast of Patagonia reveal!

At Port St. Julian, in some red mud capping the gravel on the 90-feet plain, I found half the skeleton of the Macrauchenia Patachonica, a remarkable quadruped, full as large as a camel. It belongs to the same division of the Pachydermata with the rhinoceros, tapir, and palæotherium; but in the structure of the bones of its long neck it shows a clear relation to the camel, or rather to the guanaco and llama.* From recent sea-shells being found on two of the higher step-formed plains, which must have been modelled and upraised before the mud was deposited in which the Macrauchenia was intombed, it is certain that this curious quadruped lived long after the sea was inhabited by its present shells. I was at first much surprised how a large quadruped could so lately have subsisted, in lat. 49° 15′, on these wretched gravel plains with their stunted vegetation; but the relationship of the Macrauchenia to the guanaco, now an inhabitant of the most sterile parts, partly explains this difficulty.

The relationship, though distant, between the Macrauchenia and the Guanaco, between the Toxodon and the Capybara,*—the closer relationship between the many extinct Edentata* and the living sloths, ant-eaters, and armadillos, now so eminently characteristic of South American zoology,—and the still closer relationship between the fossil and living species of Ctenomys and Hydrochærus,* are most interesting facts. This relationship is shown wonderfully—as wonderfully as between the fossil and extinct Marsupial animals of Australia—by the great collection lately brought to Europe from the caves of Brazil by MM. Lund and Clausen. In this collection there are extinct species of all the thirty-two genera, excepting four, of the terrestrial quadrupeds now inhabiting the provinces in which the caves occur; and the extinct species are much more numerous than those now living: there are fossil ant-eaters, armadillos, tapirs, peccaries, guanacos, opossums, and numerous South American gnawers and monkeys, and other animals. This wonderful relationship in the same continent between the dead and the living, will, I do not doubt, hereafter throw more light on the appearance of organic beings on our earth, and their disappearance from it, than any other class of facts.

It is impossible to reflect on the changed state of the American continent without the deepest astonishment. Formerly it must have swarmed with great monsters: now we find mere pigmies, compared with the antecedent, allied races. If Buffon had known of the gigantic sloth and armadillo-like animals, and of the lost Pachydermata, he might have said with a greater semblance of truth that the creative force in America had lost its power, rather than that it had never possessed great vigour. The greater number, if not all, of these extinct quadrupeds lived at a late period, and were the contemporaries of most of the existing sea-shells. Since they lived, no very great change in the form of the land can have taken place. What, then, has exterminated so many species and whole genera? The mind at first is irresistibly hurried into the belief of some great catastrophe; but thus to destroy animals, both large and small, in Southern Patagonia, in Brazil, on the Cordillera of Peru, in North America up to Behring's Straits, we must shake the entire framework of the globe. An examination, moreover, of the geology of La Plata and Patagonia, leads to the belief that all the features of the land result from slow and gradual changes. It appears from the character of the fossils in Europe, Asia, Australia, and in North and South America, that those conditions which favour the life of the *larger* quadrupeds were lately co-extensive with the world: what those conditions were, no one has yet even conjectured. It could hardly have been a change of temperature, which at about the same time destroyed the inhabitants of tropical, temperate, and arctic latitudes on both sides of the globe. In North America we positively know from Mr. Lyell, that the large quadrupeds lived subsequently to that period, when boulders were brought into latitudes at which icebergs now never arrive: from conclusive but indirect reasons we may feel sure, that in the southern hemisphere the Macrauchenia, also, lived long subsequently to the ice-transporting boulder-period. Did man, after his first inroad into South America, destroy, as has been suggested, the unwieldy Megatherium and the other Edentata? We must at least look to some other cause for the destruction of the little tucutuco at Bahia Blanca,* and of the many fossil mice and other small quadrupeds in Brazil. No one will imagine that a drought, even far severer than those which cause such losses in the provinces of La Plata, could destroy every individual of every species from Southern Patagonia to Behring's Straits. What shall we say of the extinction of the horse? Did those plains fail of pasture, which have since been overrun by thousands and hundreds of thousands of the descendants of the stock introduced by the Spaniards? Have the subsequently introduced species consumed the food of the great antecedent races? Can we believe that the Capybara has taken the food of the Toxodon, the Guanaco of the Macrauchenia, the existing small Edentata of their numerous gigantic prototypes? Certainly, no fact in the long history of the world is so startling as the wide and repeated exterminations of its inhabitants.

Nevertheless, if we consider the subject under another point of view, it will appear less perplexing. We do not steadily bear in mind, how profoundly ignorant we are of the conditions of existence of every animal; nor do we always remember, that some check is constantly preventing the too rapid increase of every organized being left in a state of nature. The supply of food, on an average, remains constant; yet the tendency in every animal to increase by propagation is geometrical; and its surprising effects have nowhere been more astonishingly shown, than in the case of the European animals run wild during the last few centuries in America. Every animal in a state of nature regularly breeds; yet in a species long established, any *great* increase in numbers is obviously impossible, and must be checked by some means. We are, nevertheless, seldom able with certainty to tell in any given species, at what period of life, or at what period of the year, or whether only at long intervals, the check falls; or, again, what is the precise nature of the check. Hence probably it is, that we feel so little surprise at one, of two species closely allied in habits, being rare and the other abundant in the same district; or, again, that one should be abundant in one district, and another, filling the same place in the economy of nature, should be abundant in a neighbouring district, differing very little in its conditions. If asked how this is, one immediately replies that it is determined by some slight difference in climate, food, or the number of enemies: yet how rarely, if ever, we can point out the precise cause and manner of action of the check! We are, therefore, driven to the conclusion, that causes generally quite inappreciable by us, determine whether a given species shall be abundant or scanty in numbers.

In the cases where we can trace the extinction of a species through man, either wholly or in one limited district, we know that it becomes rarer and rarer, and is then lost: it would be difficult to point out any just distinction between a species destroyed by man or by the increase of its natural enemies. The evidence of rarity preceding extinction, is more striking in the successive tertiary strata, as remarked by several able observers; it has often been found that a shell very common in a tertiary stratum is now most rare, and has even long been thought to be extinct. If then, as appears probable, species first become rare and then extinct—if the too rapid increase of every species, even the most favoured, is steadily checked, as we must admit, though how and when it is hard to say—and if we see, without the smallest surprise, though unable to assign the precise reason, one species abundant and another closely-allied species rare in the same district—why should we feel such great astonishment at the rarity being carried a step further to extinction? An action going on, on every side of us, and yet barely appreciable, might surely be carried a little further, without exciting our observation. Who would feel any great surprise at hearing that the Megalonyx* was formerly rare compared with the Megatherium, or that one of the fossil monkeys was few in number compared with one of the now living monkeys? and yet in this comparative rarity, we should have the plainest evidence of less favourable conditions for their existence. To admit that species generally become rare before they become extinct—to feel no surprise at the comparative rarity of one species with another, and yet to call in some extraordinary agent and to marvel greatly when a species ceases to exist, appears to me much the same as to admit that sickness in the individual is the prelude to death—to feel no surprise at sickness—but when the sick man dies, to wonder, and to believe that he died through violence.

[*In Chapter IX Darwin explores the southern part of Patagonia and discusses the geology, natural history, and politics of the Falkland Islands, condemning the local British population as 'runaway rebels and murderers'.*]

CHAPTER X
Tierra Del Fuego

December 17*th*, 1832. — Having now finished with Patagonia and the Falkland Islands, I will describe our first arrival in Tierra del Fuego. A little after noon we doubled Cape St. Diego, and entered the famous strait of Le Maire. We kept close to the Fuegian shore, but the outline of the rugged, inhospitable Staten-land was visible amidst the clouds. In the afternoon we anchored in the Bay of Good Success. While entering we were saluted in a manner becoming the inhabitants of this savage land. A group of Fuegians partly concealed by the entangled forest, were perched on a wild point overhanging the sea; and as we passed by, they sprang up and waving their tattered cloaks sent forth a loud and sonorous shout. The savages followed the ship, and just before dark we saw their fire, and again heard their wild cry. The harbour consists of a fine piece of water half surrounded by low rounded mountains of clay-slate, which are covered to the water's edge by one dense gloomy forest. A single glance at the landscape was sufficient to show me how widely different it was from any thing I had ever beheld. At night it blew a gale of wind, and heavy squalls from the mountains swept past us. It would have been a bad time out at sea, and we, as well as others, may call this Good Success Bay.

In the morning the Captain sent a party to communicate with the Fuegians. When we came within hail, one of the four natives who were present advanced to receive us, and began to shout most vehemently, wishing to direct us where to land. When we were on shore the party looked rather alarmed, but continued talking and making gestures with great rapidity. It was without exception the most curious and interesting spectacle I ever beheld: I could not have believed how wide was the difference between savage and civilized man: it is greater than between a wild and domesticated animal, inasmuch as in man there is a greater power of improvement. The chief spokesman was old, and appeared to be the head of the family; the three others were powerful young men, about six feet high. The women and children had been sent away. These Fuegians are a very different race from the stunted, miserable wretches farther westward; and they seem closely allied to the famous Patagonians of the Strait of Magellan.* Their only garment consists of a mantle made of guanaco skin, with the wool outside; this they wear just thrown over their shoulders, leaving their persons as often exposed as covered. Their skin is of a dirty coppery red colour.

The old man had a fillet of white feathers tied round his head, which partly confined his black, coarse, and entangled hair. His face was crossed by two broad transverse bars; one, painted bright red, reached from ear to ear and included the upper lip; the other, white like chalk, extended above and parallel to the first, so that even his eyelids were thus coloured. The other two men were ornamented by streaks of black powder, made of charcoal. The party altogether closely resembled the devils which come on the stage in plays like Der Freischutz.*

Their very attitudes were abject, and the expression of their countenances distrustful, surprised, and startled. After we had presented them with some scarlet cloth, which they immediately tied round their necks, they became good friends. This was shown by the old man patting our breasts, and making a chuckling kind of noise, as people do when feeding chickens. I walked with the old man, and this demonstration of friendship was repeated several times; it was concluded by three hard slaps, which were given me on the breast and back at the same time. He then bared his bosom for me to return the compliment, which being done, he seemed highly pleased. The language of these people, according to our notions, scarcely deserves to be called articulate.* Captain Cook has compared it to a man clearing his throat, but certainly no European ever cleared his throat with so many hoarse, guttural, and clicking sounds.

They are excellent mimics: as often as we coughed or yawned, or made any odd motion, they immediately imitated us. Some of our party began to squint and look awry; but one of the young Fuegians (whose whole face was painted black, excepting a white band across his eyes) succeeded in making far more hideous grimaces. They could repeat with perfect correctness each word in any sentence we addressed them, and they remembered such words for some time. Yet we Europeans all know how difficult it is to distinguish apart the sounds in a foreign language. Which of us, for instance, could follow an American Indian through a sentence of more than three words? All savages appear to possess, to an uncommon degree, this power of mimicry. I was told, almost in the same words, of the same ludicrous habit among the Caffres:* the Australians, likewise, have long been notorious for being able to imitate and describe the gait of any man, so that he may be recognised. How can this faculty be explained? is it a consequence of the more practised habits of perception and keener senses, common to all men in a savage state, as compared with those long civilized?

When a song was struck up by our party, I thought the Fuegians would have fallen down with astonishment. With equal surprise they viewed our dancing; but one of the young men, when asked, had no objection to a little waltzing. Little accustomed to Europeans as they appeared to be, yet they knew and dreaded our fire-arms; nothing would tempt them to take a gun in their hands. They begged for knives, calling them by the Spanish word "cuchilla." They explained also what they wanted, by acting as if they had a piece of blubber in their mouth, and then pretending to cut instead of tear it.

I have not as yet noticed the Fuegians whom we had on board. During the former voyage of the Adventure and Beagle in 1826 to 1830, Captain Fitz Roy seized on a party of natives, as hostages for the loss of a boat, which had been stolen, to the great jeopardy of a party employed on the survey; and some of these natives, as well as a child whom he bought for a pearl-button, he took with him to England, determining to educate them and instruct them in religion at his own expense. To settle these natives in their own country, was one chief inducement to Captain Fitz Roy to undertake our present voyage; and before the Admiralty had resolved to send out this expedition, Captain Fitz Roy had generously chartered a vessel, and would himself have taken them back. The natives were accompanied by a missionary, R. Matthews; of whom and of the natives, Captain Fitz Roy has published a full and excellent account. Two men, one of whom died in England of the small-pox, a boy and a little girl, were originally taken; and we had now on board, York Minster, Jemmy Button (whose name expresses his purchase-money), and Fuegia Basket. York Minster was a full-grown, short, thick, powerful man: his disposition was reserved, taciturn, morose, and when excited violently passionate; his affections were very strong towards a few friends on board; his intellect good. Jemmy Button was a universal favourite, but likewise passionate; the expression of his face at once showed his nice disposition. He was merry and often laughed, and was remarkably sympathetic with any one in pain: when the water was rough, I was often a little sea-sick, and he used to come to me and say in a plaintive voice, "Poor, poor fellow!" but the notion, after his aquatic life, of a man being sea-sick, was too ludicrous, and he was generally obliged to turn on one side to hide a smile or laugh, and then he would repeat his "Poor, poor fellow!" He was of a patriotic disposition; and he liked to praise his own tribe and country, in which he truly said there were "plenty of trees," and he abused all the other tribes: he stoutly declared that there was no Devil in his land. Jemmy was short, thick, and fat, but vain of his personal appearance; he used always to wear gloves, his hair was neatly cut, and he was distressed if his well-polished shoes were dirtied. He was fond of admiring himself in a looking-glass; and a merry-faced little Indian boy from the Rio Negro, whom we had for some months on board, soon perceived this, and used to mock him: Jemmy, who was always rather jealous of the attention paid to this little boy, did not at all like this, and used to say, with rather a contemptuous twist of his head, "Too much skylark." It seems yet wonderful to me, when I think over all his many good qualities, that he should have been of the same race, and doubtless partaken of the same character, with the miserable, degraded savages whom we first met here. Lastly, Fuegia Basket was a nice, modest, reserved young girl, with a rather pleasing but sometimes sullen expression, and very quick in learning anything, especially languages. This she showed in picking up some Portuguese and Spanish, when left on shore for only a short time at Rio de Janeiro and Monte Video, and in her knowledge of English. York Minster was very jealous of any attention paid to her; for it was clear he determined to marry her as soon as they were settled on shore.

Although all three could both speak and understand a good deal of English, it was singularly difficult to obtain much information from them, concerning the habits of their countrymen: this was partly owing to their apparent difficulty in understanding the simplest alternative. Every one accustomed to very young children, knows how seldom one can get an answer even to so simple a question as whether a thing is black *or* white; the idea of black or white seems alternately to fill their minds. So it was with these Fuegians, and hence it was generally impossible to find out, by cross-questioning, whether one had rightly understood anything which they had asserted. Their sight was remarkably acute: it is well known that sailors, from long practice, can make out a distant object much better than a landsman; but both York and Jemmy were much superior to any sailor on board: several times they have declared what some distant object has been, and though doubted by every one, they have proved right, when it has been examined through a telescope. They were quite conscious of this power; and Jemmy, when he had any little quarrel with the officer on watch, would say, "Me see ship, me no tell."

It was interesting to watch the conduct of the savages, when we landed, towards Jemmy Button: they immediately perceived the difference between him and ourselves, and held much conversation one with another on the subject. The old man addressed a long harangue to Jemmy, which it seems was to invite him to stay with them. But Jemmy understood very little of their language, and was, moreover, thoroughly ashamed of his countrymen. When York Minster afterwards came on shore, they noticed him in the same way, and told him he ought to shave; yet he had not twenty dwarf hairs on his face, whilst we all wore our untrimmed beards. They examined the colour of his skin, and compared it with ours. One of our arms being bared, they expressed the liveliest surprise and admiration at its whiteness, just in the same way in which I have seen the ourangoutang do at the Zoological Gardens. We thought that they mistook two or three of the officers, who were rather shorter and fairer, though adorned with large beards, for the ladies of our party. The tallest amongst the Fuegians was evidently much pleased at his height being noticed. When placed back to back with the tallest of the boat's crew, he tried his best to edge on higher ground, and to stand on tiptoe. He opened his mouth to show his teeth, and turned his face for a side view; and all this was done with such alacrity, that I dare say he thought himself the handsomest man in Tierra del Fuego. After our first feeling of grave astonishment was over, nothing could be more ludicrous than the odd mixture of surprise and imitation which these savages every moment exhibited.

The next day I attempted to penetrate some way into the country. Tierra del Fuego may be described as a mountainous land, partly submerged in the sea, so that deep inlets and bays occupy the place where valleys should exist. The mountain sides, except on the exposed western coast, are covered from the water's edge upwards by one great forest. The trees reach to an elevation of between 1000 and 1500 feet, and are succeeded by a band of peat, with minute alpine plants; and this again is succeeded by the line of perpetual snow, which, according to Captain King, in the Strait of Magellan descends to between 3000 and 4000 feet. To find an acre of level land in any part of the country is most rare. I recollect only one little flat piece near Port Famine, and another of rather larger extent near Goeree Road. In both places, and everywhere else, the surface is covered by a thick bed of swampy peat. Even within the forest, the ground is concealed by a mass of slowly putrefying vegetable matter, which, from being soaked with water, yields to the foot.

Finding it nearly hopeless to push my way through the wood, I followed the course of a mountain torrent. At first, from the waterfalls and number of dead trees, I could hardly crawl along; but the bed of the stream soon became a little more open, from the floods having swept the sides. I continued slowly to advance for an hour along the broken and rocky banks, and was amply repaid by the grandeur of the scene. The gloomy depth of the ravine well accorded with the universal signs of violence. On every side were lying irregular masses of rock and torn-up trees; other trees, though still erect, were decayed to the heart and ready to fall. The entangled mass of the thriving and the fallen reminded me of the forests within the tropics—yet there was a difference: for in these still solitudes, Death, instead of Life, seemed the predominant spirit. I followed the watercourse till I came to a spot, where a great slip had cleared a straight space down the mountain side. By this road I ascended to a considerable elevation, and obtained a good view of the surrounding woods. The trees all belong to one kind, the Fagus betuloides; for the number of the other species of Fagus and of the Winter's Bark, is quite inconsiderable. This beech keeps its leaves throughout the year; but its foliage is of a peculiar brownish-green colour, with a tinge of yellow. As the whole landscape is thus coloured, it has a sombre, dull appearance; nor is it often enlivened by the rays of the sun.

December 20th. —One side of the harbour is formed by a hill about 1500 feet high, which Captain Fitz Roy has called after Sir J. Banks, in commemoration of his disastrous excursion, which proved fatal to two men of his party, and nearly so to Dr. Solander. The snow-storm, which was the cause of their misfortune, happened in the middle of January, corresponding to our July, and in the latitude of Durham! I was anxious to reach the summit of this mountain to collect alpine plants; for flowers of any kind in the lower parts are few in number. We followed the same watercourse as on the previous day, till it dwindled away, and we were then compelled to crawl blindly among the trees. These, from the effects of the elevation and of the impetuous winds, were low, thick, and crooked. At length we reached that which from a distance appeared like a carpet of fine green turf, but which, to our vexation, turned out to be a compact mass of little beech-trees about four or five feet high. They were as thick together as box in the border of a garden, and we were obliged to struggle over the flat but treacherous surface. After a little more trouble we gained the peat, and then the bare slate rock.

A ridge connected this hill with another, distant some miles, and more lofty, so that patches of snow were lying on it. As the day was not far advanced, I determined to walk there and collect plants along the road. It would have been very hard work, had it not been for a well-beaten and straight path made by the guanacos; for these animals, like sheep, always follow the same line. When we reached the hill we found it the highest in the immediate neighbourhood, and the waters flowed to the sea in opposite directions. We obtained a wide view over the surrounding country: to the north a swampy moorland extended, but to the south we had a scene of savage magnificence, well becoming Tierra del Fuego. There was a degree of mysterious grandeur in mountain behind mountain, with the deep intervening valleys, all covered by one thick, dusky mass of forest. The atmosphere, likewise, in this climate, where gale succeeds gale, with rain, hail, and sleet, seems blacker than anywhere else. In the Strait of Magellan, looking due southward from Port Famine, the distant channels between the mountains appeared from their gloominess to lead beyond the confines of this world.

December 21st. — The Beagle got under way: and on the succeeding day, favoured to an uncommon degree by a fine easterly breeze, we closed in with the Barnevelts, and running past Cape Deceit with its stony peaks, about three o'clock doubled the weather-beaten Cape Horn. The evening was calm and bright, and we enjoyed a fine view of the surrounding isles. Cape Horn, however, demanded his tribute, and before night sent us a gale of wind directly in our teeth. We stood out to sea, and on the second day again made the land, when we saw on our weather-bow this notorious promontory in its proper form — veiled in a mist, and its dim outline surrounded by a storm of wind and water. Great black clouds were rolling across the heavens, and squalls of rain, with hail, swept by us with such extreme violence, that the Captain determined to run into Wigwam Cove. This is a snug little harbour, not far from Cape Horn; and here, at Christmas-eve, we anchored in smooth water. The only thing which reminded us of the gale outside, was every now and then a puff from the mountains, which made the ship surge at her anchors.

December 25th. — Close by the cove, a pointed hill, called Kater's Peak, rises to the height of 1700 feet. The surrounding islands all consist of conical masses of greenstone, associated sometimes with less regular hills of baked and altered clay-slate. This part of Tierra del Fuego may be considered as the extremity of the submerged chain of mountains already alluded to. The cove takes its name of "Wigwam" from some of the Fuegian habitations; but every bay in the neighbourhood might be so called with equal propriety. The inhabitants, living chiefly upon shell-fish, are obliged constantly to change their place of residence; but they return at intervals to the same spots, as is evident from the piles of old shells, which must often amount to many tons in weight. These heaps can be distinguished at a long distance by the bright green colour of certain plants, which invariably grow on them. Among these may be enumerated the wild celery and scurvy grass, two very serviceable plants, the use of which has not been discovered by the natives.

The Fuegian wigwam resembles, in size and dimensions, a haycock. It merely consists of a few broken branches stuck in the ground, and very imperfectly thatched on one side with a few tufts of grass and rushes. The whole cannot be the work of an hour, and it is only used for a few days. At Goeree Roads I saw a place where one of these naked men had slept, which absolutely offered no more cover than the form of a hare. The man was evidently living by himself, and York Minster said he was "very bad man," and that probably he had stolen something. On the west coast, however, the wigwams are rather better, for they are covered with seal-skins. We were de-tained here several days by the bad weather. The climate is certainly wretched: the summer solstice was now passed, yet every day snow fell on the hills, and in the valleys there was rain, accompanied by sleet.

The thermometer generally stood about 45°, but in the night fell to 38° or 40°. From the damp and boisterous state of the atmosphere, not cheered by a gleam of sunshine, one fancied the climate even worse than it really was.

While going one day on shore near Wollaston Island, we pulled alongside a canoe with six Fuegians. These were the most abject and miserable creatures I anywhere beheld. On the east coast the natives, as we have seen, have guanaco cloaks, and on the west, they possess seal-skins. Amongst these central tribes the men generally have an otter-skin, or some small scrap about as large as a pocket-handkerchief, which is barely sufficient to cover their backs as low down as their loins. It is laced across the breast by strings, and according as the wind blows, it is shifted from side to side. But these Fuegians in the canoe were quite naked, and even one full-grown woman was absolutely so. It was raining heavily, and the fresh water, together with the spray, trickled down her body. In another harbour not far distant, a woman, who was suckling a recently-born child, came one day alongside the vessel, and remained there out of mere curiosity, whilst the sleet fell and thawed on her naked bosom, and on the skin of her naked baby!* These poor wretches were stunted in their growth, their hideous faces bedaubed with white paint, their skins filthy and greasy, their hair entangled, their voices discordant, and their gestures violent. Viewing such men, one can hardly make oneself believe that they are fellow-creatures, and inhabitants of the same world. It is a common subject of conjecture what pleasure in life some of the lower animals can enjoy: how much more reasonably the same question may be asked with respect to these barbarians! At night, five or six human beings, naked and scarcely protected from the wind and rain of this tempestuous climate, sleep on the wet ground coiled up like animals. Whenever it is low water, winter or summer, night or day, they must rise to pick shellfish from the rocks; and the women either dive to collect sea-eggs, or sit patiently in their canoes, and with a baited hair-line without any hook, jerk out little fish. If a seal is killed, or the floating carcass of a putrid whale discovered, it is a feast; and such miserable food is assisted by a few tasteless berries and fungi.

They often suffer from famine: I heard Mr. Low, a sealing-master intimately acquainted with the natives of this country, give a curious account of the state of a party of one hundred and fifty natives on the west coast, who were very thin and in great distress. A succession of gales prevented the women from getting shell-fish on the rocks, and they could not go out in their canoes to catch seal. A small party of these men one morning set out, and the other Indians explained to him, that they were going a four days' journey for food: on their return, Low went to meet them, and he found them excessively tired, each man carrying a great square piece of putrid whales-blubber with a hole in the middle, through which they put their heads, like the Gauchos do through their ponchos or cloaks. As soon as the blubber was brought into a wigwam, an old man cut off thin slices, and muttering over them, broiled them for a minute, and distributed them to the famished party, who during this time preserved a profound silence. Mr. Low believes that whenever a whale is cast on shore, the natives bury large pieces of it in the sand, as a resource in time of famine; and a native boy, whom he had on board, once found a stock thus buried. The different tribes when at war are cannibals. From the concurrent, but quite independent evidence of the boy taken by Mr. Low, and of Jemmy Button, it is certainly true, that when pressed in winter by hunger, they kill and devour their old women before they kill their dogs: the boy, being asked by Mr. Low why they did this, answered, "Doggies catch otters, old women no." This boy described the manner in which they are killed by being held over smoke and thus choked; he imitated their screams as a joke, and described the parts of their bodies which are considered best to eat. Horrid as such a death by the hands of their friends and relatives must be, the fears of the old women, when hunger begins to press, are more painful to think of; we were told that they then often run away into the mountains, but that they are pursued by the men and brought back to the slaughter-house at their own fire-sides!*

Captain Fitz Roy could never ascertain that the Fuegians have any distinct belief in a future life. They sometimes bury their dead in caves, and sometimes in the mountain forests; we do not know what ceremonies they perform. Jemmy Button would not eat land-birds, because "eat dead men:" they are unwilling even to mention their dead friends. We have no reason to believe that they perform any sort of religious worship; though perhaps the muttering of the old man before he distributed the putrid blubber to his famished party, may be of this nature. Each family or tribe has a wizard or conjuring doctor,* whose office we could never clearly ascertain.

Jemmy believed in dreams, though not, as I have said, in the devil: I do not think that our Fuegians were much more superstitious than some of the sailors; for an old quarter-master firmly believed that the successive heavy gales, which we encountered off Cape Horn, were caused by our having the Fuegians on board. The nearest approach to a religious feeling which I heard of, was shown by York Minster, who, when Mr. Bynoe shot some very young ducklings as specimens, declared in the most solemn manner, "Oh Mr. Bynoe, much rain, snow, blow much." This was evidently a retributive punishment for wasting human food. In a wild and excited manner he also related, that his brother, one day whilst returning to pick up some dead birds which he had left on the coast, observed some feathers blown by the wind. His brother said (York imitating his manner), "What that?" and crawling onwards, he peeped over the cliff, and saw "wild man" picking his birds; he crawled a little nearer, and then hurled down a great stone and killed him. York declared for a long time afterwards storms raged, and much rain and snow fell. As far as we could make out, he seemed to consider the elements themselves as the avenging agents: it is evident in this case, how naturally, in a race a little more advanced in culture, the elements would become personified. What the "bad wild men" were, has always appeared to me most mysterious: from what York said, when we found the place like the form of a hare, where a single man had slept the night before, I should have thought that they were thieves who had been driven from their tribes; but other obscure speeches made me doubt this; I have sometimes imagined that the most probable explanation was that they were insane.

The different tribes have no government or chief; yet each is surrounded by other hostile tribes, speaking different dialects, and separated from each other only by a deserted border or neutral territory: the cause of their warfare appears to be the means of subsistence. Their country is a broken mass of wild rocks, lofty hills, and useless forests: and these are viewed through mists and endless storms. The habitable land is reduced to the stones on the beach; in search of food they are compelled unceasingly to wander from spot to spot, and so steep is the coast, that they can only move about in their wretched canoes. They cannot know the feeling of having a home, and still less that of domestic affection; for the husband is to the wife a brutal master to a laborious slave. Was a more horrid deed ever perpetrated, than that witnessed on the west coast by Byron, who saw a wretched mother pick up her bleeding dying infant-boy, whom her husband had mercilessly dashed on the stones for dropping a basket of sea-eggs! How little can the higher powers of the mind be brought into play: what is there for imagination to picture, for reason to compare, for judgment to decide upon? to knock a limpet from the rock does not require even cunning, that lowest power of the mind. Their skill in some respects may be compared to the instinct of animals; for it is not improved by experience: the canoe, their most ingenious work, poor as it is, has remained the same, as we know from Drake, for the last two hundred and fifty years.

Whilst beholding these savages, one asks, whence have they come? What could have tempted, or what change compelled a tribe of men, to leave the fine regions of the north, to travel down the Cordillera or backbone of America, to invent and build canoes, which are not used by the tribes of Chile, Peru, and Brazil, and then to enter on one of the most inhospitable countries within the limits of the globe? Although such reflections must at first seize on the mind, yet we may feel sure that they are partly erroneous. There is no reason to believe that the Fuegians decrease in number; therefore we must suppose that they enjoy a sufficient share of happiness, of whatever kind it may be, to render life worth having. Nature by making habit omnipotent, and its effects hereditary, has fitted the Fuegian to the climate and the productions of his miserable country.

After having been detained six days in Wigwam Cove by very bad weather, we put to sea on the 30th of December. Captain Fitz Roy wished to get westward to land York and Fuegia in their own country. When at sea we had a constant succession of gales, and the current was against us: we drifted to 57° 23′ south. On the 11th of January, 1833, by carrying a press of sail, we fetched within a few miles of the great rugged mountain of York Minster (so called by Captain Cook, and the origin of the name of the elder Fuegian), when a violent squall compelled us to shorten sail and stand out to sea. The surf was breaking fearfully on the coast, and the spray was carried over a cliff estimated at 200 feet in height. On the 12th the gale was very heavy, and we did not know exactly where we were: it was a most unpleasant sound to hear constantly repeated, "keep a good look-out to leeward." On the 13th the storm raged with its full fury: our horizon was narrowly limited by the sheets of spray borne by the wind. The sea looked ominous, like a dreary waving plain with patches of drifted snow: whilst the ship laboured heavily, the albatross glided with its expanded wings right up the wind. At noon a great sea broke over us, and filled one of the whale-boats, which was obliged to be instantly cut away. The poor Beagle trembled at the shock, and for a few minutes would not obey her helm; but soon, like a good ship that she was, she righted and came up to the wind again. Had another sea followed the first, our fate would have been decided soon, and for ever. We had now been twenty-four days trying in vain to get westward; the men were worn out with fatigue, and they had not had for many nights or days a dry thing to put on. Captain Fitz Roy gave up the attempt to get westward by the outside coast. In the evening we ran in behind False Cape Horn, and dropped our anchor in forty-seven fathoms, fire flashing from the windlass as the chain rushed round it. How delightful was that still night, after having been so long involved in the din of the warring elements!

January 15*th*, 1833. — The Beagle anchored in Goeree Roads. Captain Fitz Roy having resolved to settle the Fuegians, according to their wishes, in Ponsonby Sound, four boats were equipped to carry them there through the Beagle Channel. This channel, which was discovered by Captain Fitz Roy during the last voyage, is a most remarkable feature in the geography of this, or indeed of any other country: it may be compared to the valley of Lochness in Scotland, with its chain of lakes and friths. It is about one hundred and twenty miles long, with an average breadth, not subject to any very great variation, of about two miles; and is throughout the greater part so perfectly straight, that the view, bounded on each side by a line of mountains, gradually becomes indistinct in the long distance. It crosses the southern part of Tierra del Fuego in an east and west line, and in the middle is joined at right angles on the south side by an irregular channel, which has been called Ponsonby Sound. This is the residence of Jemmy Button's tribe and family.

19*th*. — Three whale-boats and the yawl, with a party of twenty-eight, started under the command of Captain Fitz Roy. In the afternoon we entered the eastern mouth of the channel, and shortly afterwards found a snug little cove concealed by some surrounding islets. Here we pitched our tents and lighted our fires. Nothing could look more comfortable than this scene. The glassy water of the little harbour, with the branches of the trees hanging over the rocky beach, the boats at anchor, the tents supported by the crossed oars, and the smoke curling up the wooded valley, formed a picture of quiet retirement. The next day (20th) we smoothly glided onwards in our little fleet, and came to a more inhabited district. Few if any of these natives could ever have seen a white man; certainly nothing could exceed their astonishment at the apparition of the four boats. Fires were lighted on every point (hence the name of Tierra del Fuego, or the land of fire), both to attract our attention and to spread far and wide the news. Some of the men ran for miles along the shore. I shall never forget how wild and savage one group appeared: suddenly four or five men came to the edge of an overhanging cliff; they were absolutely naked, and their long hair streamed about their faces; they held rugged staffs in their hands, and, springing from the ground, they waved their arms round their heads, and sent forth the most hideous yells.

At dinner-time we landed among a party of Fuegians. At first they were not inclined to be friendly; for until the Captain pulled in a-head of the other boats, they kept their slings in their hands. We soon, however, delighted them by trifling presents, such as tying red tape round their heads. They liked our biscuit: but one of the savages touched with his finger some of the meat preserved in tin cases which I was eating, and feeling it soft and cold, showed as much disgust at it, as I should have done at putrid blubber. Jemmy was thoroughly ashamed of his countrymen, and declared his own tribe were quite different, in which he was wofully mistaken. It was as easy to please as it was difficult to satisfy these savages. Young and old, men and children, never ceased repeating the word "yammer-schooner," which means "give me."* After pointing to almost every object, one after the other, even to the buttons on our coats, and saying their favourite word in as many intonations as possible, they would then use it in a neuter sense, and vacantly repeat "yammer-schooner." After yammerschoonering for any article very eagerly, they would by a simple artifice point to their young women or little children, as much as to say, "If you will not give it me, surely you will to such as these."

At night we endeavoured in vain to find an uninhabited cove; and at last were obliged to bivouac not far from a party of natives. They were very inoffensive as long as they were few in numbers, but in the morning (21st) being joined by others they showed symptoms of hostility, and we thought that we should have come to a skirmish. An European labours under great disadvantages when treating with savages like these, who have not the least idea of the power of fire-arms. In the very act of levelling his musket he appears to the savage far inferior to a man armed with a bow and arrow, a spear, or even a sling. Nor is it easy to teach them our superiority except by striking a fatal blow. Like wild beasts, they do not appear to compare numbers; for each individual, if attacked, instead of retiring, will endeavour to dash your brains out with a stone, as certainly as a tiger under similar circumstances would tear you. Captain Fitz Roy on one occasion being very anxious, from good reasons, to frighten away a small party, first flourished a cutlass near them, at which they only laughed; he then twice fired his pistol close to a native. The man both times looked astounded, and carefully but quickly rubbed his head; he then stared awhile, and gabbled to his companions, but he never seemed to think of running away. We can hardly put ourselves in the position of these savages, and understand their actions. In the case of this Fuegian, the possibility of such a sound as the report of a gun close to his ear could never have entered his mind. He perhaps literally did not for a second know whether it was a sound or a blow, and therefore very naturally rubbed his head. In a similar manner, when a savage sees a mark struck by a bullet, it may be some time before he is able at all to understand how it is effected; for the fact of a body being invisible from its velocity would perhaps be to him an idea totally inconceivable. Moreover, the extreme force of a bullet, that penetrates a hard substance without tearing it, may convince the savage that it has no force at all. Certainly I believe that many savages of the lowest grade, such as these of Tierra del Fuego, have seen objects struck, and even small animals killed by the musket, without being in the least aware how deadly an instrument it is.

22d.—After having passed an unmolested night, in what would appear to be neutral territory between Jemmy's tribe and the people whom we saw yesterday, we sailed pleasantly along. I do not know anything which shows more clearly the hostile state of the different tribes, than these wide border or neutral tracts. Although Jemmy Button well knew the force of our party, he was, at first, unwilling to land amidst the hostile tribe nearest to his own. He often told us how the savage Oens men* "when the leaf red," crossed the mountains from the eastern coast of Tierra del Fuego, and made inroads on the natives of this part of the country. It was most curious to watch him when thus talking, and see his eyes gleaming and his whole face assume a new and wild expression. As we proceeded along the Beagle Channel, the scenery assumed a peculiar and very magnificent character; but the effect was much lessened from the lowness of the point of view in a boat, and from looking along the valley, and thus losing all the beauty of a succession of ridges. The mountains were here about three thousand feet high, and terminated in sharp and jagged points. They rose in one unbroken sweep from the water's edge, and were covered to the height of fourteen or fifteen hundred feet by the dusky-coloured forest. It was most curious to observe, as far as the eye could range, how level and truly horizontal the line on the mountain side was, at which trees ceased to grow: it precisely resembled the high-water mark of drift-weed on a sea-beach.

At night we slept close to the junction of Ponsonby Sound with the Beagle Channel. A small family of Fuegians, who were living in the cove, were quiet and inoffensive, and soon joined our party round a blazing fire. We were well clothed, and though sitting close to the fire were far from too warm; yet these naked savages, though further off, were observed, to our great surprise, to be streaming with per-spiration at undergoing such a roasting. They seemed, however, very well pleased, and all joined in the chorus of the seamen's songs: but the manner in which they were invariably a little behindhand was quite ludicrous.

During the night the news had spread, and early in the morning (23d) a fresh party arrived, belonging to the Tekenika,* or Jemmy's tribe. Several of them had run so fast that their noses were bleeding, and their mouths frothed from the rapidity with which they talked; and with their naked bodies all bedaubed with black, white, and red, they looked like so many demoniacs who had been fighting. We then proceeded (accompanied by twelve canoes, each holding four or five people) down Ponsonby Sound to the spot where poor Jemmy expected to find his mother and relatives. He had already heard that his father was dead; but as he had had a "dream in his head" to that effect, he did not seem to care much about it, and repeatedly com-forted himself with the very natural reflection — "Me no help it." He was not able to learn any particulars regarding his father's death, as his relations would not speak about it.

Jemmy was now in a district well known to him, and guided the boats to a quiet pretty cove named Woollya, surrounded by islets, every one of which and every point had its proper native name. We found here a family of Jemmy's tribe, but not his relations: we made friends with them; and in the evening they sent a canoe to inform Jemmy's mother and brothers. The cove was bordered by some acres of good sloping land, not covered (as elsewhere) either by peat or by forest-trees. Captain Fitz Roy originally intended, as before stated, to have taken York Minster and Fuegia to their own tribe on the west coast; but as they expressed a wish to remain here, and as the spot was singularly favourable, Captain Fitz Roy determined to settle here the whole party, including Matthews, the missionary. Five days were spent in building for them three large wigwams, in landing their goods, in digging two gardens, and sowing seeds.

The next morning after our arrival (the 24th) the Fuegians began to pour in, and Jemmy's mother and brothers arrived. Jemmy recognised the stentorian voice of one of his brothers at a prodigious distance. The meeting was less interesting than that between a horse, turned out into a field, when he joins an old companion. There was no demonstration of affection; they simply stared for a short time at each other; and the mother immediately went to look after her canoe. We heard, however, through York that the mother had been inconsolable for the loss of Jemmy, and had searched everywhere for him, thinking that he might have been left after having been taken in the boat. The women took much notice of and were very kind to Fuegia. We had already perceived that Jemmy had almost forgotten his own language. I should think there was scarcely another human being with so small a stock of language, for his English was very imperfect. It was laughable, but almost pitiable, to hear him speak to his wild brother in English, and then ask him in Spanish ("no sabe?") whether he did not understand him.

Everything went on peaceably during the three next days, whilst the gardens were digging and wigwams building. We estimated the number of natives at about one hundred and twenty. The women worked hard, whilst the men lounged about all day long, watching us. They asked for everything they saw, and stole what they could. They were delighted at our dancing and singing, and were particularly interested at seeing us wash in a neighbouring brook; they did not pay much attention to anything else, not even to our boats.

Of all the things which York saw, during his absence from his country, nothing seems more to have astonished him than an ostrich, near Maldonado: breathless with astonishment he came running to Mr. Bynoe, with whom he was out walking—"Oh, Mr. Bynoe, oh, bird all same horse!" Much as our white skins surprised the natives, by Mr. Low's account a negro-cook to a sealing vessel, did so more effectually; and the poor fellow was so mobbed and shouted at that he would never go on shore again. Everything went on so quietly, that some of the officers and myself took long walks in the surrounding hills and woods. Suddenly, however, on the 27th, every woman and child disappeared. We were all uneasy at this, as neither York nor Jemmy could make out the cause. It was thought by some that they had been frightened by our cleaning and firing off our muskets on the previous evening: by others, that it was owing to offence taken by an old savage, who, when told to keep further off, had coolly spit in the sentry's face, and had then, by gestures acted over a sleeping Fuegian, plainly showed, as it was said, that he should like to cut up and eat our man. Captain Fitz Roy, to avoid the chance of an encounter, which would have been fatal to so many of the Fuegians, thought it advisable for us to sleep at a cove a few miles distant. Matthews, with his usual quiet fortitude (remarkable in a man apparently possessing little energy of character), determined to stay with the Fuegians, who evinced no alarm for themselves; and so we left them to pass their first awful night.

On our return in the morning (28th) we were delighted to find all quiet, and the men employed in their canoes spearing fish. Captain Fitz Roy determined to send the yawl and one whale-boat back to the ship; and to proceed with the two other boats, one under his own command (in which he most kindly allowed me to accompany him), and one under Mr. Hammond, to survey the western parts of the Beagle Channel, and afterwards to return and visit the settlement. The day to our astonishment was overpoweringly hot, so that our skins were scorched: with this beautiful weather, the view in the middle of the Beagle Channel was very remarkable. Looking towards either hand, no object intercepted the vanishing points of this long canal between the mountains. The circumstance of its being an arm of the sea was rendered very evident by several huge whales* spouting in different directions. On one occasion I saw two of these monsters, probably male and female, slowly swimming one after the other, within less than a stone's throw of the shore, over which the beech-tree extended its branches.

We sailed on till it was dark, and then pitched our tents in a quiet creek. The greatest luxury was to find for our beds a beach of pebbles, for they were dry and yielded to the body. Peaty soil is damp; rock is uneven and hard; sand gets into one's meat, when cooked and eaten boat-fashion; but when lying in our blanket-bags, on a good bed of smooth pebbles, we passed most comfortable nights.

It was my watch till one o'clock. There is something very solemn in these scenes. At no time does the consciousness in what a remote corner of the world you are then standing, come so strongly before the mind. Everything tends to this effect; the stillness of the night is interrupted only by the heavy breathing of the seamen beneath the tents, and sometimes by the cry of a night-bird. The occasional barking of a dog, heard in the distance, reminds one that it is the land of the savage.

January 29th. — Early in the morning we arrived at the point where the Beagle Channel divides into two arms; and we entered the northern one. The scenery here becomes even grander than before. The lofty mountains on the north side compose the granitic axis, or backbone of the country, and boldly rise to a height of between three and four thousand feet, with one peak above six thousand feet. They are covered by a wide mantle of perpetual snow, and numerous cascades pour their waters, through the woods, into the narrow channel below. In many parts, magnificent glaciers extend from the mountain side to the water's edge. It is scarcely possible to imagine any thing more beautiful than the beryl-like blue of these glaciers, and especially as contrasted with the dead white of the upper expanse of snow. The fragments which had fallen from the glacier into the water, were floating away, and the channel with its icebergs presented, for the space of a mile, a miniature likeness of the Polar Sea. The boats being hauled on shore at our dinner-hour, we were admiring from the distance of half a mile a perpendicular cliff of ice, and were wishing that some more fragments would fall. At last, down came a mass with a roaring noise, and immediately we saw the smooth outline of a wave travelling towards us. The men ran down as quickly as they could to the boats; for the chance of their being dashed to pieces was evident. One of the seamen just caught hold of the bows, as the curling breaker reached it: he was knocked over and over, but not hurt; and the boats, though thrice lifted on high and let fall again, received no damage.

This was most fortunate for us, for we were a hundred miles distant from the ship, and we should have been left without provisions or fire-arms. I had previously observed that some large fragments of rock on the beach had been lately displaced; but until seeing this wave, I did not understand the cause. One side of the creek was formed by a spur of mica-slate; the head by a cliff of ice about forty feet high; and the other side by a promontory fifty feet high, built up of huge rounded fragments of granite and mica-slate, out of which old trees were growing. This promontory was evidently a moraine, heaped up at a period when the glacier had greater dimensions.

When we reached the western mouth of this northern branch of the Beagle Channel, we sailed amongst many unknown desolate islands, and the weather was wretchedly bad. We met with no natives. The coast was almost everywhere so steep, that we had several times to pull many miles before we could find space enough to pitch our two tents: one night we slept on large round boulders, with putrefying sea-weed between them; and when the tide rose, we had to get up and move our blanket-bags. The farthest point westward which we reached was Stewart Island, a distance of about one hundred and fifty miles from our ship. We returned into the Beagle Channel by the southern arm, and thence proceeded, with no adventure, back to Ponsonby Sound.

February 6th.—We arrived at Woollya. Matthews gave so bad an account of the conduct of the Fuegians, that Captain Fitz Roy determined to take him back to the Beagle; and ultimately he was left at New Zealand, where his brother was a missionary. From the time of our leaving, a regular system of plunder commenced; fresh parties of the natives kept arriving: York and Jemmy lost many things, and Matthews almost every thing which had not been concealed underground. Every article seemed to have been torn up and divided by the natives. Matthews described the watch he was obliged always to keep as most harassing; night and day he was surrounded by the natives, who tried to tire him out by making an incessant noise close to his head. One day an old man, whom Matthews asked to leave his wigwam, immediately returned with a large stone in his hand: another day a whole party came armed with stones and stakes, and some of the younger men and Jemmy's brother were crying: Matthews met them with presents. Another party showed by signs that they wished to strip him naked and pluck all the hairs out of his face and body. I think we arrived just in time to save his life. Jemmy's relatives had been so vain and foolish, that they had showed to strangers their plunder, and their manner of obtaining it. It was quite melancholy leaving the three Fuegians with their savage countrymen; but it was a great comfort that they had no personal fears. York, being a powerful resolute man, was pretty sure to get on well, together with his wife Fuegia. Poor Jemmy looked rather disconsolate, and would then, I have little doubt, have been glad to have returned with us. His own brother had stolen many things from him; and as he remarked, 'what fashion call that:' he abused his countrymen, 'all bad men, no sabe (know) nothing' and, though I never heard him swear before, 'damned fools.' Our three Fuegians, though they had been only three years with civilized men, would, I am sure, have been glad to have retained their new habits; but this was obviously impossible. I fear it is more than doubtful, whether their visit will have been of any use to them.

In the evening, with Matthews on board, we made sail back to the ship, not by the Beagle Channel, but by the southern coast. The boats were heavily laden and the sea rough, and we had a dangerous passage. By the evening of the 7th we were on board the Beagle after an absence of twenty days, during which time we had gone three hundred miles in the open boats. On the 11th, Captain Fitz Roy paid a visit by himself to the Fuegians and found them going on well; and that they had lost very few more things.

On the last day of February in the succeeding year (1834), the Beagle anchored in a beautiful little cove at the eastern entrance of the Beagle Channel. Captain Fitz Roy determined on the bold, and as it proved successful, attempt to beat against the westerly winds by the same route, which we had followed in the boats to the settlement at Woollya. We did not see many natives until we were near Ponsonby Sound, where we were followed by ten or twelve canoes. The natives did not at all understand the reason of our tacking, and, instead of meeting us at each tack, vainly strove to follow us in our zig-zag course. I was amused at finding what a difference the circumstance of being quite superior in force made, in the interest of beholding these savages. While in the boats I got to hate the very sound of their voices, so much trouble did they give us. The first and last word was "yammerschooner." When, entering some quiet little cove, we have looked round and thought to pass a quiet night, the odious word "yammerschooner" has shrilly sounded from some gloomy nook, and then the little signal-smoke has curled up to spread the news far and wide. On leaving some place we have said to each other, 'Thank Heaven, we have at last fairly left these wretches!' when one more faint halloo from an all-powerful voice, heard at a prodigious distance, would reach our ears, and clearly could we distinguish—"yammerschooner." But now, the more Fuegians the merrier; and very merry work it was. Both parties laughing, wondering, gaping at each other; we pitying them, for giving us good fish and crabs for rags, &c.; they grasping at the chance of finding people so foolish as to exchange such splendid ornaments for a good supper. It was most amusing to see the undisguised smile of satisfaction with which one young woman with her face painted black, tied several bits of scarlet cloth round her head with rushes. Her husband, who enjoyed the very universal privilege in this country of possessing two wives, evidently became jealous of all the attention paid to his young wife; and, after a consultation with his naked beauties, was paddled away by them.

Some of the Fuegians plainly showed that they had a fair notion of barter. I gave one man a large nail (a most valuable present) without making any signs for a return; but he immediately picked out two fish, and handed them up on the point of his spear. If any present was designed for one canoe, and it fell near another, it was invariably given to the right owner. The Fuegian boy, whom Mr. Low had on board, showed, by going into the most violent passion, that he quite understood the reproach of being called a liar, which in truth he was. We were this time, as on all former occasions, much surprised at the little notice, or rather none whatever, which was taken of many things, the use of which must have been evident to the natives. Simple circumstances—such as the beauty of scarlet cloth or blue beads, the absence of women, our care in washing ourselves,—excited their admiration far more than any grand or complicated object, such as our ship. Bougainville has well remarked concerning these people, that they treat the "chef-d'œuvres de l'industrie humaine, comme ils traitent les loix de la nature et ses phénomènes."*

On the 5th of March, we anchored in the cove at Woollya, but we saw not a soul there. We were alarmed at this, for the natives in Ponsonby Sound showed by gestures, that there had been fighting; and we afterwards heard that the dreaded Oens men had made a descent. Soon a canoe, with a little flag flying, was seen approaching, with one of the men in it washing the paint off his face. This man was poor

Jemmy,—now a thin haggard savage, with long disordered hair, and naked, except a bit of a blanket round his waist. We did not recognise him till he was close to us; for he was ashamed of himself, and turned his back to the ship. We had left him plump, fat, clean, and well dressed;—I never saw so complete and grievous a change. As soon however as he was clothed, and the first flurry was over, things wore a good appearance. He dined with Captain Fitz Roy, and ate his dinner as tidily as formerly. He told us he had 'too much' (meaning enough) to eat, that he was not cold, that his relations were very good people, and that he did not wish to go back to England: in the evening we found out the cause of this great change in Jemmy's feelings, in the arrival of his young and nice-looking wife.* With his usual good feeling, he brought two beautiful otter-skins for two of his best friends, and some spear-heads and arrows made with his own hands for the Captain. He said he had built a canoe for himself, and he boasted that he could talk a little of his own language! But it is a most singular fact, that he appears to have taught all his tribe some English: an old man spontaneously announced 'Jemmy Button's wife.' Jemmy had lost all his property. He told us that York Minster had built a large canoe, and with his wife Fuegia,* had several months since gone to his own country, and had taken farewell by an act of consummate villainy; he persuaded Jemmy and his mother to come with him, and then on the way deserted them by night, stealing every article of their property.

Jemmy went to sleep on shore, and in the morning returned, and remained on board till the ship got under weigh, which frightened his wife, who continued crying violently till he got into his canoe. He returned loaded with valuable property. Every soul on board was heartily sorry to shake hands with him for the last time. I do not now doubt that he will be as happy as, perhaps happier than, if he had never left his own country. Every one must sincerely hope that Captain Fitz Roy's noble hope may be fulfilled, of being rewarded for the many generous sacrifices which he made for these Fuegians, by some shipwrecked sailor being protected by the descendants of Jemmy Button and his tribe! When Jemmy reached the shore, he lighted a signal fire, and the smoke curled up, bidding us a last and long farewell, as the ship stood on her course into the open sea.

The perfect equality among the individuals composing the Fuegian tribes, must for a long time retard their civilization. As we see those animals, whose instinct compels them to live in society and obey a chief, are most capable of improvement, so is it with the races of mankind. Whether we look at it as a cause or a consequence, the more civilized always have the most artificial governments. For instance, the inhabitants of Otaheite,* who, when first discovered, were governed by hereditary kings, had arrived at a far higher grade than another branch of the same people, the New Zealanders, — who, although benefited by being compelled to turn their attention to agriculture, were republicans in the most absolute sense. In Tierra del Fuego, until some chief shall arise with power sufficient to secure any acquired advantage, such as the domesticated animals, it seems scarcely possible that the political state of the country can be improved. At present, even a piece of cloth given to one is torn into shreds and distributed; and no one individual becomes richer than another. On the other hand, it is difficult to understand how a chief can arise till there is property of some sort by which he might manifest his superiority and increase his power.

I believe, in this extreme part of South America, man exists in a lower state of improvement than in any other part of the world. The South Sea Islanders of the two races inhabiting the Pacific, are comparatively civilized. The Esquimaux, in his subterranean hut, enjoys some of the comforts of life, and in his canoe, when fully equipped, manifests much skill. Some of the tribes of Southern Africa, prowling about in search of roots, and living concealed on the wild and arid plains, are sufficiently wretched. The Australian, in the simplicity of the arts of life, comes nearest the Fuegian: he can, however, boast of his boomerang, his spear and throwing-stick, his method of climbing trees, of tracking animals, and of hunting. Although the Australian may be superior in acquirements, it by no means follows that he is likewise superior in mental capacity: indeed, from what I saw of the Fuegians when on board, and from what I have read of the Australians, I should think the case was exactly the reverse.

[On the Beagle *as it charted the Straits of Magellan, Darwin reflects in Chapter XI on the climate of the Antarctic region and its unique flora and fauna. Turning up the western coast of South America, the ship anchors at the Chilean seaport of Valparaiso, which allows Darwin to explore the lower reaches of the Andes or Cordillera (Chapter XII). Doubling back to the south, the ship lands at Chiloe and the Chronos Islands, which offer further*

opportunities for observing and collecting (Chapter XIII). At Valdivia, Darwin witnesses an earthquake, and Chapter XIV describes in graphic detail the destruction at nearby Concepción. Travelling deeper into north-ern Chile and Peru, he is even more impressed by the forces which have elevated the mountains in geologically recent times. The Beagle leaves South America in early September 1835, and after a week lands on the Galapagos.]

CHAPTER XVII

Galapagos Archipelago

September 15*th.*—This archipelago consists of ten principal islands, of which five exceed the others in size. They are situated under the Equator, and between five and six hundred miles westward of the coast of America. They are all formed of volcanic rocks; a few fragments of granite curiously glazed and altered by the heat, can hardly be considered as an exception. Some of the craters, surmounting the larger islands, are of immense size, and they rise to a height of between

three and four thousand feet. Their flanks are studded by innumerable smaller orifices. I scarcely hesitate to affirm, that there must be in the whole archipelago at least two thousand craters. These consist either of lava and scoriæ, or of finely-stratified, sandstone-like tuff. Most of the latter are beautifully symmetrical; they owe their origin to eruptions of volcanic mud without any lava: it is a remarkable circumstance that every one of the twenty-eight tuff-craters which were examined, had their southern sides either much lower than the other sides, or quite broken down and removed. As all these craters apparently have been formed when standing in the sea, and as the waves from the trade wind and the swell from the open Pacific here unite their forces on the southern coasts of all the islands, this singular uniformity in the broken state of the craters, composed of the soft and yielding tuff, is easily explained.

Considering that these islands are placed directly under the equator, the climate is far from being excessively hot; this seems chiefly caused by the singularly low temperature of the surrounding water, brought here by the great southern Polar current. Excepting during one short season, very little rain falls, and even then it is irregular; but the clouds generally hang low. Hence, whilst the lower parts of the islands are very sterile, the upper parts, at a height of a thousand feet and upwards, possess a damp climate and a tolerably luxuriant vegetation. This is especially the case on the windward sides of the islands, which first receive and condense the moisture from the atmosphere.

In the morning (17th) we landed on Chatham Island, which, like the others, rises with a tame and rounded outline, broken here and there by scattered hillocks, the remains of former craters. Nothing could be less inviting than the first appearance. A broken field of black basaltic lava, thrown into the most rugged waves, and crossed by great fissures, is every where covered by stunted, sun-burnt brushwood, which shows little signs of life. The dry and parched surface, being heated by the noon-day sun, gave to the air a close and sultry feeling, like that from a stove: we fancied even that the bushes smelt unpleasantly. Although I diligently tried to collect as many plants as possible, I succeeded in getting very few; and such wretched-looking little weeds would have better become an arctic than an equatorial Flora. The brushwood appears, from a short distance, as leafless as our trees during winter; and it was some time before I discovered that not only almost every plant was now in full leaf, but that the greater number were in flower. The commonest bush is one of the Euphorbiaceæ:* an acacia and a great odd-looking cactus are the only trees which afford any shade. After the season of heavy rains, the islands are said to appear for a short time partially green. The volcanic island of Fernando Noronha, placed in many respects under nearly similar conditions, is the only other country where I have seen a vegetation at all like this of the Galapagos islands.

The Beagle sailed round Chatham Island, and anchored in several bays. One night I slept on shore on a part of the island, where black truncated cones were extraordinarily numerous: from one small eminence I counted sixty of them, all surmounted by craters more or less perfect. The greater number consisted merely of a ring of red scoriæ or slags, cemented together: and their height above the plain of lava was not more than from fifty to a hundred feet: none had been very lately active. The entire surface of this part of the island seems to have been permeated, like a sieve, by the subterranean vapours: here and there the lava, whilst soft, has been blown into great bubbles; and in other parts, the tops of caverns similarly formed have fallen in, leaving circular pits with steep sides. From the regular form of the many craters, they gave to the country an artificial appearance, which vividly reminded me of those parts of Staffordshire, where the great iron-foundries are most numerous. The day was glowing hot, and the scrambling over the rough surface and through the intricate thickets, was very fatiguing; but I was well repaid by the strange Cyclopean scene. As I was walking along I met two large tortoises, each of which must have weighed at least two hundred pounds: one was eating a piece of cactus, and as I approached, it stared at me and slowly stalked away; the other gave a deep hiss, and drew in its head. These huge reptiles, surrounded by the black lava, the leafless shrubs, and large cacti, seemed to my fancy like some antediluvian animals.* The few dull-coloured birds cared no more for me, than they did for the great tortoises.

23rd. — The Beagle proceeded to Charles Island. This archipelago has long been frequented, first by the Bucaniers, and latterly by whalers, but it is only within the last six years, that a small colony has been established here. The inhabitants are between two and three hundred in number: they are nearly all people of colour, who have been banished for political crimes from the Republic of the Equator, of which Quito is the capital. The settlement is placed about four and a half miles inland, and at a height probably of a thousand feet. In the first part of the road we passed through leafless thickets, as in Chatham Island. Higher up, the woods gradually became greener; and as soon as we crossed the ridge of the island, we were cooled by a fine southerly breeze, and our sight refreshed by a green and thriving vegetation. In this upper region coarse grasses and ferns abound; but there are no tree-ferns: I saw nowhere any member of the Palm family, which is the more singular, as 360 miles northward, Cocos Island takes its name from the number of cocoa-nuts. The houses are irregularly scattered over a flat space of ground, which is cultivated with sweet potatoes and bananas. It will not easily be imagined how pleasant the sight of black mud was to us, after having been so long accustomed to the parched soil of Peru and northern Chile. The inhabitants, although complaining of poverty, obtain, without much trouble, the means of subsistence. In the woods there are many wild pigs and goats; but the staple article of animal food is supplied by the tortoises. Their numbers have of course been greatly reduced in this island, but the people yet count on two days' hunting giving them food for the rest of the week. It is said that formerly single vessels have taken away as many as seven hundred, and that the ship's company of a frigate some years since brought down in one day two hundred tortoises to the beach.

September 29th. — We doubled the south-west extremity of Albemarle Island, and the next day were nearly becalmed between it and Narborough Island. Both are covered with immense deluges of black naked lava, which have flowed either over the rims of the great caldrons, like pitch over the rim of a pot in which it has been boiled, or have burst forth from smaller orifices on the flanks; in their descent they have spread over miles of the sea-coast. On both of these islands, eruptions are known to have taken place; and in Albemarle, we saw a small jet of smoke curling from the summit of one of the great craters. In the evening we anchored in Bank's Cove, in Albemarle Island. The next morning I went out walking. To the south of the broken tuff-crater, in which the Beagle was anchored, there was another beautifully symmetrical one of an elliptic form; its longer axis was a little less than a mile, and its depth about 500 feet. At its bottom there was a shallow lake, in the middle of which a tiny crater formed an islet. The day was overpoweringly hot, and the lake looked clear and blue: I hurried down the cindery slope, and choked with dust eagerly tasted the water — but, to my sorrow, I found it salt as brine.

The rocks on the coast abounded with great black lizards, between three and four feet long; and on the hills, an ugly yellowish-brown species was equally common. We saw many of this latter kind, some clumsily running out of our way, and others shuffling into their burrows. I shall presently describe in more detail the habits of both these reptiles. The whole of this northern part of Albemarle Island is miserably sterile.

October 8*th*.— We arrived at James Island: this island, as well as Charles Island, were long since thus named after our kings of the Stuart line. Mr. Bynoe, myself, and our servants were left here for a week, with provisions and a tent, whilst the Beagle went for water. We found here a party of Spaniards, who had been sent from Charles Island to dry fish, and to salt tortoise-meat. About six miles inland, and at the height of nearly 2000 feet, a hovel had been built in which two men lived, who were employed in catching tortoises, whilst the others were fishing on the coast. I paid this party two visits, and slept there one night. As in the other islands, the lower region was covered by nearly leafless bushes, but the trees were here of a larger growth than elsewhere, several being two feet and some even two feet nine inches in diameter. The upper region being kept damp by the clouds, supports a green and flourishing vegetation. So damp was the ground, that there were large beds of a coarse cyperus, in which great numbers of a very small water-rail lived and bred. While staying in this upper region, we lived entirely upon tortoise-meat: the breast-plate roasted (as the Gauchos do *carne con cuero*), with the flesh on it, is very good; and the young tortoises make excellent soup; but otherwise the meat to my taste is indifferent.

One day we accompanied a party of the Spaniards in their whale-boat to a salina, or lake from which salt is procured. After landing, we had a very rough walk over a rugged field of recent lava, which has almost surrounded a tuff-crater, at the bottom of which the salt-lake lies. The water is only three or four inches deep, and rests on a layer of beautifully crystallized, white salt. The lake is quite circular, and is fringed with a border of bright green succulent plants; the almost precipitous walls of the crater are clothed with wood, so that the scene was altogether both picturesque and curious. A few years since, the sailors belonging to a sealing-vessel murdered their captain in this quiet spot; and we saw his skull lying among the bushes.

During the greater part of our stay of a week, the sky was cloudless, and if the trade-wind failed for an hour, the heat became very oppressive. On two days, the thermometer within the tent stood for some hours at 93°; but in the open air, in the wind and sun, at only 85°. The sand was extremely hot; the thermometer placed in some of a brown colour immediately rose to 137°, and how much above that it would have risen, I do not know, for it was not graduated any higher. The black sand felt much hotter, so that even in thick boots it was quite disagreeable to walk over it.

The natural history of these islands is eminently curious, and well deserves attention. Most of the organic productions are aboriginal creations, found nowhere else; there is even a difference between the inhabitants of the different islands; yet all show a marked relationship with those of America, though separated from that continent by an open space of ocean, between 500 and 600 miles in width. The archipelago is a little world within itself, or rather a satellite attached to America, whence it has derived a few stray colonists, and has received the general character of its indigenous productions. Considering the small size of these islands, we feel the more astonished at the number of their aboriginal beings, and at their confined range. Seeing every height crowned with its crater, and the boundaries of most of the lava-streams still distinct, we are led to believe that within a period, geologically recent, the unbroken ocean was here spread out. Hence, both in space and time, we seem to be brought somewhat near to that great fact—that mystery of mysteries—the first appearance of new beings on this earth.

Of terrestrial mammals, there is only one which must be considered as indigenous, namely, a mouse (Mus Galapagoensis), and this is confined, as far as I could ascertain, to Chatham island, the most easterly island of the group. It belongs, as I am informed by Mr. Waterhouse, to a division of the family of mice characteristic of America. At James island, there is a rat sufficiently distinct from the common kind to have been named and described by Mr. Waterhouse; but as it belongs to the old-world division of the family, and as this island has been frequented by ships for the last hundred and fifty years, I can hardly doubt that this rat is merely a variety, produced by the new and peculiar climate, food, and soil, to which it has been subjected. Although no one has a right to speculate without distinct facts, yet even with respect to the Chatham island mouse, it should be borne in mind, that it may possibly be an American species imported here; for I have seen, in a most unfrequented part of the Pampas, a native mouse living in the roof of a newly-built hovel, and therefore its transportation in a vessel is not improbable: analogous facts have been observed by Dr. Richardson in North America.

Of land-birds I obtained twenty-six kinds, all peculiar to the group and found nowhere else, with the exception of one lark-like finch from North America (Dolichonyx oryzivorus), which ranges on that continent as far north as 54°, and generally frequents marshes. The other twenty-five birds consist, firstly, of a hawk, curiously intermediate in structure between a Buzzard and the American group of carrion-feeding Polybori;* and with these latter birds it agrees most closely in every habit and even tone of voice. Secondly, there are two owls, representing the short-eared and white barn-owls of Europe. Thirdly, a wren, three tyrant fly-catchers (two of them species of Pyrocephalus, one or both of which would be ranked by some ornithologists as only varieties), and a dove—all analogous to, but distinct from, American species. Fourthly, a swallow, which though differing from the Progne purpurea* of both Americas, only in being rather duller coloured, smaller, and slenderer, is considered by Mr. Gould as specifically distinct. Fifthly, there are three species of mocking-thrush—a form highly characteristic of America. The remaining land-birds form a most singular group of finches, related to each other in the structure of their beaks, short tails, form of body, and plumage: there are thirteen species, which Mr. Gould has divided into four subgroups. All these species are peculiar to this archipelago; and so is the whole group, with the exception of one species of the sub-group Cactornis, lately brought from Bow island, in the Low Archipelago. Of Cactornis, the two species may be often seen climbing about the flowers of the great cactus-trees; but all the other species of this group of finches, mingled together in flocks, feed on the dry and sterile ground of the lower districts. The males of all, or certainly of the greater number, are jet black; and the females (with perhaps one or two exceptions) are brown. The most curious fact is the perfect gradation in the size of the beaks in the different species of Geospiza, from one as large as that of a hawfinch to that of a chaffinch, and (if Mr. Gould is right in including his sub-group, Certhidea, in the main group), even to that of a warbler. The largest

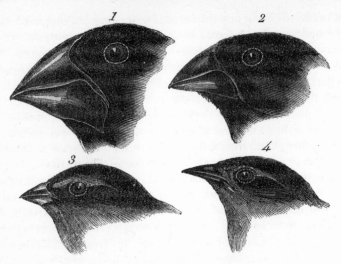

1. *Geospiza magnirostris.* 2. *Geospiza fortis.*
3. *Geospiza parvula.* 4. *Certhidea olivacea.*

beak in the genus Geospiza is shown in Fig. 1, and the smallest in Fig. 3; but instead of there being only one intermediate species, with a beak of the size shown in Fig. 2, there are no less than six species with insensibly graduated beaks. The beak of the sub-group Certhidea, is shown in Fig. 4. The beak of Cactornis is somewhat like that of a starling: and that of the fourth sub-group, Camarhynchus, is slightly parrot-shaped. Seeing this gradation and diversity of structure in one small, intimately related group of birds, one might really fancy that from an original paucity of birds in this archipelago, one species had been taken and modified for different ends. In a like manner it might be fancied that a bird originally a buzzard, had been induced here to undertake the office of the carrion-feeding Polybori of the American continent.

Of waders and water-birds I was able to get only eleven kinds, and of these only three (including a rail confined to the damp summits of the islands) are new species. Considering the wandering habits of the gulls, I was surprised to find that the species inhabiting these islands is peculiar, but allied to one from the southern parts of South America. The far greater peculiarity of the land-birds, namely, twenty-five out of twenty-six being new species or at least new races, compared with the waders and web-footed birds, is in accordance with the greater range which these latter orders have in all parts of the world. We shall hereafter see this law of aquatic forms, whether marine or fresh-water, being less peculiar at any given point of the earth's surface than the terrestrial forms of the same classes, strikingly illustrated in the shells, and in a lesser degree in the insects of this archipelago.

Two of the waders are rather smaller than the same species brought from other places: the swallow is also smaller, though it is doubtful whether or not it is distinct from its analogue. The two owls, the two tyrant fly-catchers (Pyrocephalus) and the dove, are also smaller than the analogous but distinct species, to which they are most nearly related; on the other hand, the gull is rather larger. The two owls, the swallow, all three species of mocking-thrush, the dove in its separate colours though not in its whole plumage, the Totanus,* and the gull, are likewise duskier coloured than their analogous species; and in the case of the mocking-thrush and Totanus, than any other species of the two genera. With the exception of a wren with a fine yellow breast, and of a tyrant fly-catcher with a scarlet tuft and breast, none of the birds are brilliantly coloured, as might have been expected in an equatorial district. Hence it would appear probable, that the same causes which here make the immigrants of some species smaller, make most of the peculiar Galapageian species also smaller, as well as very generally more dusky coloured. All the plants have a wretched, weedy appearance, and I did not see one beautiful flower. The insects, again, are small sized and dull coloured, and, as Mr. Waterhouse informs me, there is nothing in their general appearance which would have led him to imagine that they had come from under the equator. The birds, plants, and insects have a desert character, and are not more brilliantly coloured than those from southern Patagonia; we may, therefore, conclude that the usual gaudy colouring of the inter-tropical productions, is not related either to the heat or light of those zones, but to some other cause, perhaps to the conditions of existence being generally favourable to life.

We will now turn to the order of reptiles, which gives the most striking character to the zoology of these islands. The species are not numerous, but the numbers of individuals of each species are extraordinarily great. There is one small lizard belonging to a

South American genus, and two species (and probably more) of the Amblyrhynchus—a genus confined to the Galapagos islands. There is one snake which is numerous; it is identical, as I am informed by M. Bibron, with the Psammophis Temminckii from Chile. Of sea-turtle I believe there is more than one species; and of tortoises there are, as we shall presently show, two or three species or races. Of toads and frogs there are none: I was surprised at this, considering how well suited for them the temperate and damp upper woods appeared to be. It recalled to my mind the remark made by Bory St. Vincent, namely, that none of this family are found on any of the volcanic islands in the great oceans. As far as I can ascertain from various works, this seems to hold good throughout the Pacific, and even in the large islands of the Sandwich archipelago. Mauritius offers an apparent exception, where I saw the Rana Mascariensis in abundance: this frog is said now to inhabit the Seychelles, Madagascar, and Bourbon; but on the other hand, Du Bois, in his voyage in 1669, states that there were no reptiles in Bourbon except tortoises; and the Officier du Roi asserts that before 1768 it had been attempted, without success, to introduce frogs into Mauritius—I presume, for the purpose of eating: hence it may be well doubted whether this frog is an aboriginal of these islands. The absence of the frog family in the oceanic islands is the more remarkable, when contrasted with the case of lizards, which swarm on most of the smallest islands. May this difference not be caused, by the greater facility with which the eggs of lizards, protected by calcareous shells, might be transported through salt-water, than could the slimy spawn of frogs?

I will first describe the habits of the tortoise (Testudo nigra, formerly called Indica), which has been so frequently alluded to. These animals are found, I believe, on all the islands of the Archipelago; certainly on the greater number. They frequent in preference the high damp parts, but they likewise live in the lower and arid districts. I have already shown, from the numbers which have been caught in a single day, how very numerous they must be. Some grow to an immense size: Mr. Lawson, an Englishman, and vice-governor of the colony, told us that he had seen several so large, that it required six or eight men to lift them from the ground; and that some had afforded as much as two hundred pounds of meat. The old males are the largest, the females rarely growing to so great a size: the male can readily be distinguished from the female by the greater length of its tail.

The tortoises which live on those islands where there is no water, or in the lower and arid parts of the others, feed chiefly on the succulent cactus. Those which frequent the higher and damp regions, eat the leaves of various trees, a kind of berry (called guayavita)* which is acid and austere, and likewise a pale green filamentous lichen (Usnera plicata), that hangs in tresses from the boughs of the trees.

The tortoise is very fond of water, drinking large quantities, and wallowing in the mud. The larger islands alone possess springs, and these are always situated towards the central parts, and at a considerable height. The tortoises, therefore, which frequent the lower districts, when thirsty, are obliged to travel from a long distance. Hence broad and well-beaten paths branch off in every direction from the wells down to the sea-coast; and the Spaniards by following them up, first discovered the watering-places. When I landed at Chatham Island, I could not imagine what animal travelled so methodically along well-chosen tracks. Near the springs it was a curious spectacle to behold many of these huge creatures, one set eagerly travelling onwards with outstretched necks, and another set returning, after having drunk their fill. When the tortoise arrives at the spring, quite regardless of any spectator, he buries his head in the water above his eyes, and greedily swallows great mouthfulls, at the rate of about ten in a minute. The inhabitants say each animal stays three or four days in the neighbourhood of the water, and then returns to the lower country; but they differed respecting the frequency of these visits. The animal probably regulates them according to the nature of the food on which it has lived. It is, however, certain, that tortoises can subsist even on those islands, where there is no other water than what falls during a few rainy days in the year.

I believe it is well ascertained, that the bladder of the frog acts as a reservoir for the moisture necessary to its existence: such seems to be the case with the tortoise. For some time after a visit to the springs, their urinary bladders are distended with fluid, which is said gradually to decrease in volume, and to become less pure. The inhabitants, when walking in the lower district, and overcome with thirst, often take advantage of this circumstance, and drink the contents of the bladder if full: in one I saw killed, the fluid was quite limpid, and had only a very slightly bitter taste. The inhabitants, however, always first drink the water in the pericardium,* which is described as being best.

The tortoises, when purposely moving towards any point, travel by night and day, and arrive at their journey's end much sooner than would be expected. The inhabitants, from observing marked individuals, consider that they travel a distance of about eight miles in two or three days. One large tortoise, which I watched, walked at the rate of sixty yards in ten minutes, that is 360 yards in the hour, or four miles a day,—allowing a little time for it to eat on the road. During the breeding season, when the male and female are together, the male utters a hoarse roar or bellowing, which, it is said, can be heard at the distance of more than a hundred yards. The female never uses her voice, and the male only at these times; so that when the people hear this noise, they know that the two are together. They were at this time (October) laying their eggs. The female, where the soil is sandy, deposits them together, and covers them up with sand; but where the ground is rocky she drops them indiscriminately in any hole: Mr. Bynoe found seven placed in a fissure. The egg is white and spherical; one which I measured was seven inches and three-eighths in circumference, and therefore larger than a hen's egg. The young tortoises, as soon as they are hatched, fall a prey in great numbers to the carrion-feeding buzzard. The old ones seem generally to die from accidents, as from falling down precipices: at least, several of the inhabitants told me, that they had never found one dead without some evident cause.

The inhabitants believe that these animals are absolutely deaf; certainly they do not overhear a person walking close behind them. I was always amused when overtaking one of these great monsters, as it was quietly pacing along, to see how suddenly, the instant I passed, it would draw in its head and legs, and uttering a deep hiss fall to the ground with a heavy sound, as if struck dead. I frequently got on their backs, and then giving a few raps on the hinder part of their shells, they would rise up and walk away;—but I found it very difficult to keep my balance. The flesh of this animal is largely employed, both fresh and salted; and a beautifully clear oil is prepared from the fat. When a tortoise is caught, the man makes a slit in the skin near its tail, so as to see inside its body, whether the fat under the dorsal plate is thick. If it is not, the animal is liberated; and it is said to recover soon from this strange operation. In order to secure the tortoises, it is not sufficient to turn them like turtle, for they are often able to get on their legs again.

There can be little doubt that this tortoise is an aboriginal inhabitant of the Galapagos; for it is found on all, or nearly all, the islands, even on some of the smaller ones where there is no water; had it been an imported species, this would hardly have been the case in a group which has been so little frequented. Moreover, the old Bucaniers found this tortoise in greater numbers even than at present: Wood and Rogers* also, in 1708, say that it is the opinion of the Spaniards, that it is found nowhere else in this quarter of the world. It is now widely distributed; but it may be questioned whether it is in any other place an aboriginal. The bones of a tortoise at Mauritius, associated with those of the extinct Dodo, have generally been considered as belonging to this tortoise: if this had been so, undoubtedly it must have been there indigenous; but M. Bibron informs me that he believes that it was distinct, as the species now living there certainly is.

The Amblyrhynchus, a remarkable genus of lizards, is confined to this archipelago: there are two species, resembling each other in general form, one being terrestrial and the other aquatic. This latter species (A. cristatus) was first characterised by Mr. Bell, who well foresaw, from its short, broad head, and strong claws of equal length, that its habits of life would turn out very peculiar, and different from those of its nearest ally, the Iguana. It is extremely common on all the islands throughout the group, and lives exclusively on the rocky sea-beaches, being never found, at least I never saw one, even ten yards in-shore. It is a hideous-looking creature, of a dirty black colour, stupid, and sluggish in its movements. The usual length of a full-grown one is about a yard, but there are some even four feet long; a large one

Amblyrhynchus cristatus. a, Tooth of natural size, and likewise magnified.

weighed twenty pounds: on the island of Albemarle they seem to grow to a greater size than elsewhere. Their tails are flattened sideways, and all four feet partially webbed. They are occasionally seen some hundred yards from the shore, swimming about; and Captain Collnett, in his Voyage, says, "They go to sea in herds a-fishing, and sun themselves on the rocks; and may be called alligators in miniature." It must not, however, be supposed that they live on fish. When in the water this lizard swims with perfect ease and quickness, by a serpentine movement of its body and flattened tail—the legs being motionless and closely collapsed on its sides. A seaman on board sank one, with a heavy weight attached to it, thinking thus to kill it directly; but when, an hour afterwards, he drew up the line, it was quite active. Their limbs and strong claws are admirably adapted for crawling over the rugged and fissured masses of lava, which everywhere form the coast. In such situations, a group of six or seven of these hideous reptiles may oftentimes be seen on the black rocks, a few feet above the surf, basking in the sun with outstretched legs.

I opened the stomachs of several, and found them largely distended with minced sea-weed (Ulvæ), which grows in thin foliaceous expansions of a bright green or a dull red colour. I do not recollect having observed this sea-weed in any quantity on the tidal rocks; and I have reason to believe it grows at the bottom of the sea, at some little distance from the coast. If such be the case, the object of these animals occasionally going out to sea is explained. The stomach contained nothing but the sea-weed. Mr. Bynoe, however, found a piece of a crab in one; but this might have got in accidentally, in the same manner as I have seen a caterpillar, in the midst of some lichen, in the paunch of a tortoise. The intestines were large, as in other herbivorous animals. The nature of this lizard's food, as well as the structure of its tail and feet, and the fact of its having been seen voluntarily swimming out at sea, absolutely prove its aquatic habits; yet there is in this respect one strange anomaly, namely, that when frightened it will not enter the water. Hence it is easy to drive these lizards down to any little point overhanging the sea, where they will sooner allow a person to catch hold of their tails than jump into the water. They do not seem to have any notion of biting; but when much frightened they squirt a drop of fluid from each nostril. I threw one several times as far as I could, into a deep pool left by the retiring tide; but it invariably returned in a direct line to the spot where

I stood. It swam near the bottom, with a very graceful and rapid movement, and occasionally aided itself over the uneven ground with its feet. As soon as it arrived near the edge, but still being under water, it tried to conceal itself in the tufts of sea-weed, or it entered some crevice. As soon as it thought the danger was past, it crawled out on the dry rocks, and shuffled away as quickly as it could. I several times caught this same lizard, by driving it down to a point, and though possessed of such perfect powers of diving and swimming, nothing would induce it to enter the water; and as often as I threw it in, it returned in the manner above described. Perhaps this singular piece of apparent stupidity may be accounted for by the circumstance, that this reptile has no enemy whatever on shore, whereas at sea it must often fall a prey to the numerous sharks. Hence, probably, urged by a fixed and hereditary instinct that the shore is its place of safety, whatever the emergency may be, it there takes refuge.

During our visit (in October), I saw extremely few small individuals of this species, and none I should think under a year old. From this circumstance it seems probable that the breeding season had not then commenced. I asked several of the inhabitants if they knew where it laid its eggs: they said that they knew nothing of its propagation, although well acquainted with the eggs of the land kind—a fact, considering how very common this lizard is, not a little extraordinary.

We will now turn to the terrestrial species (A. Demarlii), with a round tail, and toes without webs. This lizard, instead of being found like the other on all the islands, is confined to the central part of the archipelago, namely to Albemarle, James, Barrington, and Indefatigable islands. To the southward, in Charles, Hood, and Chatham islands, and to the northward, in Towers, Bindloes, and Abingdon, I neither saw nor heard of any. It would appear as if it had been created in the centre of the archipelago, and thence had been dispersed only to a certain distance. Some of these lizards inhabit the high and damp parts of the islands, but they are much more numerous in the lower and sterile districts near the coast. I cannot give a more forcible proof of their numbers, than by stating that when we were left at James Island, we could not for some time find a spot free from their burrows on which to pitch our single tent. Like their brothers the sea-kind, they are ugly animals, of a yellowish orange beneath, and of a brownish red colour above: from their low facial angle they have a singularly stupid appearance. They are, perhaps, of a rather less size than the marine species; but several of them weighed between ten and fifteen pounds. In their movements they are lazy and half torpid. When not frightened, they slowly crawl along with their tails and bellies dragging on the ground. They often stop, and doze for a minute or two, with closed eyes and hind legs spread out on the parched soil.

They inhabit burrows, which they sometimes make between fragments of lava, but more generally on level patches of the soft sandstone-like tuff. The holes do not appear to be very deep, and they enter the ground at a small angle; so that when walking over these lizard-warrens, the soil is constantly giving way, much to the annoyance of the tired walker. This animal, when making its burrow, works alternately the opposite sides of its body. One front leg for a short time scratches up the soil, and throws it towards the hind foot, which is well placed so as to heave it beyond the mouth of the hole. That side of the body being tired, the other takes up the task, and so on alternately. I watched one for a long time, till half its body was buried; I then walked up and pulled it by the tail; at this it was greatly astonished, and soon shuffled up to see what was the matter; and then stared me in the face, as much as to say, "What made you pull my tail?"

They feed by day, and do not wander far from their burrows; if frightened, they rush to them with a most awkward gait. Except when running down hill, they cannot move very fast, apparently from the lateral position of their legs. They are not at all timorous: when attentively watching any one, they curl their tails, and, raising themselves on their front legs, nod their heads vertically, with a quick movement, and try to look very fierce: but in reality they are not at all so; if one just stamps on the ground, down go their tails, and off they shuffle as quickly as they can. I have frequently observed small fly-eating lizards, when watching anything, nod their heads in precisely the same manner; but I do not at all know for what purpose. If this Amblyrhynchus is held and plagued with a stick, it will bite it very severely; but I caught many by the tail, and they never tried to bite me. If two are placed on the ground and held together, they will fight, and bite each other till blood is drawn.

The individuals, and they are the greater number, which inhabit the lower country, can scarcely taste a drop of water throughout the year; but they consume much of the succulent cactus, the branches of which are occasionally broken off by the wind. I several times threw a piece to two or three of them when together; and it was amusing enough to see them trying to seize and carry it away in their mouths, like so many hungry dogs with a bone. They eat very deliberately, but do not chew their food. The little birds are aware how harmless these creatures are: I have seen one of the thick-billed finches picking at one end of a piece of cactus (which is much relished by all the animals of the lower region), whilst a lizard was eating at the other end; and afterwards the little bird with the utmost indifference hopped on the back of the reptile.

I opened the stomachs of several, and found them full of vegetable fibres and leaves of different trees, especially of an acacia. In the upper region they live chiefly on the acid and astringent berries of the guayavita, under which trees I have seen these lizards and the huge tortoises feeding together. To obtain the acacia-leaves they crawl up the low stunted trees; and it is not uncommon to see a pair quietly browsing, whilst seated on a branch several feet above the ground. These lizards, when cooked, yield a white meat, which is liked by those whose stomachs soar above all prejudices. Humboldt has remarked that in intertropical South America, all lizards which inhabit dry regions are esteemed delicacies for the table. The inhabitants state that those which inhabit the upper damp parts drink water, but that the others do not, like the tortoises, travel up for it from the lower sterile country. At the time of our visit, the females had within their bodies numerous, large, elongated eggs, which they lay in their burrows: the inhabitants seek them for food.

These two species of Amblyrhynchus agree, as I have already stated, in their general structure, and in many of their habits. Neither have that rapid movement, so characteristic of the genera Lacerta and Iguana. They are both herbivorous, although the kind of vegetation on which they feed is so very different. Mr. Bell has given the name to the genus from the shortness of the snout; indeed, the form of the mouth may almost be compared to that of the tortoise: one is led to suppose that this is an adaptation to their herbivorous appetites. It is very interesting thus to find a well-characterized genus, having its marine and terrestrial species, belonging to so confined a portion of the world. The aquatic species is by far the most remarkable, because it is the only existing lizard which lives on marine vegetable productions. As I at first observed, these islands are not so remarkable for the number of the species of reptiles, as for that of the individuals; when we remember the well-beaten paths made

by the thousands of huge tortoises—the many turtles—the great war-
rens of the terrestrial Amblyrhynchus—and the groups of the marine
species basking on the coast-rocks of every island—we must admit that
there is no other quarter of the world where this Order replaces the
herbivorous mammalia in so extraordinary a manner. The geologist on
hearing this will probably refer back in his mind to the Secondary
epochs,* when lizards, some herbivorous, some carnivorous, and of
dimensions comparable only with our existing whales, swarmed on the
land and in the sea. It is, therefore, worthy of his observation, that this
archipelago, instead of possessing a humid climate and rank vegetation,
cannot be considered otherwise than extremely arid, and, for an equatorial region, remarkably temperate.

To finish with the zoology: the fifteen kinds of sea-fish which
I procured here are all new species; they belong to twelve genera, all
widely distributed, with the exception of Prionotus, of which the
four previously known species live on the eastern side of America.*
Of land-shells I collected sixteen kinds (and two marked varieties), of
which, with the exception of one Helix found at Tahiti, all are peculiar to this archipelago: a single fresh-water shell (Paludina) is common to Tahiti and Van Diemen's Land. Mr. Cuming, before our
voyage, procured here ninety species of sea-shells, and this does not
include several species not yet specifically examined, of Trochus,
Turbo, Monodonta, and Nassa. He has been kind enough to give me
the following interesting results: of the ninety shells, no less than
forty-seven are unknown elsewhere—a wonderful fact, considering
how widely distributed sea-shells generally are. Of the forty-three
shells found in other parts of the world, twenty-five inhabit the
western coast of America, and of these eight are distinguishable as
varieties; the remaining eighteen (including one variety) were found
by Mr. Cuming in the Low archipelago, and some of them also at the
Philippines. This fact of shells from islands in the central parts of the
Pacific occurring here, deserves notice, for not one single sea-shell is
known to be common to the islands of that ocean and to the west
coast of America. The space of open sea running north and south off
the west coast, separates two quite distinct conchological provinces;
but at the Galapagos Archipelago we have a halting-place, where many
new forms have been created, and whither these two great conchological provinces have each sent several colonists. The American province has also sent here representative species; for there is a Galapageian

species of Monoceros, a genus only found on the west coast of America; and there are Galapageian species of Fissurella and Cancellaria, genera common on the west coast, but not found (as I am informed by Mr. Cuming) in the central islands of the Pacific. On the other hand, there are Galapageian species of Oniscia and Stylifer, genera common to the West Indies and to the Chinese and Indian seas, but not found either on the west coast of America or in the central Pacific. I may here add, that after the comparison by Messrs. Cuming and Hinds of about 2000 shells from the eastern and western coasts of America, only one single shell was found in common, namely, the Purpura patula, which inhabits the West Indies, the coast of Panama, and the Galapagos. We have, therefore, in this quarter of the world, three great conchological sea-provinces, quite distinct, though surprisingly near each other, being separated by long north and south spaces either of land or of open sea.

I took great pains in collecting the insects, but, excepting Tierra del Fuego, I never saw in this respect so poor a country. Even in the upper and damp region I procured very few, excepting some minute Diptera and Hymenoptera,* mostly of common mundane forms. As before remarked, the insects, for a tropical region, are of very small size and dull colours. Of beetles I collected twenty-five species (excluding a Dermestes and Corynetes imported, wherever a ship touches); of these, two belong to the Harpalidæ, two to the Hydrophilidæ, nine to three families of the Heteromera, and the remaining twelve to as many different families. This circumstance of insects (and I may add plants), where few in number, belonging to many different families, is, I believe, very general. Mr. Waterhouse, who has published an account of the insects of this archipelago, and to whom I am indebted for the above details, informs me that there are several new genera; and that of the genera not new, one or two are American, and the rest of mundane distribution. With the exception of a wood-feeding Apate, and of one or probably two water-beetles from the American continent, all the species appear to be new.

The botany of this group is fully as interesting as the zoology. Dr. J. Hooker will soon publish in the 'Linnean Transactions' a full account of the Flora, and I am much indebted to him for the following details. Of flowering plants there are, as far as at present is known, 185 species, and 40 cryptogamic species, making together 225; of this number I was fortunate enough to bring home 193. Of the flowering plants, 100 are new species, and are probably confined to this archipelago. Dr. Hooker conceives that, of the plants not so confined, at least 10 species found near the cultivated ground at Charles Island, have been imported. It is, I think, surprising that more American species have not been introduced naturally, considering that the distance is only between 500 and 600 miles from the continent; and that (according to Collnet, p. 58) drift-wood, bamboos, canes, and the nuts of a palm, are often washed on the south-eastern shores. The proportion of 100 flowering plants out of 185 (or 175 excluding the imported weeds) being new, is sufficient, I conceive, to make the Galapagos Archipelago a distinct botanical province; but this Flora is not nearly so peculiar as that of St. Helena, nor, as I am informed by Dr. Hooker, of Juan Fernandez. The peculiarity of the Galapageian Flora is best shown in certain families;—thus there are 21 species of Compositæ, of which 20 are peculiar to this archipelago; these belong to twelve genera, and of these genera no less than ten are confined to the archipelago! Dr. Hooker informs me that the Flora has an undoubted Western American character; nor can he detect in it any affinity with that of the Pacific. If, therefore, we except the eighteen marine, the one fresh-water, and one land-shell, which have apparently come here as colonists from the central islands of the Pacific, and likewise the one distinct Pacific species of the Galapageian group of finches, we see that this archipelago, though standing in the Pacific Ocean, is zoologically part of America.

If this character were owing merely to immigrants from America, there would be little remarkable in it; but we see that a vast majority of all the land animals, and that more than half of the flowering plants, are aboriginal productions. It was most striking to be surrounded by new birds, new reptiles, new shells, new insects, new plants, and yet by innumerable trifling details of structure, and even by the tones of voice and plumage of the birds, to have the temperate plains of Patagonia, or the hot dry deserts of Northern Chile, vividly brought before my eyes. Why, on these small points of land, which within a late geological period must have been covered by the ocean, which are formed of basaltic lava, and therefore differ in geological character from the American continent, and which are placed under a peculiar climate,—why were their aboriginal inhabitants, associated, I may add, in different proportions both in kind and number from those on the continent, and therefore acting on each other in a different manner — why were they created on American types of organization? It is probable that the islands of the Cape de Verd group resemble, in all their physical conditions, far more closely the Galapagos Islands than these latter physically resemble the coast of America; yet the aboriginal inhabitants of the two groups are totally unlike; those of the Cape de Verd Islands bearing the impress of Africa, as the inhabitants of the Galapagos Archipelago are stamped with that of America.

I have not as yet noticed by far the most remarkable feature in the natural history of this archipelago; it is, that the different islands to a considerable extent are inhabited by a different set of beings. My attention was first called to this fact by the Vice-Governor, Mr. Lawson, declaring that the tortoises differed from the different islands, and that he could with certainty tell from which island any one was brought. I did not for some time pay sufficient attention to this statement, and I had already partially mingled together the collections from two of the islands. I never dreamed that islands, about fifty or sixty miles apart, and most of them in sight of each other, formed of precisely the same rocks, placed under a quite similar climate, rising to a nearly equal height, would have been differently tenanted; but we shall soon see that this is the case. It is the fate of most voyagers, no sooner to discover what is most interesting in any locality, than they are hurried from it; but I ought, perhaps, to be thankful that I obtained sufficient materials to establish this most remarkable fact in the distribution of organic beings.

The inhabitants, as I have said, state that they can distinguish the tortoises from the different islands; and that they differ not only in size, but in other characters. Captain Porter has described those from Charles and from the nearest island to it, namely, Hood Island, as having their shells in front thick and turned up like a Spanish saddle, whilst the tortoises from James Island are rounder, blacker, and have a better taste when cooked. M. Bibron, moreover, informs me that he has seen what he considers two distinct species of tortoise from the Galapagos, but he does not know from which islands. The specimens that I brought from three islands were young ones; and probably owing to this cause, neither Mr. Gray nor myself could find in them any specific differences. I have remarked that the marine Amblyrhynchus was larger at Albemarle Island than elsewhere; and M. Bibron informs me that he has seen two distinct aquatic species of this genus; so that the different islands probably have their representative species or races of the Amblyrhynchus, as well as of the tortoise. My attention was first thoroughly aroused, by comparing together the numerous specimens, shot by myself and several other parties on board, of the mocking-thrushes, when, to my astonishment, I discovered that all those from Charles Island belonged to one species (Mimus trifasciatus); all from Albemarle Island to M. parvulus; and all from James and Chatham Islands (between which two other islands are situated, as connecting links) belonged to M. melanotis. These two latter species are closely allied, and would by some ornithologists be considered as only well-marked races or varieties; but the Mimus trifasciatus is very distinct. Unfortunately most of the specimens of the finch tribe were mingled together; but I have strong reasons to suspect that some of the species of the sub-group Geospiza are confined to separate islands. If the different islands have their representatives of Geospiza, it may help to explain the singularly large number of the species of this sub-group in this one small archipelago, and as a probable consequence of their numbers, the perfectly graduated series in the size of their beaks. Two species of the sub-group Cactornis, and two of Camarhynchus, were procured in the archipelago; and of the numerous specimens of these two sub-groups shot by four collectors at James Island, all were found to belong to one species of each; whereas the numerous specimens shot either on Chatham or Charles Island (for the two sets were mingled together) all belonged to the two other species: hence we may feel almost sure that these islands possess their representative species of these two sub-groups. In land-shells this law of distribution does not appear to hold good. In my very small collection of insects, Mr. Waterhouse remarks, that of those which were ticketed with their locality, not one was common to any two of the islands.

If we now turn to the Flora, we shall find the aboriginal plants of the different islands wonderfully different. I give all the following results on the high authority of my friend Dr. J. Hooker. I may premise that I indiscriminately collected everything in flower on the different islands, and fortunately kept my collections separate. Too much confidence, however, must not be placed in the proportional results, as the small collections brought home by some other naturalists, though in some respects confirming the results, plainly show that much

remains to be done in the botany of this group: the Leguminosæ,*
moreover, have as yet been only approximately worked out: —

Name of Island	Total No. of Species	No. of Species found in other parts of the world	No. of Species confined to the Galapagos Archipelago	No. confined to the one Island	No. of Species confined to the Galapagos Archipelago, but found on more than the one Island
James Island	71	33	38	30	8
Albemarle Island	46	18	26	22	4
Chatham Island	32	16	16	12	4
Charles Island	68	39 (or 29, if the probably imported plants be subtracted)	29	21	8

Hence we have the truly wonderful fact, that in James Island, of
the thirty-eight Galapageian plants, or those found in no other part
of the world, thirty are exclusively confined to this one island; and in
Albemarle Island, of the twenty-six aboriginal Galapageian plants,
twenty-two are confined to this one island, that is, only four are
at present known to grow in the other islands of the archipelago;
and so on, as shown in the above table, with the plants from Chatham
and Charles Islands. This fact will, perhaps, be rendered even more
striking, by giving a few illustrations: — thus, Scalesia, a remarkable
arborescent genus of the Compositæ,* is confined to the archipelago:
it has six species; one from Chatham, one from Albemarle, one from
Charles Island, two from James Island, and the sixth from one of the
three latter islands, but it is not known from which: not one of these
six species grows on any two islands. Again, Euphorbia, a mundane
or widely distributed genus, has here eight species, of which seven are
confined to the archipelago, and not one found on any two islands:
Acalypha and Borreria, both mundane genera, have respectively six

and seven species, none of which have the same species on two islands, with the exception of one Borreria, which does occur on two islands. The species of the Compositæ are particularly local; and Dr. Hooker has furnished me with several other most striking illustrations of the difference of the species on the different islands. He remarks that this law of distribution holds good both with those genera confined to the archipelago, and those distributed in other quarters of the world: in like manner we have seen that the different islands have their proper species of the mundane genus of tortoise, and of the widely distributed American genus of the mocking-thrush, as well as of two of the Galapageian sub-groups of finches, and almost certainly of the Galapageian genus Amblyrhynchus.

The distribution of the tenants of this archipelago would not be nearly so wonderful, if, for instance, one island had a mocking-thrush, and a second island some other quite distinct genus;—if one island had its genus of lizard, and a second island another distinct genus, or none whatever;—or if the different islands were inhabited, not by representative species of the same genera of plants, but by totally different genera, as does to a certain extent hold good; for, to give one instance, a large berry-bearing tree at James Island has no representative species in Charles Island. But it is the circumstance, that several of the islands possess their own species of the tortoise, mocking-thrush, finches, and numerous plants, these species having the same general habits, occupying analogous situations, and obviously filling the same place in the natural economy of this archipelago, that strikes me with wonder. It may be suspected that some of these representative species, at least in the case of the tortoise and of some of the birds, may hereafter prove to be only well-marked races; but this would be of equally great interest to the philosophical naturalist. I have said that most of the islands are in sight of each other: I may specify that Charles Island is fifty miles from the nearest part of Chatham Island, and thirty-three miles from the nearest part of Albemarle Island. Chatham Island is sixty miles from the nearest part of James Island, but there are two intermediate islands between them which were not visited by me. James Island is only ten miles from the nearest part of Albemarle Island, but the two points where the collections were made are thirty-two miles apart. I must repeat, that neither the nature of the soil, nor height of the land, nor the climate, nor the general character of the associated beings, and therefore their action one on another, can differ much in the different islands. If there be any sensible difference in their climates, it must be between the windward group (namely Charles and Chatham Islands), and that to leeward; but there seems to be no corresponding difference in the productions of these two halves of the archipelago.

The only light which I can throw on this remarkable difference in the inhabitants of the different islands, is, that very strong currents of the sea running in a westerly and W.N.W. direction must separate, as far as transportal by the sea is concerned, the southern islands from the northern ones; and between these northern islands a strong N.W. current was observed, which must effectually separate James and Albemarle Islands. As the archipelago is free to a most remarkable degree from gales of wind, neither the birds, insects, nor lighter seeds, would be blown from island to island. And lastly, the profound depth of the ocean between the islands, and their apparently recent (in a geological sense) volcanic origin, render it highly unlikely that they were ever united; and this, probably, is a far more important consideration than any other, with respect to the geographical distribution of their inhabitants. Reviewing the facts here given, one is astonished at the amount of creative force, if such an expression may be used, displayed on these small, barren, and rocky islands; and still more so, at its diverse yet analogous action on points so near each other. I have said that the Galapagos Archipelago might be called a satellite attached to America, but it should rather be called a group of satellites, physically similar, organically distinct, yet intimately related to each other, and all related in a marked, though much lesser degree, to the great American continent.

I will conclude my description of the natural history of these islands, by giving an account of the extreme tameness of the birds.

This disposition is common to all the terrestrial species; namely, to the mocking-thrushes, the finches, wrens, tyrant-flycatchers, the dove, and carrion-buzzard. All of them are often approached sufficiently near to be killed with a switch, and sometimes, as I myself tried, with a cap or hat. A gun is here almost superfluous; for with the muzzle I pushed a hawk off the branch of a tree. One day, whilst lying down, a mocking-thrush alighted on the edge of a pitcher, made of the shell of a tortoise, which I held in my hand, and began very quietly to sip the water; it allowed me to lift it from the ground whilst seated on the vessel: I often tried, and very nearly succeeded, in catching these birds by their legs. Formerly the birds appear to have been even tamer than at present. Cowley (in the year 1684) says that the "Turtle-doves were so tame, that they would often alight upon our hats and arms, so as that we could take them alive: they not fearing man, until such time as some of our company did fire at them, whereby they were rendered more shy." Dampier also, in the same year, says that a man in a morning's walk might kill six or seven dozen of these doves. At present, although certainly very tame, they do not alight on people's arms, nor do they suffer themselves to be killed in such large numbers. It is surprising that they have not become wilder; for these islands during the last hundred and fifty years have been frequently visited by bucaniers and whalers; and the sailors, wandering through the woods in search of tortoises, always take cruel delight in knocking down the little birds.

These birds, although now still more persecuted, do not readily become wild: in Charles Island, which had then been colonized about six years, I saw a boy sitting by a well with a switch in his hand, with which he killed the doves and finches as they came to drink. He had already procured a little heap of them for his dinner; and he said that he had constantly been in the habit of waiting by this well for the same purpose. It would appear that the birds of this archipelago, not having as yet learnt that man is a more dangerous animal than the tortoise or the Amblyrhynchus, disregard him, in the same manner as in England shy birds, such as magpies, disregard the cows and horses grazing in our fields.

The Falkland Islands offer a second instance of birds with a similar disposition. The extraordinary tameness of the little Opetiorhynchus* has been remarked by Pernety, Lesson, and other voyagers. It is not, however, peculiar to that bird: the Polyborus, snipe, upland and lowland goose, thrush, bunting, and even some true hawks, are all more or less tame. As the birds are so tame there, where foxes, hawks, and owls occur, we may infer that the absence of all rapacious animals at the Galapagos, is not the cause of their tameness here. The upland geese at the Falklands show, by the precaution they take in building on the islets, that they are aware of their danger from the foxes; but they are not by this rendered wild towards man. This tameness of the birds, especially of the waterfowl, is strongly contrasted with the habits of the same species in Tierra del Fuego, where for ages past they have been persecuted by the wild inhabitants. In the Falklands, the sportsman may sometimes kill more of the upland geese in one day than he can carry home; whereas in Tierra del Fuego, it is nearly as difficult to kill one, as it is in England to shoot the common wild goose.

In the time of Pernety (1763), all the birds there appear to have been much tamer than at present; he states that the Opetiorhynchus would almost perch on his finger; and that with a wand he killed ten in half an hour. At that period the birds must have been about as tame, as they now are at the Galapagos. They appear to have learnt caution more slowly at these latter islands than at the Falklands, where they have had proportionate means of experience; for besides frequent visits from vessels, those islands have been at intervals colonized during the entire period. Even formerly, when all the birds were so tame, it was impossible by Pernety's account to kill the black-necked swan — a bird of passage, which probably brought with it the wisdom learnt in foreign countries.

I may add that, according to Du Bois, all the birds at Bourbon in 1571–72, with the exception of the flamingoes and geese, were so extremely tame, that they could be caught by the hand, or killed in any number with a stick. Again, at Tristan d'Acunha in the Atlantic, Carmichael states that the only two land-birds, a thrush and a bunting, were "so tame as to suffer themselves to be caught with a hand-net." From these several facts we may, I think, conclude, first, that the wildness of birds with regard to man, is a particular instinct directed against *him*, and not dependent on any general degree of caution arising from other sources of danger; secondly, that it is not acquired by individual birds in a short time, even when much persecuted; but that in the course of successive generations it becomes hereditary. With domesticated animals we are accustomed to see new mental habits or instincts acquired and rendered hereditary; but with animals in a state of nature, it must always be most difficult to discover instances of acquired hereditary knowledge. In regard to the wildness of birds towards man, there is no way of accounting for it, except as an inherited habit: comparatively few young birds, in any one year, have been injured by man in England, yet almost all, even nestlings, are afraid of him; many individuals, on the other hand, both at the Galapagos and at the Falklands, have been pursued and injured by man, but yet have not learned a salutary dread of him. We may infer from these facts, what havoc the introduction of any new beast of prey must cause in a country, before the instincts of the indigenous inhabitants have become adapted to the stranger's craft or power.

[*Crossing the Pacific, Darwin compares in Chapter XVIII the civilized Tahitians with the perceived savagery of the inhabitants of New Zealand. In the following chapter the* Beagle *visits Australia, stopping at New South Wales, Tasmania, and King George's Sound. 'Farewell, Australia!', Darwin writes on leaving the penal colony; 'you are a rising child, and doubtless some day will reign a great princess in the South: but you are too great and ambitious for affection, yet not great enough for respect. I leave your shores without sorrow or regret'.*]

CHAPTER XX

Keeling Island: — Coral Formations

April 1st. [1836]—We arrived in view of the Keeling or Cocos Islands, situated in the Indian Ocean, and about six hundred miles distant from the coast of Sumatra. This is one of the lagoon-islands (or atolls) of coral formation, similar to those in the Low Archipelago which we passed near. When the ship was in the channel at the entrance, Mr. Liesk, an English resident, came off in his boat. The history of the inhabitants of this place, in as few words as possible, is as follows. About nine years ago, Mr. Hare, a worthless character, brought from the East Indian archipelago a number of Malay slaves, which now, including children, amount to more than a hundred. Shortly afterwards, Captain Ross, who had before visited these islands in his merchant-ship, arrived from England, bringing with him his family and goods for settlement: along with him came Mr. Liesk, who had been a mate in his vessel. The Malay slaves soon ran away from the islet on which Mr. Hare was settled, and joined Captain Ross's party. Mr. Hare upon this was ultimately obliged to leave the place.

The Malays are now nominally in a state of freedom, and certainly are so, as far as regards their personal treatment; but in most other points they are considered as slaves. From their discontented state, from the repeated removals from islet to islet, and perhaps also from a little mismanagement, things are not very prosperous. The island has no domestic quadruped, excepting the pig, and the main vegetable production is the cocoa-nut. The whole prosperity of the place depends on this tree: the only exports being oil from the nut, and the nuts themselves, which are taken to Singapore and Mauritius, where they are chiefly used, when grated, in making curries. On the cocoa-nut, also, the pigs, which are loaded with fat, almost entirely subsist, as do the ducks and poultry. Even a huge land-crab is furnished by nature with the means to open and feed on this most useful production.

The ring-formed reef of the lagoon-island is surmounted in the greater part of its length by linear islets. On the northern or leeward side, there is an opening through which vessels can pass to the anchorage within. On entering, the scene was very curious and rather pretty; its beauty, however, entirely depended on the brilliancy of the surrounding colours. The shallow, clear, and still water of the lagoon, resting in its greater part on white sand, is, when illumined by a vertical sun, of the most vivid green. This brilliant expanse, several miles in width, is on all sides divided, either by a line of snow-white breakers from the dark heaving waters of the ocean, or from the blue vault of heaven by the strips of land, crowned by the level tops of the cocoa-nut trees. As a white cloud here and there affords a pleasing contrast with the azure sky, so in the lagoon, bands of living coral darken the emerald green water.

[*Darwin describes the range of animals, plants, and people found on the islands. A few days later, he joins FitzRoy in visiting a Malay settlement.*]

After dinner we stayed to see a curious half superstitious scene acted by the Malay women. A large wooden spoon dressed in garments, and which had been carried to the grave of a dead man, they pretend becomes inspired at the full of the moon, and will dance and jump about. After the proper preparations, the spoon, held by two women, became convulsed, and danced in good time to the song of the surrounding children and women. It was a most foolish spectacle; but Mr. Liesk maintained that many of the Malays believed in its spiritual movements. The dance did not commence till the moon had risen, and it was well worth remaining to behold her bright orb so quietly shining through the long arms of the cocoa-nut trees as they waved in the evening breeze. These scenes of the tropics are in themselves so delicious, that they almost equal those dearer ones at home, to which we are bound by each best feeling of the mind.

The next day I employed myself in examining the very interesting, yet simple structure and origin of these islands. The water being unusually smooth, I waded over the outer flat of dead rock as far as the living mounds of coral, on which the swell of the open sea breaks. In some of the gullies and hollows there were beautiful green and other coloured fishes, and the forms and tints of many of the zoo-phytes were admirable. It is excusable to grow enthusiastic over the infinite numbers of organic beings with which the sea of the tropics, so prodigal of life, teems; yet I must confess I think those naturalists who have described, in well-known words, the submarine grottoes decked with a thousand beauties, have indulged in rather exuberant language.

April 6th. — I accompanied Captain Fitz Roy to an island at the head of the lagoon: the channel was exceedingly intricate, winding through fields of delicately branched corals. We saw several turtle, and two boats were then employed in catching them. The water was so clear and shallow, that although at first a turtle quickly dives out of sight, yet in a canoe or boat under sail, the pursuers after no very long chase come up to it. A man standing ready in the bow, at this moment dashes through the water upon the turtle's back; then cling-ing with both hands by the shell of its neck, he is carried away till the animal becomes exhausted and is secured. It was quite an interesting chase to see the two boats thus doubling about, and the men dashing head foremost into the water trying to seize their prey. Captain Moresby informs me that in the Chagos archipelago in this same ocean, the natives, by a horrible process, take the shell from the back of the living turtle. "It is covered with burning charcoal, which causes the outer shell to curl upwards; it is then forced off with a knife, and before it becomes cold flattened between boards. After this barbarous process the animal is suffered to regain its native element, where, after a certain time, a new shell is formed; it is, however, too thin to be of any service, and the animal always appears languishing and sickly."

When we arrived at the head of the lagoon, we crossed a narrow islet, and found a great surf breaking on the windward coast. I can hardly explain the reason, but there is to my mind much grandeur in the view of the outer shores of these lagoon-islands. There is a sim-plicity in the barrier-like beach, the margin of green bushes and tall cocoa-nuts, the solid flat of dead coral-rock, strewed here and there with great loose fragments, and the line of furious breakers, all rounding away towards either hand. The ocean throwing its waters over the broad reef appears an invincible, all-powerful enemy; yet we see it resisted, and even conquered, by means which at first seem most weak and inefficient. It is not that the ocean spares the rock of coral; the great fragments scattered over the reef, and heaped on the beach, whence the tall cocoa-nut springs, plainly bespeak the unrelenting power of the waves. Nor are any periods of repose granted. The long swell caused by the gentle but steady action of the trade-wind, always blowing in one direction over a wide area, causes breakers, almost equalling in force those during a gale of wind in the temperate regions, and which never cease to rage. It is impossible to behold these waves without feeling a conviction that an island, though built of the hardest rock, let it be porphyry, granite, or quartz, would ultimately yield and be demolished by such an irresistible power. Yet these low, insignificant coral-islets stand and are victorious: for here another power, as an antagonist, takes part in the contest. The organic forces separate the atoms of carbonate of lime, one by one, from the foaming breakers, and unite them into a symmetrical structure. Let the hurricane tear up its thousand huge fragments; yet what will that tell against the accumulated labour of myriads of architects at work night and day, month after month? Thus do we see the soft and gelatinous body of a polypus, through the agency of the vital laws, conquering the great mechanical power of the waves of an ocean which neither the art of man nor the inanimate works of nature could successfully resist.

We did not return on board till late in the evening, for we staid a long time in the lagoon, examining the fields of coral and the gigantic shells of the chama,* into which, if a man were to put his hand, he would not, as long as the animal lived, be able to withdraw it. Near the head of the lagoon, I was much surprised to find a wide area, considerably more than a mile square, covered with a forest of delicately branching corals, which, though standing upright, were all dead and rotten. At first I was quite at a loss to understand the cause; afterwards it occurred to me that it was owing to the following rather curious combination of circumstances. It should, however, first be stated, that corals are not able to survive even a short exposure in the air to the sun's rays, so that their upward limit of growth is determined by that of lowest water at spring tides. It appears, from some old charts, that the long island to windward was formerly separated by wide channels into several islets; this fact is likewise indicated by the trees being younger on these portions. Under the former condition of the reef, a strong breeze, by throwing more water over the barrier, would tend to raise the level of the lagoon. Now it acts in a directly contrary manner; for the water within the lagoon not only is not increased by currents from the outside, but is itself blown outwards by the force of the wind. Hence it is observed, that the tide near the head of the lagoon does not rise so high during a strong breeze as it does when it is calm. This difference of level, although no doubt very small, has, I believe, caused the death of those coral-groves, which under the former and more open condition of the outer reef had attained the utmost possible limit of upward growth.

[*After discussing the transport of non-coralline rocks to the islands in the roots of trees, Darwin gives further details of the zoology and botany of the islands.*]

April 12*th.* — In the morning we stood out of the lagoon on our passage to the Isle of France. I am glad we have visited these islands: such formations surely rank high amongst the wonderful objects of this world. Captain Fitz Roy found no bottom with a line 7200 feet in length, at the distance of only 2200 yards from the shore; hence this island forms a lofty submarine mountain, with sides steeper even than those of the most abrupt volcanic cone. The saucer-shaped summit is nearly ten miles across; and every single atom, from the least particle to the largest fragment of rock, in this great pile, which however is small compared with very many other lagoon-islands, bears the stamp of having been subjected to organic arrangement. We feel surprise when travellers tell us of the vast dimensions of the Pyramids and other great ruins, but how utterly insignificant are the greatest of these, when compared to these mountains of stone accumulated by the agency of various minute and tender animals! This is a wonder which does not at first strike the eye of the body, but, after reflection, the eye of reason.

I will now give a very brief account of the three great classes of coral-reefs; namely, Atolls, Barrier, and Fringing-reefs, and will explain my views on their formation. Almost every voyager who has crossed the Pacific has expressed his unbounded astonishment at the lagoon-islands, or as I shall for the future call them by their Indian name of atolls, and has attempted some explanation. Even as long ago as the year 1605, Pyrard de Laval well exclaimed, "C'est une meruille de voir chacun de ces atollons, enuironné d'un grand banc de pierre tout autour, n'y ayant point d'artifice humain."* The accompanying sketch of Whitsunday Island in the Pacific, copied from Capt. Beechey's admirable Voyage, gives but a faint idea of the singular aspect of an atoll: it is one of the smallest size, and has its narrow islets united together in a ring. The immensity of the ocean, the fury of the breakers, contrasted with the lowness of the land and the smoothness of the bright green water within the lagoon, can hardly be imagined without having been seen.

The earlier voyagers fancied that the coral-building animals instinctively built up their great circles to afford themselves protection in the inner parts; but so far is this from the truth, that those massive kinds, to whose growth on the exposed outer shores the very existence of the reef depends, cannot live within the lagoon, where other delicately-branching kinds flourish. Moreover, on this view, many species of distinct genera and families are supposed to combine for one end; and of such a combination, not a single instance can be found in the whole of nature. The theory that has been most generally received is, that atolls are based on submarine craters; but when we consider the form and size of some, the number, proximity, and relative positions of others, this idea loses its plausible character: thus, Suadiva atoll is 44 geographical miles in diameter in one line, by 34 miles in another line; Rimsky is 54 by 20 miles across, and it has a strangely sinuous margin; Bow atoll is 30 miles long, and on an average only 6 in width; Menchicoff atoll consists of three atolls

united or tied together. This theory, moreover, is totally inapplicable to the northern Maldiva atolls in the Indian Ocean (one of which is 88 miles in length, and between 10 and 20 in breadth), for they are not bounded like ordinary atolls by narrow reefs, but by a vast number of separate little atolls; other little atolls rising out of the great central lagoon-like spaces. A third and better theory was advanced by Chamisso, who thought that from the corals growing more vigorously where exposed to the open sea, as undoubtedly is the case, the outer edges would grow up from the general foundation before any other part, and that this would account for the ring or cup-shaped structure. But we shall immediately see, that in this, as well as in the crater-theory, a most important consideration has been overlooked, namely, on what have the reef-building corals, which cannot live at a great depth, based their massive structures?

Numerous soundings were carefully taken by Captain Fitz Roy on the steep outside of Keeling atoll, and it was found that within ten fathoms, the prepared tallow at the bottom of the lead, invariably came up marked with the impressions of living corals, but as perfectly clean as if it had been dropped on a carpet of turf; as the depth increased, the impressions became less numerous, but the adhering particles of sand more and more numerous, until at last it was evident that the bottom consisted of a smooth sandy layer: to carry on the analogy of the turf, the blades of grass grew thinner and thinner, till at last the soil was so sterile, that nothing sprang from it. From these observations, confirmed by many others, it may be safely inferred that the utmost depth at which corals can construct reefs is between 20 and 30 fathoms. Now there are enormous areas in the Pacific and Indian Oceans, in which every single island is of coral formation, and is raised only to that height to which the waves can throw up fragments, and the winds pile up sand. Thus the Radack group of atolls is an irregular square, 520 miles long and 240 broad; the Low archipelago is elliptic-formed, 840 miles in its longer, and 420 in its shorter axis: there are other small groups and single low islands between these two archipelagoes, making a linear space of ocean actually more than 4000 miles in length, in which not one single island rises above the specified height. Again, in the Indian Ocean there is a space of ocean 1500 miles in length, including three archipelagoes, in which every island is low and of coral formation. From the fact of the reef-building corals not living at great depths, it is absolutely certain that throughout these vast areas, wherever there is now an atoll, a foundation must have originally existed within a depth of from 20 to 30 fathoms from the surface. It is improbable in the highest degree that broad, lofty, isolated, steep-sided banks of sediment, arranged in groups and lines hundreds of leagues in length, could have been deposited in the central and profoundest parts of the Pacific and Indian Oceans, at an immense distance from any continent, and where the water is perfectly limpid. It is equally improbable that the elevatory forces should have uplifted throughout the above vast areas, innumerable great rocky banks within 20 to 30 fathoms, or 120 to 180 feet, of the surface of the sea, and not one single point above that level; for where on the whole face of the globe can we find a single chain of mountains, even a few hundred miles in length, with their many summits rising within a few feet of a given level, and not one pinnacle above it? If then the foundations, whence the atoll-building corals sprang, were not formed of sediment, and if they were not lifted up to the required level, they must of necessity have subsided into it; and this at once solves the difficulty. For as mountain after mountain, and island after island, slowly sank beneath the water, fresh bases would be successively afforded for the growth of the corals. It is impossible here to enter into all the necessary details, but I venture to defy any one to explain in any other manner how it is possible that numerous islands should be distributed throughout vast areas — all the islands being low — all being built of corals, absolutely requiring a foundation within a limited depth from the surface.

Before explaining how atoll-formed reefs acquire their peculiar structure, we must turn to the second great class, namely, Barrier-reefs. These either extend in straight lines in front of the shores of a continent or of a large island, or they encircle smaller islands; in both cases, being separated from the land by a broad and rather deep channel of water, analogous to the lagoon within an atoll. It is remarkable how little attention has been paid to encircling barrier-reefs; yet they are truly wonderful structures. The following sketch represents part of the barrier encircling the island of Bolabola in the Pacific, as seen from one of the central peaks. In this instance the whole line of reef has been converted into land; but usually a snow-white line of great breakers, with only here and there a single low islet crowned with cocoa-nut trees, divides the dark heaving waters of the ocean from the light-green expanse of the lagoon-channel.

And the quiet waters of this channel generally bathe a fringe of low alluvial soil, loaded with the most beautiful productions of the tropics, and lying at the foot of the wild, abrupt, central mountains.

Encircling barrier-reefs are of all sizes, from three miles to no less than forty-four miles in diameter; and that which fronts one side, and encircles both ends, of New Caledonia, is 400 miles long. Each reef includes one, two, or several rocky islands of various heights; and in one instance, even as many as twelve separate islands. The reef runs at a greater or less distance from the included land; in the Society archipelago generally from one to three or four miles; but at Hogoleu the reef is 20 miles on the southern side, and 14 miles on the opposite or northern side, from the included islands. The depth within the lagoon-channel also varies much; from 10 to 30 fathoms may be taken as an average; but at Vanikoro there are spaces no less than 56 fathoms or 336 feet deep. Internally the reef either slopes gently into the lagoon-channel, or ends in a perpendicular wall sometimes between two and three hundred feet under water in height: externally the reef rises, like an atoll, with extreme abruptness out of the profound depths of the ocean. What can be more singular than these structures? We see an island, which may be compared to a castle situated on the summit of a lofty submarine mountain, protected by a great wall of coral-rock, always steep externally and sometimes internally, with a broad level summit, here and there breached by narrow gateways, through which the largest ships can enter the wide and deep encircling moat.

As far as the actual reef of coral is concerned, there is not the smallest difference, in general size, outline, grouping, and even in quite trifling details of structure, between a barrier and an atoll. The geographer Balbi has well remarked, that an encircled island is an atoll with high land rising out of its lagoon; remove the land from within, and a perfect atoll is left.

But what has caused these reefs to spring up at such great distances from the shores of the included islands? It cannot be that the corals will not grow close to the land; for the shores within the lagoon-channel, when not surrounded by alluvial soil, are often fringed by living reefs; and we shall presently see that there is a whole class, which I have called Fringing Reefs from their close attachment to the shores both of continents and of islands. Again, on what have the reef-building corals, which cannot live at great depths, based their encircling structures? This is a great apparent difficulty, analogous to that in the case of atolls, which has generally been overlooked. It will be perceived more clearly by inspecting the following sections, which are real ones, taken in north and south lines, through the islands with their barrier-reefs, of Vanikoro, Gambier, and Maurua; and they are laid down, both vertically and horizontally, on the same scale of a quarter of an inch to a mile.

It should be observed that the sections might have been taken in any direction through these islands, or through many other encircled islands, and the general features would have been the same. Now

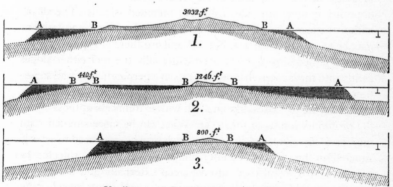

1. Vanikoro. 2. Gambier Islands. 3. Maurus

The horizontal shading shows the barrier-reefs and lagoon-channels. The inclined shading above the level of the sea (**AA**), shows the actual form of the land; the inclined shading below this line, shows its probable prolongation under water.

bearing in mind that reef-building coral cannot live at a greater depth than from 20 to 30 fathoms, and that the scale is so small that the plummets on the right hand show a depth of 200 fathoms, on what are these barrier-reefs based? Are we to suppose that each island is surrounded by a collar-like submarine ledge of rock, or by a great bank of sediment, ending abruptly where the reef ends? If the sea had formerly eaten deeply into the islands, before they were protected by the reefs, thus having left a shallow ledge round them under water, the present shores would have been invariably bounded by great precipices; but this is most rarely the case. Moreover, on this notion, it is not possible to explain why the corals should have sprung up, like a wall, from the extreme outer margin of the ledge, often leaving a broad space of water within, too deep for the growth of corals. The accumulation of a wide bank of sediment all round these islands, and generally widest where the included islands are smallest, is highly improbable, considering their exposed positions in the central and deepest parts of the ocean. In the case of the barrier-reef of New Caledonia, which extends for 150 miles beyond the northern point of the island, in the same straight line with which it fronts the west coast, it is hardly possible to believe, that a bank of sediment could thus have been straightly deposited in front of a lofty island, and so far beyond its termination in the open sea. Finally, if we look to other oceanic islands of about the same height and of similar geological constitution, but not encircled by coral-reefs, we may in vain search for so trifling a circum-ambient depth as 30 fathoms, except quite near to their shores; for usually land that rises abruptly out of water, as do most of the encircled and non-encircled oceanic islands, plunges abruptly under it. On what then, I repeat, are these barrier-reefs based? Why, with their wide and deep moat-like channels, do they stand so far from the included land? We shall soon see how easily these difficulties disappear.

We come now to our third class of Fringing Reefs, which will require a very short notice. Where the land slopes abruptly under water, these reefs are only a few yards in width, forming a mere ribbon or fringe round the shores: where the land slopes gently under the water the reef extends further, sometimes even as much as a mile from the land; but in such cases the soundings outside the reef, always show that the submarine prolongation of the land is gently inclined. In fact the reefs extend only to that distance from the shore, at which a foundation within the requisite depth from 20 to 30 fathoms is found. As far as the actual reef is concerned, there is no essential difference between it and that forming a barrier or an atoll: it is, however, generally of less width, and consequently few islets have been formed on it. From the corals growing more vigorously on the outside, and from the noxious effect of the sediment washed inwards, the outer edge of the reef is the highest part, and between it and the land there is generally a shallow sandy channel a few feet in depth. Where banks of sediment have accumulated near to the surface, as in parts of the West Indies, they sometimes become fringed with corals, and hence in some degree resemble lagoon-islands or atolls; in the same manner as fringing-reefs, surrounding gently-sloping islands, in some degree resemble barrier-reefs.

No theory on the formation of coral-reefs can be considered satisfactory which does not include the three great classes. We have seen that we are driven to believe in the subsidence of those vast areas, interspersed with low islands, of which not one rises above the height to which the wind and waves can throw up matter, and yet are constructed by animals requiring a foundation, and that foundation to lie at no great depth. Let us then take an island surrounded by fringing-reefs, which offer no difficulty in their structure; and let this island with its reef, represented by the unbroken lines in the woodcut, slowly subside. Now as the island sinks down, either a few feet at a time or quite insensibly, we may safely infer, from what is known of the conditions favourable to the growth of coral, that the living masses,

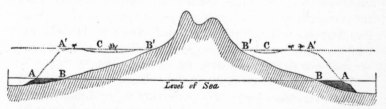

AA. Outer edges of the fringing-reef, at the level of the sea. **BB.** The shores of the fringed island.

A′A′. Outer edges of the reef, after its upward growth during a period of subsidence, now converted into a barrier, with islets on it. **B′B′.** The shores of the now encircled island. **CC.** Lagoon-channel.

N.B. In this and the following woodcut, the subsidence of the land could be represented only by an apparent rise in the level of the sea.

bathed by the surf on the margin of the reef, will soon regain the surface. The water, however, will encroach little by little on the shore, the island becoming lower and smaller, and the space between the inner edge of the reef and the beach proportionally broader. A section of the reef and island in this state, after a subsidence of several hundred feet, is given by the dotted lines. Coral islets are supposed to have been formed on the reef; and a ship is anchored in the lagoon-channel. This channel will be more or less deep, according to the rate of subsidence, to the amount of sediment accumulated in it, and to the growth of the delicately branched corals which can live there. The section in this state resembles in every respect one drawn through an encircled island: in fact, it is a real section (on the scale of .517 of an inch to a mile) through Bolabola in the Pacific. We can now at once see why encircling barrier-reefs stand so far from the shores which they front. We can also perceive, that a line drawn perpendicularly down from the outer edge of the new reef, to the foundation of solid rock beneath the old fringing-reef, will exceed by as many feet as there have been feet of subsidence, that small limit of depth at which the effective corals can live:—the little architects having built up their great wall-like mass, as the whole sank down, upon a basis formed of other corals and their consolidated fragments. Thus the difficulty on this head, which appeared so great, disappears.

If, instead of an island, we had taken the shore of a continent fringed with reefs, and had imagined it to have subsided, a great straight barrier, like that of Australia or New Caledonia, separated from the land by a wide and deep channel, would evidently have been the result.

Let us take our new encircling barrier-reef, of which the section is now represented by unbroken lines, and which, as I have said, is a real section through Bolabola, and let it go on subsiding. As the barrier-reef slowly sinks down, the corals will go on vigorously growing upwards; but as the island sinks, the water will gain inch by inch on the shore—the separate mountains first forming separate islands within one great reef—and finally, the last and highest pinnacle disappearing. The instant this takes place, a perfect atoll is formed: I have said, remove the high land from within an encircling barrier-reef, and an atoll is left, and the land has been removed. We can now perceive how it comes that atolls, having sprung from encircling barrier-reefs, resemble them in general size, form, in the manner in which they are grouped together, and in their arrangement in single or double lines;

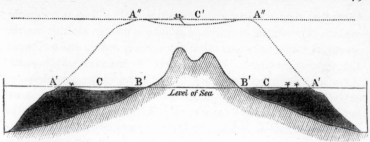

A′A′. Outer edges of the barrier-reef at the level of the sea, with islets on it.
B′B′. The shores of the included island. **CC.** The lagoon-channel.

A″A″. Outer edges of the reef, now converted into an atoll. **C′.** The lagoon of the new atoll.

N.B. According to the true scale, the depths of the lagoon-channel and lagoon are much exaggerated.

for they may be called rude outline charts of the sunken islands over which they stand. We can further see how it arises that the atolls in the Pacific and Indian oceans extend in lines parallel to the generally prevailing strike of the high islands and great coast-lines of those oceans. I venture, therefore, to affirm, that on the theory of the upward growth of the corals during the sinking of the land, all the leading features in those wonderful structures, the lagoon-islands or atolls, which have so long excited the attention of voyagers, as well as in the no less wonderful barrier-reefs, whether encircling small islands or stretching for hundreds of miles along the shores of a continent, are simply explained.

It may be asked, whether I can offer any direct evidence of the sub-sidence of barrier-reefs or atolls; but it must be borne in mind how difficult it must ever be to detect a movement, the tendency of which is to hide under water the part affected. Nevertheless, at Keeling atoll I observed on all sides of the lagoon old cocoa-nut trees undermined and falling; and in one place the foundation-posts of a shed, which the inhabitants asserted had stood seven years before just above high-water mark, but now was daily washed by every tide: on inquiry I found that three earthquakes, one of them severe, had been felt here during the last ten years. At Vanikoro, the lagoon-channel is remark-ably deep, scarcely any alluvial soil has accumulated at the foot of the lofty included mountains, and remarkably few islets have been formed by the heaping of fragments and sand on the wall-like barrier-reef; these facts, and some analogous ones, led me to believe that this island must lately have subsided and the reef grown upwards: here again earthquakes are frequent and very severe. In the Society archipelago, on the other hand, where the lagoon-channels are almost choked up, where much low alluvial land has accumulated, and where in some cases long islets have been formed on the barrier-reefs—facts all showing that the islands have not very lately subsided—only feeble shocks are most rarely felt. In these coral formations, where the land and water seem struggling for mastery, it must be ever difficult to decide between the effects of a change in the set of the tides and of a slight subsidence: that many of these reefs and atolls are subject to changes of some kind is certain; on some atolls the islets appear to have increased greatly within a late period; on others they have been partially or wholly washed away. The inhabitants of parts of the Maldiva archipelago know the date of the first formation of some islets; in other parts, the corals are now flourishing on water-washed reefs, where holes made for graves attest the former existence of inhabited land. It is difficult to believe in frequent changes in the tidal currents of an open ocean; whereas, we have in the earthquakes recorded by the natives on some atolls, and in the great fissures observed on other atolls, plain evidence of changes and disturbances in progress in the subterranean regions.

It is evident, on our theory, that coasts merely fringed by reefs cannot have subsided to any perceptible amount; and therefore they must, since the growth of their corals, either have remained stationary or have been upheaved. Now it is remarkable how generally it can be shown, by the presence of upraised organic remains, that the fringed islands have been elevated: and so far, this is indirect evidence in favour of our theory. I was particularly struck with this fact, when I found to my surprise, that the descriptions given by MM. Quoy and Gaimard were applicable, not to reefs in general as implied by them, but only to those of the fringing-class; my surprise, however, ceased when I afterwards found that, by a strange chance, all the several islands visited by these eminent naturalists, could be shown by their own statements to have been elevated within a recent geological era.

[*Further examples are given of the explanatory power of the theory.*]

One more point in detail: as in two neighbouring archipelagoes corals flourish in one and not in the other, and as so many conditions before enumerated must affect their existence, it would be an inexplicable fact if, during the changes to which earth, air, and water are subjected, the reef-building corals were to keep alive for perpetuity on any one spot or area. And as by our theory the areas including atolls and barrier-reefs are subsiding, we ought occasionally to find reefs both dead and submerged. In all reefs, owing to the sediment being washed out of the lagoon or lagoon-channel to leeward, that side is least favourable to the long-continued vigorous growth of the corals; hence dead portions of reef not unfrequently occur on the leeward side; and these, though still retaining their proper wall-like form, are now in several instances sunk several fathoms beneath the surface. The Chagos group appears from some cause, possibly from the subsidence having been too rapid, at present to be much less favourably circumstanced for the growth of reefs than formerly: one atoll has a portion of its marginal reef, nine miles in length, dead and submerged; a second has only a few quite small living points which rise to the surface; a third and fourth are entirely dead and submerged; a fifth is a mere wreck, with its structure almost obliterated. It is remarkable that in all these cases, the dead reefs and portions of reef lie at nearly the same depth, namely, from six to eight fathoms beneath the surface, as if they had been carried down by one uniform movement. One of these "half-drowned atolls," so called by Capt. Moresby (to whom I am indebted for much invaluable information), is of vast size, namely, ninety nautical miles across in one direction, and seventy miles in another line; and is in many respects eminently curious. As by our theory it follows that new atolls will generally be formed in each new area of subsidence, two weighty objections might have been raised, namely, that atolls must be increasing indefinitely in number; and secondly, that in old areas of subsidence each separate atoll must be increasing indefinitely in thickness, if proofs of their occasional destruction could not have been adduced. Thus have we traced the history of these great rings of coral-rock, from their first origin through their normal changes, and through the occasional accidents of their existence, to their death and final obliteration.

In my volume on 'Coral Formations' I have published a map, in which I have coloured all the atolls dark-blue, the barrier-reefs pale-blue, and the fringing-reefs red. These latter reefs have been formed whilst the land has been stationary, or, as appears from the frequent presence of upraised organic remains, whilst it has been slowly rising: atolls and barrier-reefs, on the other hand, have grown up during the directly opposite movement of subsidence, which movement must have been very gradual, and in the case of atolls so vast in amount as to have buried every mountain-summit over wide ocean-spaces. Now in this map we see that the reefs tinted pale and dark-blue, which have been produced by the same order of movement, as a general rule manifestly stand near each other. Again we see, that the areas with the two blue tints are of wide extent; and that they lie separate from extensive lines of coast coloured red, both of which circumstances might naturally have been inferred, on the theory of the nature of the reefs having been governed by the nature of the earth's movement. It deserves notice, that in more than one instance where single red and blue circles approach near each other, I can show that there have been oscillations of level; for in such cases the red or fringed circles consist of atolls, originally by our theory formed during subsidence, but subsequently upheaved; and on the other hand, some of the pale-blue or encircled islands are composed of coral-rock, which must have been uplifted to its present height before that subsidence took place, during which the existing barrier-reefs grew upwards.

Authors have noticed with surprise, that although atolls are the commonest coral-structures throughout some enormous oceanic tracts, they are entirely absent in other seas, as in the West Indies: we can now at once perceive the cause, for where there has not been subsidence, atolls cannot have been formed; and in the case of the West Indies and parts of the East Indies, these tracts are known to have been rising within the recent period. The larger areas, coloured red and blue, are all elongated; and between the two colours there is a degree of rude alternation, as if the rising of one had balanced the sinking of the other. Taking into consideration the proofs of recent elevation both on the fringed coasts and on some others (for instance, in South America) where there are no reefs, we are led to conclude that the great continents are for the most part rising areas; and from the nature of the coral-reefs, that the central parts of the great oceans are sinking areas. The East Indian archipelago, the most broken land in the world, is in most parts an area of elevation, but surrounded and penetrated, probably in more lines than one, by narrow areas of subsidence.

I have marked with vermilion spots all the many known active volcanos within the limits of this same map. Their entire absence from every one of the great subsiding areas, coloured either pale or dark blue, is most striking; and not less so is the coincidence of the chief volcanic chains with the parts coloured red, which we are led to conclude have either long remained stationary, or more generally have been recently upraised. Although a few of the vermilion spots occur within no great distance of single circles tinted blue, yet not one single active volcano is situated within several hundred miles of an archipelago, or even small group of atolls. It is, therefore, a striking fact that in the Friendly Archipelago, which consists of a group of atolls upheaved and since partially worn down, two volcanos, and perhaps more, are historically known to have been in action. On the other hand, although most of the islands in the Pacific which are encircled by barrier-reefs, are of volcanic origin, often with the remnants of craters still distinguishable, not one of them is known to have ever been in eruption. Hence in these cases it would appear, that volcanos burst forth into action and become extinguished on the same spots, accordingly as elevatory or subsiding movements prevail there. Numberless facts could be adduced to prove that upraised organic remains are common wherever there are active volcanos; but until it could be shown that in areas of subsidence, volcanos were either absent or inactive, the inference, however probable in itself, that their distribution depended on the rising or falling of the earth's surface, would have been hazardous. But now, I think, we may freely admit this important deduction.

Taking a final view of the map, and bearing in mind the statements made with respect to the upraised organic remains, we must feel astonished at the vastness of the areas, which have suffered changes in level either downwards or upwards, within a period not geologically remote. It would appear, also, that the elevatory and subsiding movements follow nearly the same laws. Throughout the spaces interspersed with atolls, where not a single peak of high land has been left above the level of the sea, the sinking must have been immense in amount. The sinking, moreover, whether continuous, or recurrent with intervals sufficiently long for the corals again to bring up their living edifices to the surface, must necessarily have been extremely slow. This conclusion is probably the most important one, which can be deduced from the study of coral formations;—and it is one which it is difficult to imagine, how otherwise could ever have been arrived at. Nor can I quite pass over the probability of the former existence of large archipelagoes of lofty islands, where now only rings of coral-rock scarcely break the open expanse of the sea, throwing some light on the distribution of the inhabitants of the other high islands, now left standing so immensely remote from each other in the midst of the great oceans. The reef-constructing corals have indeed reared and preserved wonderful memorials of the subterranean oscillations of level; we see in each barrier-reef a proof that the land has there subsided, and in each atoll a monument over an island now lost. We may thus, like unto a geologist who had lived his ten thousand years and kept a record of the passing changes, gain some insight into the great system by which the surface of this globe has been broken up, and land and water interchanged.

CHAPTER XXI

Mauritius to England

[*Sailing home, the* Beagle *stops at Mauritius, where Darwin finds a 'most interesting spectacle' in the remarkable range of human racial types. After stopping in at the Cape of Good Hope, the ship lands on the mid-Atlantic island of St Helena (site of Napoleon's tomb) and the volcanic island of Ascension.*]

ON leaving Ascension we sailed for Bahia, on the coast of Brazil, in order to complete the chronometrical measurement of the world. We arrived there on August 1st, and stayed four days, during which I took several long walks. I was glad to find my enjoyment in tropical scenery had not decreased from the want of novelty, even in the slightest degree. The elements of the scenery are so simple, that they are worth mentioning, as a proof on what trifling circumstances exquisite natural beauty depends.

The country may be described as a level plain of about three hundred feet in elevation, which in all parts has been worn into flat-bottomed valleys. This structure is remarkable in a granitic land, but is nearly universal in all those softer formations of which plains are usually composed. The whole surface is covered by various kinds of stately trees, interspersed with patches of cultivated ground, out of which houses, convents, and chapels arise. It must be remembered that within the tropics, the wild luxuriance of nature is not lost even in the vicinity of large cities; for the natural vegetation of the hedges and hill-sides overpowers in picturesque effect the artificial labour of man. Hence, there are only a few spots where the bright red soil affords a strong contrast with the universal clothing of green. From the edges of the plain there are distant views either of the ocean, or of the great Bay with its low-wooded shores, and on which numerous boats and canoes show their white sails. Excepting from these points, the scene is extremely limited; following the level pathways, on each hand, only glimpses into the wooded valleys below can be obtained. The houses, I may add, and especially the sacred edifices, are built in a peculiar and rather fantastic style of architecture. They are all whitewashed; so that when illumined by the brilliant sun of midday, and as seen against the pale blue sky of the horizon, they stand out more like shadows than real buildings.

Such are the elements of the scenery, but it is a hopeless attempt to paint the general effect. Learned naturalists describe these scenes of the tropics by naming a multitude of objects, and mentioning some characteristic feature of each. To a learned traveller this possibly may communicate some definite ideas: but who else from seeing a plant in an herbarium can imagine its appearance when growing in its native soil? Who from seeing choice plants in a hothouse, can magnify some into the dimensions of forest trees, and crowd others into an entangled jungle? Who when examining in the cabinet of the entomologist the gay exotic butterflies, and singular cicadas, will associate with these lifeless objects, the ceaseless harsh music of the latter, and the lazy flight of the former, — the sure accompaniments of the still, glowing noonday of the tropics? It is when the sun has attained its greatest height, that such scenes should be viewed: then the dense splendid foliage of the mango hides the ground with its darkest shade, whilst the upper branches are rendered from the profusion of light of the most brilliant green. In the temperate zones the case is different — the vegetation there is not so dark or so rich, and hence the rays of the declining sun, tinged of a red, purple, or bright yellow colour, add most to the beauties of those climes.

When quietly walking along the shady pathways, and admiring each successive view, I wished to find language to express my ideas. Epithet after epithet was found too weak to convey to those who have not visited the intertropical regions, the sensation of delight which the mind experiences. I have said that the plants in a hothouse fail to communicate a just idea of the vegetation, yet I must recur to it. The land is one great wild, untidy, luxuriant hothouse, made by Nature for herself, but taken possession of by man, who has studded it with gay houses and formal gardens. How great would be the desire in every admirer of nature to behold, if such were possible, the scenery of another planet! yet to every person in Europe, it may be truly said, that at the distance of only a few degrees from his native soil, the glories of another world are opened to him. In my last walk I stopped again and again to gaze on these beauties, and endeavoured to fix in my mind for ever, an impression which at the time I knew sooner or later must fail. The form of the orange-tree, the cocoa-nut, the palm, the mango, the tree-fern, the banana, will remain clear and separate; but the thousand beauties which unite these into one perfect scene must fade away; yet they will leave, like a tale heard in childhood, a picture full of indistinct, but most beautiful figures.

August 6th.—In the afternoon we stood out to sea, with the intention of making a direct course to the Cape de Verd Islands. Unfavourable winds, however, delayed us, and on the 12th we ran into Pernambuco,—a large city on the coast of Brazil, in latitude 8° south. We anchored outside the reef; but in a short time a pilot came on board and took us into the inner harbour, where we lay close to the town.

Pernambuco is built on some narrow and low sand-banks, which are separated from each other by shoal channels of salt water. The three parts of the town are connected together by two long bridges built on wooden piles. The town is in all parts disgusting, the streets being narrow, ill-paved, and filthy; the houses, tall and gloomy. The season of heavy rains had hardly come to an end, and hence the surrounding country, which is scarcely raised above the level of the sea, was flooded with water; and I failed in all my attempts to take long walks.

The flat swampy land on which Pernambuco stands is surrounded, at the distance of a few miles, by a semicircle of low hills, or rather by the edge of a country elevated perhaps two hundred feet above the sea. The old city of Olinda stands on one extremity of this range. One day I took a canoe, and proceeded up one of the channels to visit it; I found the old town from its situation both sweeter and cleaner than that of Pernambuco. I must here commemorate what happened for the first time during our nearly five years' wandering, namely, having met with a want of politeness: I was refused in a sullen manner at two different houses, and obtained with difficulty from a third, permission to pass through their gardens to an uncultivated hill, for the purpose of viewing the country. I feel glad that this happened in the land of the Brazilians, for I bear them no good will—a land also of slavery, and therefore of moral debasement. A Spaniard would have felt ashamed at the very thought of refusing such a request, or of behaving to a stranger with rudeness. The channel by which we went to and returned from Olinda, was bordered on each side by mangroves, which sprang like a miniature forest out of the greasy mud-banks. The bright green colour of these bushes always reminded me of the rank grass in a churchyard: both are nourished by putrid exhalations; the one speaks of death past, and the other too often of death to come.

The most curious object which I saw in this neighbourhood, was the reef that forms the harbour. I doubt whether in the whole world any other natural structure has so artificial an appearance. It runs for a length of several miles in an absolutely straight line, parallel to, and not far distant from, the shore. It varies in width from thirty to sixty yards, and its surface is level and smooth; it is composed of obscurely-stratified hard sandstone. At high water the waves break over it; at low water its summit is left dry, and it might then be mistaken for a breakwater erected by Cyclopean workmen. On this coast the currents of the sea tend to throw up in front of the land, long spits and bars of loose sand, and on one of these, part of the town of Pernambuco stands. In former times a long spit of this nature seems to have become consolidated by the percolation of calcareous matter, and afterwards to have been gradually upheaved; the outer and loose parts during this process having been worn away by the action of the sea, and the solid nucleus left as we now see it. Although night and day the waves of the open Atlantic, turbid with sediment, are driven against the steep outside edges of this wall of stone, yet the oldest pilots know of no tradition of any change in its appearance. This durability is much the most curious fact in its history: it is due to a tough layer, a few inches thick, of calcareous matter, wholly formed by the successive growth and death of the small shells of Serpulæ, together with some few barnacles and nulliporæ.* These nulliporæ, which are hard, very simply-organized sea-plants, play an analogous and important part in protecting the upper surfaces of coral-reefs, behind and within the breakers, where the true corals, during the outward growth of the mass, become killed by exposure to the sun and air. These insignificant organic beings, especially the Serpulæ, have done good service to the people of Pernambuco; for without their protective aid the bar of sandstone would inevitably have been long ago worn away, and without the bar, there would have been no harbour.

On the 19th of August we finally left the shores of Brazil. I thank God, I shall never again visit a slave-country. To this day, if I hear a distant scream, it recalls with painful vividness my feelings, when passing a house near Pernambuco, I heard the most pitiable moans, and could not but suspect that some poor slave was being tortured, yet knew that I was as powerless as a child even to remonstrate. I suspected that these moans were from a tortured slave, for I was told that this was the case in another instance. Near Rio de Janeiro I lived opposite to an old lady, who kept screws to crush the fingers of her female slaves. I have staid in a house where a young household mulatto, daily and hourly, was reviled, beaten, and persecuted enough to break the spirit of the lowest animal. I have seen a little boy, six or seven years old, struck thrice with a horse-whip (before I could interfere) on his naked head, for having handed me a glass of water not quite clean; I saw his father tremble at a mere glance from his master's eye. These latter cruelties were witnessed by me in a Spanish colony, in which it has always been said, that slaves are better treated than by the Portuguese, English, or other European nations. I have seen at Rio Janeiro a powerful negro afraid to ward off a blow directed, as he thought, at his face. I was present when a kind-hearted man was on the point of separating for ever the men, women, and little children of a large number of families who had long lived together. I will not even allude to the many heart-sickening atrocities which I authentically heard of;—nor would I have mentioned the above revolting details, had I not met with several people, so blinded by the constitutional gaiety of the negro, as to speak of slavery as a tolerable evil. Such people have generally visited at the houses of the upper classes, where the domestic slaves are usually well treated; and they have not, like myself, lived amongst the lower classes. Such enquirers will ask slaves about their condition; they forget that the slave must indeed be dull, who does not calculate on the chance of his answer reaching his master's ears.

It is argued that self-interest will prevent excessive cruelty; as if self-interest protected our domestic animals, which are far less likely than degraded slaves, to stir up the rage of their savage masters. It is an argument long since protested against with noble feeling, and strikingly exemplified, by the ever illustrious Humboldt. It is often attempted to palliate slavery by comparing the state of slaves with our poorer countrymen: if the misery of our poor be caused not by the laws of nature, but by our institutions, great is our sin; but how this bears on slavery, I cannot see; as well might the use of the thumb-screw be defended in one land, by showing that men in another land suffered from some dreadful disease. Those who look tenderly at the slave-owner, and with a cold heart at the slave, never seem to put themselves into the position of the latter;—what a cheerless prospect, with not even a hope of change! picture to yourself the chance, ever hanging over you, of your wife and your little children—those objects which nature urges even the slave to call his own—being torn from you and sold like beasts to the first bidder! And these deeds are done and palliated by men, who profess to love their neighbours as themselves, who believe in God, and pray that his Will be done on earth! It makes one's blood boil, yet heart tremble, to think that we Englishmen and our American descendants, with their boastful cry of liberty, have been and are so guilty: but it is a consolation to reflect, that we at least have made a greater sacrifice, than ever made by any nation, to expiate our sin.

On the last day of August we anchored for the second time at Porto Praya in the Cape de Verd archipelago; thence we proceeded to the Azores, where we staid six days. On the 2nd of October we made the shores of England; and at Falmouth I left the Beagle, having lived on board the good little vessel nearly five years.

Our Voyage having come to an end, I will take a short retrospect of the advantages and disadvantages, the pains and pleasures, of our circumnavigation of the world. If a person asked my advice, before undertaking a long voyage, my answer would depend upon his possessing a decided taste for some branch of knowledge, which could by this means be advanced. No doubt it is a high satisfaction to behold various countries and the many races of mankind, but the pleasures gained at the time do not counterbalance the evils. It is necessary to look forward to a harvest, however distant that may be, when some fruit will be reaped, some good effected.

Many of the losses which must be experienced are obvious; such as that of the society of every old friend, and of the sight of those places with which every dearest remembrance is so intimately connected. These losses, however, are at the time partly relieved by the exhaustless delight of anticipating the long wished-for day of return. If, as poets say, life is a dream, I am sure in a voyage these are the visions which best serve to pass away the long night. Other losses, although not at first felt, tell heavily after a period: these are the want of room, of seclusion, of rest; the jading feeling of constant hurry; the privation of small luxuries, the loss of domestic society, and even of music and the other pleasures of imagination. When such trifles are mentioned, it is evident that the real grievances, excepting from accidents, of a sea-life are at an end. The short space of sixty years has made an astonishing difference in the facility of distant navigation. Even in the time of Cook, a man who left his fireside for such expeditions underwent severe privations. A yacht now, with every luxury of life, can circumnavigate the globe. Besides the vast improvements in ships and naval resources, the whole western shores of America are thrown open, and Australia has become the capital of a rising continent. How different are the circumstances to a man shipwrecked at the present day in the Pacific, to what they were in the time of Cook! Since his voyage a hemisphere has been added to the civilized world.

If a person suffer much from sea-sickness, let him weigh it heavily in the balance. I speak from experience: it is no trifling evil, cured in a week. If, on the other hand, he take pleasure in naval tactics, he will assuredly have full scope for his taste. But it must be borne in mind, how large a proportion of the time, during a long voyage, is spent on the water, as compared with the days in harbour. And what are the boasted glories of the illimitable ocean? A tedious waste, a desert of water, as the Arabian calls it. No doubt there are some delightful scenes. A moonlight night, with the clear heavens and the dark glittering sea, and the white sails filled by the soft air of a gently blowing trade-wind; a dead calm, with the heaving surface polished like a mirror, and all still except the occasional flapping of the canvas. It is well once to behold a squall with its rising arch and coming fury, or the heavy gale of wind and mountainous waves. I confess, however, my imagination had painted something more grand, more terrific in the full-grown storm. It is an incomparably finer spectacle when beheld on shore, where the waving trees, the wild flight of the birds, the dark shadows and bright lights, the rushing of the torrents, all proclaim the strife of the unloosed elements. At sea the albatross and little petrel fly as if the storm were their proper sphere, the water rises and sinks as if fulfilling its usual task, the ship alone and its inhabitants seem the objects of wrath. On a forlorn and weather-beaten coast, the scene is indeed different, but the feelings partake more of horror than of wild delight.

Let us now look at the brighter side of the past time. The pleasure derived from beholding the scenery and the general aspect of the various countries we have visited, has decidedly been the most constant and highest source of enjoyment. It is probable that the picturesque beauty of many parts of Europe exceeds anything which we beheld. But there is a growing pleasure in comparing the character of the scenery in different countries, which to a certain degree is distinct from merely admiring its beauty. It depends chiefly on an acquaintance with the individual parts of each view: I am strongly induced to believe that, as in music, the person who understands every note will, if he also possesses a proper taste, more thoroughly enjoy the whole, so he who examines each part of a fine view, may also thoroughly comprehend the full and combined effect. Hence, a traveller should be a botanist, for in all views plants form the chief embellishment. Group masses of naked rock even in the wildest forms, and they may for a time afford a sublime spectacle, but they will soon grow monotonous. Paint them with bright and varied colours, as in Northern Chile, they will become fantastic; clothe them with vegetation, they must form a decent, if not a beautiful picture.

When I say that the scenery of parts of Europe is probably superior to anything which we beheld, I except, as a class by itself, that of the intertropical zones. The two classes cannot be compared together; but I have already often enlarged on the grandeur of those regions. As the force of impressions generally depends on preconceived ideas, I may add, that mine were taken from the vivid descriptions in the

Personal Narrative of Humboldt, which far exceed in merit anything else which I have read. Yet with these high-wrought ideas, my feelings were far from partaking of a tinge of disappointment on my first and final landing on the shores of Brazil.

Among the scenes which are deeply impressed on my mind, none exceed in sublimity the primeval forests undefaced by the hand of man; whether those of Brazil, where the powers of Life are predominant, or those of Tierra del Fuego, where Death and Decay prevail. Both are temples filled with the varied productions of the God of Nature:—no one can stand in these solitudes unmoved, and not feel that there is more in man than the mere breath of his body. In calling up images of the past, I find that the plains of Patagonia frequently cross before my eyes; yet these plains are pronounced by all wretched and useless. They can be described only by negative characters; without habitations, without water, without trees, without mountains, they support merely a few dwarf plants. Why then, and the case is not peculiar to myself, have these arid wastes taken so firm a hold on my memory? Why have not the still more level, the greener and more fertile Pampas, which are serviceable to mankind, produced an equal impression? I can scarcely analyze these feelings: but it must be partly owing to the free scope given to the imagination. The plains of Patagonia are boundless, for they are scarcely passable, and hence unknown: they bear the stamp of having lasted, as they are now, for ages, and there appears no limit to their duration through future time. If, as the ancients supposed, the flat earth was surrounded by an impassable breadth of water, or by deserts heated to an intolerable excess, who would not look at these last boundaries to man's knowledge with deep but ill-defined sensations?

Lastly, of natural scenery, the views from lofty mountains, though certainly in one sense not beautiful, are very memorable. When looking down from the highest crest of the Cordillera, the mind, undisturbed by minute details, was filled with the stupendous dimensions of the surrounding masses.

Of individual objects, perhaps nothing is more certain to create astonishment than the first sight in his native haunt of a barbarian,—of man in his lowest and most savage state. One's mind hurries back over past centuries, and then asks, could our progenitors have been men like these?—men, whose very signs and expressions are less intelligible to us than those of the domesticated animals; men, who do not possess the instinct of those animals, nor yet appear to boast of human reason, or at least of arts consequent on that reason. I do not believe it is possible to describe or paint the difference between savage and civilized man. It is the difference between a wild and tame animal: and part of the interest in beholding a savage, is the same which would lead every one to desire to see the lion in his desert, the tiger tearing his prey in the jungle, or the rhinoceros wandering over the wild plains of Africa.

Among the other most remarkable spectacles which we have beheld, may be ranked the Southern Cross, the cloud of Magellan, and the other constellations of the southern hemisphere—the water-spout—the glacier leading its blue stream of ice, over-hanging the sea in a bold precipice—a lagoon-island raised by the reef-building corals—an active volcano—and the overwhelming effects of a violent earthquake. These latter phenomena, perhaps, possess for me a peculiar interest, from their intimate connexion with the geological structure of the world. The earthquake, however, must be to every one a most impressive event: the earth, considered from our earliest childhood as the type of solidity, has oscillated like a thin crust beneath our feet; and in seeing the laboured works of man in a moment overthrown, we feel the insignificance of his boasted power.

It has been said, that the love of the chase is an inherent delight in man—a relic of an instinctive passion. If so, I am sure the pleasure of living in the open air, with the sky for a roof and the ground for a table, is part of the same feeling; it is the savage returning to his wild and native habits. I always look back to our boat cruises, and my land journeys, when through unfrequented countries, with an extreme delight, which no scenes of civilization could have created. I do not doubt that every traveller must remember the glowing sense of happiness which he experienced, when he first breathed in a foreign clime, where the civilized man had seldom or never trod.

There are several other sources of enjoyment in a long voyage, which are of a more reasonable nature. The map of the world ceases to be a blank; it becomes a picture full of the most varied and animated figures. Each part assumes its proper dimensions: continents are not looked at in the light of islands, or islands considered as mere specks, which are, in truth, larger than many kingdoms of Europe. Africa, or North and South America, are well-sounding names, and easily pronounced; but it is not until having sailed for weeks along small portions of their shores, that one is thoroughly convinced what vast spaces on our immense world these names imply.

From seeing the present state, it is impossible not to look forward with high expectations to the future progress of nearly an entire hemisphere. The march of improvement, consequent on the introduction of Christianity throughout the South Sea, probably stands by itself in the records of history. It is the more striking when we remember that only sixty years since, Cook, whose excellent judgment none will dispute, could foresee no prospect of a change. Yet these changes have now been effected by the philanthropic spirit of the British nation.

In the same quarter of the globe Australia is rising, or indeed may be said to have risen, into a grand centre of civilization, which, at some not very remote period, will rule as empress over the southern hemisphere. It is impossible for an Englishman to behold these distant colonies, without a high pride and satisfaction. To hoist the British flag, seems to draw with it as a certain consequence, wealth, prosperity, and civilization.

In conclusion, it appears to me that nothing can be more improving to a young naturalist, than a journey in distant countries. It both sharpens, and partly allays that want and craving, which, as Sir J. Herschel remarks, a man experiences although every corporeal sense be fully satisfied. The excitement from the novelty of objects, and the chance of success, stimulate him to increased activity. Moreover, as a number of isolated facts soon become uninteresting, the habit of comparison leads to generalization. On the other hand, as the traveller stays but a short time in each place, his descriptions must generally consist of mere sketches, instead of detailed observations. Hence arises, as I have found to my cost, a constant tendency to fill up the wide gaps of knowledge, by inaccurate and superficial hypotheses.

But I have too deeply enjoyed the voyage, not to recommend any naturalist, although he must not expect to be so fortunate in his companions as I have been, to take all chances, and to start, on travels by land if possible, if otherwise on a long voyage. He may feel assured, he will meet with no difficulties or dangers, excepting in rare cases, nearly so bad as he beforehand anticipates. In a moral point of view, the effect ought to be, to teach him good-humoured patience, freedom from selfishness, the habit of acting for himself, and of making the best of every occurrence. In short, he ought to partake of the characteristic qualities of most sailors. Travelling ought also to teach him distrust; but at the same time he will discover, how many truly kind-hearted people there are, with whom he never before had, or ever again will have any further communication, who yet are ready to offer him the most disinterested assistance.

REVIEWS AND RESPONSES

My first meeting with Mr. Darwin was in 1839, in Trafalgar Square. I was walking with an officer who had been his shipmate for a short time in the *Beagle* seven years before, but who had not, I believe, since met him. I was introduced; the interview was of course brief, and the memory of him that I carried away and still retain was that of a rather tall and rather broad-shouldered man, with a slight stoop, an agreeable and animated expression when talking, beetle brows, and a hollow but mellow voice; and that his greeting of his old acquaintance was sailor-like—that is, delightfully frank and cordial. I observed him well, for I was already aware of his attainments and labours, derived from having read various proof-sheets of his then unpublished 'Journal.' These had been submitted to Mr. (afterwards Sir Charles) Lyell by Mr. Darwin, and by him sent to his father, Ch. Lyell, Esq., of Kinnordy, who (being a very old friend of my father and taking a kind interest in my projected career as a naturalist) had allowed me to peruse them. At this time I was hurrying on my studies, so as to take my degree before volunteering to accompany Sir James Ross in the Antarctic Expedition, which had just been determined on by the Admiralty; and so pressed for time was I, that I used to sleep with the sheets of the 'Journal' under my pillow, that I might read them between waking and rising. They impressed me profoundly, I might say despairingly, with the variety of acquirements, mental and physical, required in a naturalist who should follow in Darwin's footsteps, whilst they stimulated me to enthusiasm in the desire to travel and observe.

English botanist Joseph Hooker, recalling his first meeting with Darwin, in F. Darwin (ed.), *The Life and Letters of Charles Darwin* (London, 1887), ii, 19–20

Looking at the general mass of his results, the account of which he has been kind enough to place in my hands, I cannot help considering his voyage round the world as one of the most important events for geology which has occurred for many years. We may think ourselves fortunate that Capt. Fitz Roy, who conducted the expedition, was led, by his enlightened zeal for science, to take out a naturalist with him. And we have further reason to rejoice that this lot fell to a gentleman like Mr. Darwin, who possessed the genuine spirit and zeal, as well as the knowledge of a naturalist; who had pursued the studies which fitted him for this employment, under the friendly guidance of Dr. Grant at Edinburgh, and Professor Henslow and Professor Sedgwick at Cambridge; and whose powers of reason and application had been braced and disciplined by the other studies of the University of which the latter two gentlemen are such distinguished ornaments. . . .

Guided by the principles which he learned from my distinguished predecessor in this chair [Charles Lyell], Mr. Darwin has presented this subject under an aspect which cannot but have the most powerful influence on the speculations concerning the history of our globe, to which you, gentlemen, may hereafter be led. I might say the same of the large and philosophical views which you will find illustrated in his work, on the laws of change of climate, of diffusion, duration and extinction of species, and other great problems of our science which this voyage has suggested. I know that I only express your feeling when I say, that we look with impatience to the period when this portion of the results of Captain Fitz Roy's voyage shall be published, as the scientific world in general looks eagerly for the whole record of that important expedition.

> Revd William Whewell of Cambridge in his presidential address of 16 Feb. 1838 to the Geological Society of London, *Proceedings of the Geological Society of London*, 2 (1838), 624–49, at 643–5

I have read far enough into your Journal to feel that I have to thank you for the most delightful book in my collection. It is as full of good original wholesome food as an egg, & if what I have enjoyed has not been duly digested it is because it has been too hastily devoured. I leave it reluctantly — tired eyes compelling — at night, and greet it as a new luxury at the breakfast table. . . .

> Comparative anatomist Richard Owen, in a letter to Darwin from the College of Surgeons on 11 June 1839, *Correspondence*, ii (1987), 199

I have dipped into it here and there — but have reserved its steady perusal for the ensuing fortnight in the country. My wife has it now with her — and from what we have seen — in various glimpses — I have no doubt whatever that I shall be deeply interested by reading it *attentively* (much of it requires **close** *thinking* I apprehend) when undisturbed by daily, or rather hourly calls upon one's time. I cannot think that there is an expression in it—referring to me personally—which I could wish were not in it—at all events neither I nor my wife have yet lighted upon anything that induces me to doubt in the smallest degree that I shall not be thoroughly at ease in that respect.

Captain Robert FitzRoy, in a letter to Darwin of June 1839, *Correspondence*, ii (1987), 197

Mankind like to feel convinced that the utterer of theories has spared no pains in assaying the metal to which he seeks to give general currency. They require him to be a minute and patient observer, not biased in his observations by preconceived notions. He must not seem disposed to reject the consideration of dry facts, and to rear his goodly edifice on assumed premises. If he would inspire the world with confidence, he must allow every stone to be examined, from the lowest foundations to the crown of the pediment. To us, Mr. Darwin's remarks seem to display no common sagacity and power of observation, but, denuded as his journal is in general of elemental facts, we doubt not that many of the theories therein contained will meet with a less general concurrence than they are really entitled to.

English traveller and author William Desborough Cooley, in an unsigned review in the *Athenaeum* (15 June 1839), 446–50

If I have delayed so long, Sir, in expressing to you my deep and affectionate gratitude, it is because I have had your excellent and admirable book in my possession for only a fortnight, and I did not want to answer your letter which arrived two months earlier without being able to tell you all I have learned and enjoyed in what you so modestly call 'The Journal of a Naturalist'. . . .

You told me in your kind letter that, when you were young, the manner in which I studied and depicted nature in the torrid zones contributed toward exciting in you the ardour and desire to travel in distant lands. Considering the importance of your work, Sir, this may be the greatest success that my humble work could bring. Works are of value only if they give rise to better ones. Moreover, Sir, with the illustrious name you bear, what inspiration you can draw from the reminder of scientific and literary achievements that make up a family's finest patrimony. My antediluvian piece 'on the excitation of nervous fiber' frequently attests how much I owe to the poetic author of *Zoonomia*, who proved that profound affinity with nature and an imagination that was not dreamy but powerful and productive, enlarge in superior men the realm of understanding. . . .

Being at the end of my career, enjoying without regrets, with all the purity of the love of science, the progress of intelligence and of liberty, the glory of the modern age, I do not judge my contemporaries with that austere and ill-willed severity my own works suffered for so long, but with a judgment free of national prejudices, which takes into account strength of talent, solid and wide knowledge, and a felicitous literary disposition to describe what one feels and wishes to convey to the reader. On all these counts, Sir, you rank high in my estimation. You have an excellent future ahead of you. Your work is remarkable for the number of new and ingenious observations on the geographical distribution of organisms, the physiognomy of plants, the geological structure of the earth's crust, the ancient oscillations, the influence of that unusual littoral climate which unites Cycads, hummingbirds, and parrots with forms found in Lapland. . . .

You see, Sir, that I like going over the principal points on which you have enlarged and corrected my views.

The great naturalist and traveller Alexander von Humboldt, in a letter from Potsdam dated 18 Sept. 1839, trans. from German in *Correspondence*, ii (1987), 425–6

But it is not to the scientific alone that Mr. Darwin's volume will prove highly interesting. The general reader will find in it a fund of amusement and instruction. Mr. Darwin is a first-rate landscape-painter with the pen. Even the dreariest solitudes are made to teem with interest. Nor less striking are his accounts of the state of society in South America, especially those which relate to the murderous hatred mutually felt and exercised towards each other by the aborigines and those whom they justly consider usurpers, but who look upon them more as wild beasts than fellow-men.

English lawyer and naturalist William Broderip writing anonymously in the *Quarterly Review*, 65 (Dec. 1839), 194–234, at 233

Having drawn up a brief report on the fossil of an extraordinary quadruped, found in the great alluvial layer just below the surface of our Province, which extends throughout most of the territory of

Argentina . . . I respectfully submit my account in the hopes it will merit publication in your most admirable newspaper. I am certain that the species here reported is not among those described by the estimable *Mr. Darwin* following his fascinating exploration of the Patagonian coast and other portions of our Republic in 1832–36, and so I am the first, in the account that follows, to recommend it to the attention of savants dedicated to examining these witnesses and victims of terrible, devastating catastrophes.

I recognize the skeleton in question as belonging to an individual of genus *Felis*, and resembling the *lion* in many particulars of its structure. It differs from both living and fossil forms of this formidable dictator of the quadruped tribes in precisely those features by which it is even more ferocious and destructive than others in its genus, more fearsome for the animals on which it preys.

> Francisco Javier Muñiz, letter from Luján to the editor dated 1 July 1845 describing a new vertebrate fossil in Patagonia, 'The Muñi-Felis Bonaerensis', *Gaceta Mercantil*, 9 Oct. 1845; trans. in A. Novoa and A. Levine, *¡Darwinistas!* (forthcoming)

The accuracy and novelty of the observations of so capable a naturalist as Darwin is also so generally recognized in Germany that the translator could justifiably refrain from saying anything in his [Darwin's] defence, even if he had not himself had the opportunity, through his own experience as far as Australia and New Zealand are concerned, to confirm the faithfulness of the account.

> The naturalist Ernst Dieffenbach introducing the first German translation of the *Journal of Researches* (Brunswick, 1844), i, p. vi

The scenery which he saw on his voyage was more varied and stupendous than Europe could have afforded him; but after all his tastes as an observer of nature are satisfied, he seems to be most interested in the sight of man, in those earlier stages of barbarism through which our own ancestors must have passed. Doubtless, to some future age our own civilization will appear like a variety of barbarism, or at least like a transition state; but meantime it is matter of study and reflection to observe those peculiarities out of which, or of something like which, our present social systems have sprung.

> Unitarian minister and essayist William Bourn Oliver Peabody, writing anonymously on 'Darwin's Researches in Geology and

Natural History', *North American Review*, 61 (July 1845), 181–99,
at 199

There are isolated parts of the earth, which we know to have become
dry land more recently than others. Such is the Galapagos group of
islands, situated in the Pacific, between five and six hundred miles
from the American coast. They are wholly of volcanic origin, and are
considered by Mr. Darwin as having been raised out of the sea,
"within a late geological period." Here, then, is a piece of the world
undoubtedly younger, so to speak, than most other portions. . . . What
are the organic productions of this curious archipelago? In the first
place, they are "mostly aboriginal creations, found nowhere else,"
though with an affinity to those of America. Many of them are even
peculiar to particular islands in the group. But the remarkable fact
bearing on the present inquiry is, that, excepting a rat and a mouse
on two of the islands, supposed to have been imported by foreign
vessels, *there are no mammals in the Galapagos.* The leading terrestrial
animals are reptiles, and these exist in great variety, and in some
instances of extraordinary size. Lizards and tortoises particularly
abound. There are also birds, eleven kinds of swimmers and waders,
and twenty-six purely terrestrial. All this harmonizes with our ideas
of the world at the time of the oolites. It speaks of *time* being necessary for the completion of the animal series in any scene of its development. The Galapagos have not had the full time required for the
completion of the series, and it is incomplete accordingly. The entire
harmony of this fact does, I must confess, strike my mind forcibly.
Had there been mammals and no reptiles, it would have been quite
different. We should then have said, that one decided fact against the
development theory had been ascertained.

The anonymous author of *Vestiges* using the *Journal of Researches*
to support a hypothesis of evolutionary progress towards higher
forms; from *Explanations: A Sequel to 'Vestiges of the Natural History
of Creation'* (London, 1845), 161–3

A map is always a decisive criterion of they who aspire to the rank of
geologists, every one who has not compiled a map, wants the necessary talent of combination. The spirited Darwin, with all his remarkable vivacity of mind, is for me no Geologist, only an able history
maker of what nature as he beleaves, has done, and what never she
did. Therefore the eternal repetition of protruded rocks, of elevations,

of igneous protusions, of lava streams where there are none, and all such vague assertions of very little value. This man would never make a tolerable geological *map*.

> Prussian geologist Leopold von Buch in a letter to Roderick Murchison, 20 Apr. 1846, Murchison MSS, LDGSL 838/B33/6, Geological Society of London

The volumes before us were prepared in a most acceptable form, comprising, in a condensed view, the general results of the expedition, imparting the information, collected by such an expenditure of money, time and scientific skill, to the reading world, in two elegant volumes of Harper's New Miscellany, at fifty cents each. The narrative is enlivened with the most interesting incidents of personal adventure, while it developes new and important facts in Geology, Natural History and Geography. We cannot but look upon this class of publications as the most desirable to lay before the community, and we doubt not but it will become the most popular. The appetite soon palls with the exciting verbiage which has of late deluged the market, and the perusal of which occupies the time and vitiates the taste without adding to the stock of knowledge.

> 'Notices of New Books', *United States Magazine, and Democratic Review*, 18 (May 1846), 398

First of all however permit me to state, that having some 18 months since made an excursion to the Uruguay La Plata & Parana rivers, also across the continent to Valparaiso, how much indebted I am to you for the pleasure, instruction & assistance derived from the "Journal of a Naturalist", which served me as a guide book, wherever my route was the same with yours. —

> American anatomist and ethnologist Jeffries Wyman, in a letter to Darwin from Boston, [*c*.15] Sept. 1860, *Correspondence*, viii (1993), 360

More than five years ago I adopted as my motto a passage from your book. 'A traveller should be a botanist, for in all views *plants* form the chief embellishment'.

I adopted it in my Botany of the Herald, and many people have agreed with you in thinking that without botanical knowledge it is impossible to describe scenery with any approach to correctness.

Berthold Carl Seeman, German-born naturalist on the voyage of the *Herald* (1847–51), in a letter to Darwin from London, 24 Apr. 1862, *Correspondence*, x (1997), 167

Everybody has eyes, but, as you know, some people are blind; and many of those who are not blind wear glasses, and cannot see without them. But even those whose eyes are good and strong do not all see alike. . . . So those see best who know the most, or who naturally take notice of new things. Now Charles Darwin, about whom I am going to tell you presently, is one of the best seers that ever lived, partly because he had learned so well what to look for, and partly because nothing escaped his eyes. . . . And now all the world looks at things differently from what it used to before he showed it how.

What Mr. Darwin Saw in his Voyage Round the World in the Ship 'Beagle' (New York, 1879), 13–17

At the risk of being thought extravagant, I will say that I consider this work the most exquisite and even the most important book of travels which has appeared in the world since Herodotus. I do not know which to admire most, the acuteness with which he observes, or the profound and far-reaching wisdom with which he interprets nature.

American popular writer James Parton in a biographical article on Darwin in the London-based *Young Folks; A Boys' and Girls' Paper of Instructive and Entertaining Literature* (17 Feb. 1883), 55

ORIGIN OF SPECIES

"But with regard to the material world, we can at least go so far as this — we can perceive that events are brought about not by insulated interpositions of Divine power, exerted in each particular case, but by the establishment of general laws."

W. WHEWELL: *Bridgewater Treatise*.

"To conclude, therefore, let no man out of a weak conceit of sobriety, or an ill-applied moderation, think or maintain, that a man can search too far or be too well studied in the book of God's word, or in the book of God's works; divinity or philosophy; but rather let men endeavour an endless progress or proficience in both."

BACON: *Advancement of Learning*.

Down, Bromley, Kent,
October 1st, 1859.

ON

THE ORIGIN OF SPECIES

BY MEANS OF NATURAL SELECTION, OR THE
PRESERVATION OF FAVOURED RACES
IN THE STRUGGLE FOR LIFE

Introduction

WHEN on board H.M.S. 'Beagle,' as naturalist, I was much struck with certain facts in the distribution of the inhabitants of South America, and in the geological relations of the present to the past inhabitants of that continent. These facts seemed to me to throw some light on the origin of species — that mystery of mysteries, as it has been called by one of our greatest philosophers.* On my return home, it occurred to me, in 1837, that something might perhaps be made out on this question by patiently accumulating and reflecting on all sorts of facts which could possibly have any bearing on it. After five years' work I allowed myself to speculate on the subject, and drew up some short notes; these I enlarged in 1844 into a sketch of the conclusions, which then seemed to me probable: from that period to the present day I have steadily pursued the same object. I hope that I may be excused for entering on these personal details, as I give them to show that I have not been hasty in coming to a decision.

My work is now nearly finished; but as it will take me two or three more years to complete it, and as my health is far from strong, I have been urged to publish this Abstract. I have more especially been induced to do this, as Mr. Wallace, who is now studying the natural history of the Malay archipelago, has arrived at almost exactly the same general conclusions that I have on the origin of species. Last year he sent to me a memoir on this subject, with a request that I would forward it to Sir Charles Lyell, who sent it to the Linnean Society, and it is published in the third volume of the Journal of that Society. Sir C. Lyell and Dr. Hooker, who both knew of my work — the latter having read my sketch of 1844—honoured me by thinking it advisable to publish, with Mr. Wallace's excellent memoir, some brief extracts from my manuscripts.

This Abstract, which I now publish, must necessarily be imperfect. I cannot here give references and authorities for my several statements; and I must trust to the reader reposing some confidence in my accuracy. No doubt errors will have crept in, though I hope I have always been cautious in trusting to good authorities alone. I can here give only the general conclusions at which I have arrived, with a few facts in illustration, but which, I hope, in most cases will suffice. No one can feel more sensible than I do of the necessity of hereafter publishing in detail all the facts, with references, on which my conclusions have been grounded; and I hope in a future work to do this. For I am well aware that scarcely a single point is discussed in this volume on which facts cannot be adduced, often apparently leading to conclusions directly opposite to those at which I have arrived. A fair result can be obtained only by fully stating and balancing the facts and arguments on both sides of each question; and this cannot possibly be here done.

I much regret that want of space prevents my having the satisfaction of acknowledging the generous assistance which I have received from very many naturalists, some of them personally unknown to me. I cannot, however, let this opportunity pass without expressing my deep obligations to Dr. Hooker, who for the last fifteen years has aided me in every possible way by his large stores of knowledge and his excellent judgment.

In considering the Origin of Species, it is quite conceivable that a naturalist, reflecting on the mutual affinities of organic beings, on their embryological relations, their geographical distribution, geological succession, and other such facts, might come to the conclusion that each species had not been independently created, but had descended, like varieties, from other species. Nevertheless, such a conclusion, even if well founded, would be unsatisfactory, until it could be shown how the innumerable species inhabiting this world have been modified, so as to acquire that perfection of structure and coadaptation* which most justly excites our admiration. Naturalists continually refer to external conditions, such as climate, food, &c., as the only possible cause of variation. In one very limited sense, as we shall hereafter see, this may be true; but it is preposterous to attribute to mere external conditions, the structure, for instance, of the woodpecker, with its feet, tail, beak, and tongue, so admirably adapted to catch insects under the bark of trees. In the case of the misseltoe, which draws its nourishment from certain trees, which has seeds that must be transported by certain birds, and which has flowers with separate sexes absolutely requiring the agency of certain insects to bring pollen from one flower to the other, it is equally preposterous to account for the structure of this parasite, with its relations to several distinct organic beings, by the effects of external conditions, or of habit, or of the volition of the plant itself.

The author of the 'Vestiges of Creation'* would, I presume, say that, after a certain unknown number of generations, some bird had given birth to a woodpecker, and some plant to the misseltoe, and that these had been produced perfect as we now see them; but this assumption seems to me to be no explanation, for it leaves the case of the coadaptations of organic beings to each other and to their physical conditions of life, untouched and unexplained.

It is, therefore, of the highest importance to gain a clear insight into the means of modification and coadaptation. At the commencement of my observations it seemed to me probable that a careful study of domesticated animals and of cultivated plants would offer the best chance of making out this obscure problem. Nor have I been disappointed; in this and in all other perplexing cases I have invariably found that our knowledge, imperfect though it be, of variation under domestication, afforded the best and safest clue. I may venture to express my conviction of the high value of such studies, although they have been very commonly neglected by naturalists.

From these considerations, I shall devote the first chapter of this Abstract to Variation under Domestication. We shall thus see that a large amount of hereditary modification is at least possible; and, what is equally or more important, we shall see how great is the power of man in accumulating by his Selection successive slight variations. I will then pass on to the variability of species in a state of nature; but I shall, unfortunately, be compelled to treat this subject far too briefly, as it can be treated properly only by giving long catalogues of facts. We shall, however, be enabled to discuss what circumstances are most favourable to variation. In the next chapter the Struggle for Existence amongst all organic beings throughout the world, which inevitably follows from their high geometrical powers of increase, will be treated of. This is the doctrine of Malthus, applied to the whole animal and vegetable kingdoms. As many more individuals of each species are born than can possibly survive; and as, consequently, there is a frequently recurring struggle for existence, it follows that any being, if it vary however slightly in any manner profitable to itself, under the complex and sometimes varying conditions of life, will have a better chance of surviving, and thus be *naturally selected*. From the strong principle of inheritance, any selected variety will tend to propagate its new and modified form.

This fundamental subject of Natural Selection will be treated at some length in the fourth chapter; and we shall then see how Natural Selection almost inevitably causes much Extinction of the less improved forms of life, and induces what I have called Divergence of Character. In the next chapter I shall discuss the complex and little known laws of variation and of correlation of growth. In the four succeeding chapters, the most apparent and gravest difficulties on the theory will be given: namely, first, the difficulties of transitions, or in understanding how a simple being or a simple organ can be changed and perfected into a highly developed being or elaborately constructed organ; secondly, the subject of Instinct, or the mental powers of animals; thirdly, Hybridism, or the infertility of species and the fertility of varieties when intercrossed; and fourthly, the imperfection of the Geological Record. In the next chapter I shall consider the geological succession of organic beings throughout time; in the eleventh and twelfth, their geographical distribution throughout space; in the thirteenth, their classification or mutual affinities, both when mature and in an embryonic condition. In the last chapter I shall give a brief recapitulation of the whole work, and a few concluding remarks.

No one ought to feel surprise at much remaining as yet unexplained in regard to the origin of species and varieties, if he makes due allowance for our profound ignorance in regard to the mutual relations of all the beings which live around us. Who can explain why one species ranges widely and is very numerous, and why another allied species has a narrow range and is rare? Yet these relations are of the highest importance, for they determine the present welfare, and, as I believe, the future success and modification of every inhabitant of this world. Still less do we know of the mutual relations of the innumerable inhabitants of the world during the many past geological epochs in its history. Although much remains obscure, and will long remain obscure, I can entertain no doubt, after the most deliberate study and dispassionate judgment of which I am capable, that the view which most naturalists entertain, and which I formerly entertained—namely, that each species has been independently created—is erroneous. I am fully convinced that species are not immutable; but that those belonging to what are called the same genera are lineal descendants of some other and generally extinct species, in the same manner as the acknowledged varieties of any one species are the descendants of that species. Furthermore, I am convinced that Natural Selection has been the main but not exclusive means of modification.

CHAPTER I

Variation under Domestication

WHEN we look to the individuals of the same variety or sub-variety of our older cultivated plants and animals, one of the first points which strikes us, is, that they generally differ much more from each other, than do the individuals of any one species or variety in a state of nature. When we reflect on the vast diversity of the plants and animals which have been cultivated, and which have varied during all ages under the most different climates and treatment, I think we are driven to conclude that this greater variability is simply due to our domestic productions having been raised under conditions of life not so uniform as, and somewhat different from, those to which the parent-species have been exposed under nature. There is, also, I think, some probability in the view propounded by Andrew Knight, that this variability may be partly connected with excess of food. It seems pretty clear that organic beings must be exposed during several generations to the new conditions of life to cause any appreciable amount of variation; and that when the organisation has once begun to vary, it generally continues to vary for many generations. No case is on record of a variable being ceasing to be variable under cultivation. Our oldest cultivated plants, such as wheat, still often yield new varieties: our oldest domesticated animals are still capable of rapid improvement or modification.

It has been disputed at what period of life the causes of variability, whatever they may be, generally act; whether during the early or late period of development of the embryo, or at the instant of conception. Geoffroy St. Hilaire's experiments show that unnatural treatment of the embryo causes monstrosities; and monstrosities cannot be separated by any clear line of distinction from mere variations. But I am strongly inclined to suspect that the most frequent cause of variability may be attributed to the male and female reproductive elements having been affected prior to the act of conception. Several reasons make me believe in this; but the chief one is the remarkable effect which confinement or cultivation has on the functions of the reproductive system; this system appearing to be far more susceptible than any other part of the organisation, to the action of any change in the conditions of life.

[*After further discussing the sources of inherited variations (including the effects of long-continued use and disuse), Darwin points out the danger of assuming that highly differentiated domesticated forms must have descended from several wild species.*]

The doctrine of the origin of our several domestic races from several aboriginal stocks, has been carried to an absurd extreme by some authors. They believe that every race which breeds true, let the distinctive characters be ever so slight, has had its wild prototype. At this rate there must have existed at least a score of species of wild cattle, as many sheep, and several goats in Europe alone, and several even within Great Britain. One author believes that there formerly existed in Great Britain eleven wild species of sheep peculiar to it! When we bear in mind that Britain has now hardly one peculiar mammal, and France but few distinct from those of Germany and conversely, and so with Hungary, Spain, &c., but that each of these kingdoms possesses several peculiar breeds of cattle, sheep, &c., we must admit that many domestic breeds have originated in Europe; for whence could they have been derived, as these several countries do not possess a number of peculiar species as distinct parent-stocks? So it is in India. Even in the case of the domestic dogs of the whole world, which I fully admit have probably descended from several wild species, I cannot doubt that there has been an immense amount of inherited variation. Who can believe that animals closely resembling the Italian greyhound, the bloodhound, the bull-dog, or Blenheim spaniel, &c.—so unlike all wild Canidæ—ever existed freely in a state of nature? It has often been loosely said that all our races of dogs have been produced by the crossing of a few aboriginal species; but by crossing we can get only forms in some degree intermediate between their parents; and if we account for our several domestic races by this process, we must admit the former existence of the most extreme forms, as the Italian greyhound, bloodhound, bull-dog, &c., in the wild state. Moreover, the possibility of making distinct races by crossing has been greatly exaggerated. There can be no doubt that a race may be modified by occasional crosses, if aided by the careful selection of those individual mongrels, which present any desired character; but that a race could be obtained nearly intermediate between two extremely different races or species, I can hardly believe. Sir J. Sebright expressly experimentised for this object, and failed. The offspring from the first cross between two pure breeds is tolerably and sometimes (as I have found with pigeons) extremely uniform, and everything seems simple enough; but when these mongrels are crossed one with another for several generations, hardly two of them will be alike, and then the extreme difficulty, or rather utter hopelessness, of the task becomes apparent. Certainly, a breed intermediate between *two very distinct* breeds could not be got without extreme care and long-continued selection; nor can I find a single case on record of a permanent race having been thus formed.

On the Breeds of the Domestic Pigeon.—Believing that it is always best to study some special group, I have, after deliberation, taken up domestic pigeons. I have kept every breed which I could purchase or obtain, and have been most kindly favoured with skins from several quarters of the world, more especially by the Hon. W. Elliot from India, and by the Hon. C. Murray from Persia. Many treatises in different languages have been published on pigeons, and some of them are very important, as being of considerable antiquity. I have associated with several eminent fanciers, and have been permitted to join two of the London Pigeon Clubs. The diversity of the breeds is something astonishing. Compare the English carrier and the short-faced tumbler, and see the wonderful difference in their beaks, entailing corresponding differences in their skulls. The carrier, more especially the male bird, is also remarkable from the wonderful development of the carunculated* skin about the head, and this is accompanied by greatly elongated eyelids, very large external orifices to the nostrils, and a wide gape of mouth. The short-faced tumbler has a beak in outline almost like that of a finch; and the common tumbler has the singular and strictly inherited habit of flying at a great height in a compact flock, and tumbling in the air head over heels. The runt is a bird of great size, with long, massive beak and large feet; some of the sub-breeds of runts have very long necks, others very long wings and tails, others singularly short tails. The barb is allied to the carrier, but, instead of a very long beak, has a very short and very broad one. The pouter has a much elongated body, wings, and legs; and its enormously developed crop,* which it glories in inflating, may well excite astonishment and even laughter. The turbit has a very short and conical beak, with a line of reversed feathers down the breast; and it has the habit of continually expanding slightly the upper part of the œsophagus. The Jacobin has the feathers so much reversed along the back of the neck that they form a hood, and it has, proportionally to its size, much elongated wing and tail feathers. The trumpeter and laugher, as their names express, utter a very different coo from the other breeds. The fantail has thirty or even forty tail-feathers, instead of twelve or fourteen, the normal number in all members of the great pigeon family; and these feathers are kept expanded, and are carried so erect that in good birds the head and tail touch; the oil-gland is quite aborted. Several other less distinct breeds might have been specified.

In the skeletons of the several breeds, the development of the bones of the face in length and breadth and curvature differs enormously. The shape, as well as the breadth and length of the ramus of the lower jaw,* varies in a highly remarkable manner. The number of the caudal and sacral vertebræ* vary; as does the number of the ribs, together with their relative breadth and the presence of processes. The size and shape of the apertures in the sternum are highly variable; so is the degree of divergence and relative size of the two arms of the furcula.* The proportional width of the gape of mouth, the proportional length of the eyelids, of the orifice of the nostrils, of the tongue (not always in strict correlation with the length of beak), the size of the crop and of the upper part of the œsophagus; the development and abortion of the oil-gland; the number of the primary wing and caudal feathers; the relative length of wing and tail to each other and to the body; the relative length of leg and of the feet; the number of scutellæ* on the toes, the development of skin between the toes, are all points of structure which are variable. The period at which the perfect plumage is acquired varies, as does the state of the down with which the nestling birds are clothed when hatched. The shape and size of the eggs vary. The manner of flight differs remarkably; as does in some breeds the voice and disposition. Lastly, in certain breeds, the males and females have come to differ to a slight degree from each other.

Altogether at least a score of pigeons might be chosen, which if shown to an ornithologist, and he were told that they were wild birds, would certainly, I think, be ranked by him as well-defined species. Moreover, I do not believe that any ornithologist would place the English carrier, the short-faced tumbler, the runt, the barb, pouter, and fantail in the same genus; more especially as in each of these breeds several truly-inherited sub-breeds, or species as he might have called them, could be shown him.

Great as the differences are between the breeds of pigeons, I am fully convinced that the common opinion of naturalists is correct, namely, that all have descended from the rock-pigeon (Columba livia), including under this term several geographical races or sub-species, which differ from each other in the most trifling respects. As several of the reasons which have led me to this belief are in some degree applicable in other cases, I will here briefly give them. If the several breeds are not varieties, and have not proceeded from the rock-pigeon, they must have descended from at least seven or eight aboriginal stocks; for it is impossible to make the present domestic breeds by the crossing of any lesser number: how, for instance, could a pouter be produced by crossing two breeds unless one of the parent-stocks possessed the characteristic enormous crop? The supposed aboriginal stocks must all have been rock-pigeons, that is, not breeding or willingly perching on trees. But besides C. livia, with its geographical sub-species, only two or three other species of rock-pigeons are known; and these have not any of the characters of the domestic breeds. Hence the supposed aboriginal stocks must either still exist in the countries where they were originally domesticated, and yet be unknown to ornithologists; and this, considering their size, habits, and remarkable characters, seems very improbable; or they must have become extinct in the wild state. But birds breeding on precipices, and good fliers, are unlikely to be exterminated; and the common rock-pigeon, which has the same habits with the domestic breeds, has not been exterminated even on several of the smaller British islets, or on the shores of the Mediterranean. Hence the supposed extermination of so many species having similar habits with the rock-pigeon seems to me a very rash assumption. Moreover, the several above-named domesticated breeds have been transported to all parts of the world, and, therefore, some of them must have been carried back again into their native country; but not one has ever become wild or feral, though the dovecot-pigeon, which is the rock-pigeon in a very slightly altered state, has become feral in several places. Again, all recent experience shows that it is most difficult to get any wild animal to breed freely under domestication; yet on the hypothesis of the multiple origin of our pigeons, it must be assumed that at least seven or eight species were so thoroughly domesticated in ancient times by half-civilized man, as to be quite prolific under confinement.

An argument, as it seems to me, of great weight, and applicable in several other cases, is, that the above-specified breeds, though agreeing generally in constitution, habits, voice, colouring, and in most parts of their structure, with the wild rock-pigeon, yet are certainly highly abnormal in other parts of their structure: we may look in vain throughout the whole great family of Columbidæ for a beak like that of the English carrier, or that of the short-faced tumbler, or barb; for reversed feathers like those of the jacobin; for a crop like that of the pouter; for tail-feathers like those of the fantail. Hence it must be assumed not only that half-civilized man succeeded in thoroughly domesticating several species, but that he intentionally or by chance picked out extraordinarily abnormal species; and further, that these very species have since all become extinct or unknown. So many strange contingencies seem to me improbable in the highest degree.

Some facts in regard to the colouring of pigeons well deserve consideration. The rock-pigeon is of a slaty-blue, and has a white rump (the Indian sub-species, C. intermedia of Strickland, having it bluish); the tail has a terminal dark bar, with the bases of the outer feathers externally edged with white; the wings have two black bars; some semi-domestic breeds and some apparently truly wild breeds have, besides the two black bars, the wings chequered with black. These several marks do not occur together in any other species of the whole family. Now, in every one of the domestic breeds, taking thoroughly well-bred birds, all the above marks, even to the white edging of the outer tail-feathers, sometimes concur perfectly developed. Moreover, when two birds belonging to two distinct breeds are crossed, neither of which is blue or has any of the above-specified marks, the mongrel offspring are very apt suddenly to acquire these characters; for instance, I crossed some uniformly white fantails with some uniformly black barbs, and they produced mottled brown and black birds; these I again crossed together, and one grandchild of the pure white fantail and pure black barb was of as beautiful a blue colour, with the white rump, double black wing-bar, and barred and white-edged tail-feathers, as any wild rock-pigeon! We can understand these facts, on the well-known principle of reversion to ancestral characters, if all the domestic breeds have descended from the rock-pigeon. But if we deny this, we must make one of the two following highly improbable suppositions. Either, firstly, that all the several imagined aboriginal stocks were coloured and marked like the rock-pigeon, although no other existing species is thus coloured and marked, so that in each separate breed there might be a tendency to revert to the very same colours and markings. Or, secondly, that each breed, even the purest, has within a dozen or, at most, within a score of generations, been crossed by the rock-pigeon: I say within a dozen or twenty generations, for we know of no fact countenancing the belief that the child ever reverts to some one ancestor, removed by a greater number of generations. In a breed which has been crossed only once with some distinct breed, the tendency to reversion to any character derived from such cross will naturally become less and less, as in each succeeding generation there will be less of the foreign blood; but when there has been no cross with a distinct breed, and there is a tendency in both parents to revert to a character, which has been lost during some former generation, this tendency, for all that we can see to the contrary, may be transmitted undiminished for an indefinite number of generations. These two distinct cases are often confounded in treatises on inheritance.

Lastly, the hybrids or mongrels from between all the domestic breeds of pigeons are perfectly fertile. I can state this from my own observations, purposely made on the most distinct breeds. Now, it is difficult, perhaps impossible, to bring forward one case of the hybrid offspring of two animals *clearly distinct* being themselves perfectly fertile. Some authors believe that long-continued domestication eliminates this strong tendency to sterility: from the history of the dog I think there is some probability in this hypothesis, if applied to species closely related together, though it is unsupported by a single experiment. But to extend the hypothesis so far as to suppose that species, aboriginally as distinct as carriers, tumblers, pouters, and fantails now are, should yield offspring perfectly fertile, *inter se*, seems to me rash in the extreme.

From these several reasons, namely, the improbability of man having formerly got seven or eight supposed species of pigeons to breed freely under domestication; these supposed species being quite unknown in a wild state, and their becoming nowhere feral; these species having very abnormal characters in certain respects, as compared with all other Columbidæ, though so like in most other respects to the rock-pigeon; the blue colour and various marks occasionally appearing in all the breeds, both when kept pure and when crossed; the mongrel offspring being perfectly fertile;—from these several reasons, taken together, I can feel no doubt that all our domestic breeds have descended from the Columba livia with its geographical sub-species.

In favour of this view, I may add, firstly, that C. livia, or the rock-pigeon, has been found capable of domestication in Europe and in India; and that it agrees in habits and in a great number of points of structure with all the domestic breeds. Secondly, although an English carrier or short-faced tumbler differs immensely in certain characters from the rock-pigeon, yet by comparing the several sub-breeds of these breeds, more especially those brought from distant countries, we can make an almost perfect series between the extremes of structure. Thirdly, those characters which are mainly distinctive of each breed, for instance the wattle and length of beak of the carrier, the shortness of that of the tumbler, and the number of tail-feathers in the fantail, are in each breed eminently variable; and the explanation of this fact will be obvious when we come to treat of selection. Fourthly, pigeons have been watched, and tended with the utmost care, and loved by many people. They have been domesticated for thousands of years in several quarters of the world; the earliest known record of pigeons is in the fifth Ægyptian dynasty, about 3000 B.C., as was pointed out to me by Professor Lepsius; but Mr. Birch informs me that pigeons are given in a bill of fare in the previous dynasty. In the time of the Romans, as we hear from Pliny, immense prices were given for pigeons; "nay, they are come to this pass, that they can reckon up their pedigree and race." Pigeons were much valued by Akber Khan in India, about the year 1600; never less than 20,000 pigeons were taken with the court. "The monarchs of Iran and

Turan sent him some very rare birds;" and, continues the courtly historian, "His Majesty by crossing the breeds, which method was never practised before, has improved them astonishingly." About this same period the Dutch were as eager about pigeons as were the old Romans. The paramount importance of these considerations in explaining the immense amount of variation which pigeons have undergone, will be obvious when we treat of Selection. We shall then, also, see how it is that the breeds so often have a somewhat monstrous character. It is also a most favourable circumstance for the production of distinct breeds, that male and female pigeons can be easily mated for life; and thus different breeds can be kept together in the same aviary.

I have discussed the probable origin of domestic pigeons at some, yet quite insufficient, length; because when I first kept pigeons and watched the several kinds, knowing well how true they bred, I felt fully as much difficulty in believing that they could ever have descended from a common parent, as any naturalist could in coming to a similar conclusion in regard to the many species of finches, or other large groups of birds, in nature. One circumstance has struck me much; namely, that all the breeders of the various domestic animals and the cultivators of plants, with whom I have ever conversed, or whose treatises I have read, are firmly convinced that the several breeds to which each has attended, are descended from so many aboriginally distinct species. Ask, as I have asked, a celebrated raiser of Hereford cattle, whether his cattle might not have descended from long-horns, and he will laugh you to scorn. I have never met a pigeon, or poultry, or duck, or rabbit fancier, who was not fully convinced that each main breed was descended from a distinct species. Van Mons, in his treatise on pears and apples, shows how utterly he disbelieves that the several sorts, for instance a Ribston-pippin or Codlin-apple, could ever have proceeded from the seeds of the same tree. Innumerable other examples could be given. The explanation, I think, is simple: from long-continued study they are strongly impressed with the differences between the several races; and though they well know that each race varies slightly, for they win their prizes by selecting such slight differences, yet they ignore all general arguments, and refuse to sum up in their minds slight differences accumulated during many successive generations. May not those naturalists who, knowing far less of the laws of inheritance than does the breeder, and knowing no more than he does of the intermediate links in the long lines of descent, yet admit that many of our domestic races have descended from the same parents—may they not learn a lesson of caution, when they deride the idea of species in a state of nature being lineal descendants of other species?

Selection.—Let us now briefly consider the steps by which domestic races have been produced, either from one or from several allied species. Some little effect may, perhaps, be attributed to the direct action of the external conditions of life, and some little to habit; but he would be a bold man who would account by such agencies for the differences of a dray and race horse, a greyhound and bloodhound, a carrier and tumbler pigeon. One of the most remarkable features in our domesticated races is that we see in them adaptation, not indeed to the animal's or plant's own good, but to man's use or fancy. Some variations useful to him have probably arisen suddenly, or by one step; many botanists, for instance, believe that the fuller's teazle, with its hooks, which cannot be rivalled by any mechanical contrivance, is only a variety of the wild Dipsacus;* and this amount of change may have suddenly arisen in a seedling. So it has probably been with the turnspit dog; and this is known to have been the case with the ancon sheep.* But when we compare the dray-horse and race-horse, the dromedary and camel, the various breeds of sheep fitted either for cultivated land or mountain pasture, with the wool of one breed good for one purpose, and that of another breed for another purpose; when we compare the many breeds of dogs, each good for man in very different ways; when we compare the game-cock, so pertinacious in battle, with other breeds so little quarrelsome, with "everlasting layers" which never desire to sit, and with the bantam so small and elegant; when we compare the host of agricultural, culinary, orchard, and flower-garden races of plants, most useful to man at different seasons and for different purposes, or so beautiful in his eyes, we must, I think, look further than to mere variability. We cannot suppose that all the breeds were suddenly produced as perfect and as useful as we now see them; indeed, in several cases, we know that this has not been their history. The key is man's power of accumulative selection: nature gives successive variations; man adds them up in certain directions useful to him. In this sense he may be said to make for himself useful breeds.

The great power of this principle of selection is not hypothetical. It is certain that several of our eminent breeders have, even within a single lifetime, modified to a large extent some breeds of cattle and

sheep. In order fully to realise what they have done, it is almost necessary to read several of the many treatises devoted to this subject,
and to inspect the animals. Breeders habitually speak of an animal's
organisation as something quite plastic, which they can model almost
as they please. If I had space I could quote numerous passages to this
effect from highly competent authorities. Youatt, who was probably
better acquainted with the works of agriculturalists than almost
any other individual, and who was himself a very good judge of an
animal, speaks of the principle of selection as "that which enables the
agriculturist, not only to modify the character of his flock, but to
change it altogether. It is the magician's wand, by means of which he
may summon into life whatever form and mould he pleases."

[Further testimony underlines the significance of careful selection.]

At the present time, eminent breeders try by methodical selection,
with a distinct object in view, to make a new strain or sub-breed,
superior to anything existing in the country. But, for our purpose,
a kind of Selection, which may be called Unconscious, and which
results from every one trying to possess and breed from the best
individual animals, is more important. Thus, a man who intends
keeping pointers naturally tries to get as good dogs as he can, and
afterwards breeds from his own best dogs, but he has no wish or
expectation of permanently altering the breed. Nevertheless I cannot
doubt that this process, continued during centuries, would improve
and modify any breed, in the same way as Bakewell, Collins,*
&c., by this very same process, only carried on more methodically,
did greatly modify, even during their own lifetimes, the forms and
qualities of their cattle. Slow and insensible changes of this kind
could never be recognised unless actual measurements or careful
drawings of the breeds in question had been made long ago, which
might serve for comparison. In some cases, however, unchanged or
but little changed individuals of the same breed may be found in less
civilised districts, where the breed has been less improved. There is
reason to believe that King Charles's spaniel has been unconsciously
modified to a large extent since the time of that monarch. Some
highly competent authorities are convinced that the setter is directly
derived from the spaniel, and has probably been slowly altered from it.
It is known that the English pointer has been greatly changed
within the last century, and in this case the change has, it is believed,

been chiefly effected by crosses with the fox-hound; but what concerns us is, that the change has been effected unconsciously and gradually, and yet so effectually, that, though the old Spanish pointer certainly came from Spain, Mr. Borrow has not seen, as I am informed by him, any native dog in Spain like our pointer.

By a similar process of selection, and by careful training, the whole body of English racehorses have come to surpass in fleetness and size the parent Arab stock, so that the latter, by the regulations for the Goodwood Races, are favoured in the weights they carry. Lord Spencer and others have shown how the cattle of England have increased in weight and in early maturity, compared with the stock formerly kept in this country. By comparing the accounts given in old pigeon treatises of carriers and tumblers with these breeds as now existing in Britain, India, and Persia, we can, I think, clearly trace the stages through which they have insensibly passed, and come to differ so greatly from the rock-pigeon.

Youatt gives an excellent illustration of the effects of a course of selection, which may be considered as unconsciously followed, in so far that the breeders could never have expected or even have wished to have produced the result which ensued—namely, the production of two distinct strains. The two flocks of Leicester sheep kept by Mr. Buckley and Mr. Burgess, as Mr. Youatt remarks, "have been purely bred from the original stock of Mr. Bakewell for upwards of fifty years. There is not a suspicion existing in the mind of any one at all acquainted with the subject that the owner of either of them has deviated in any one instance from the pure blood of Mr. Bakewell's flock, and yet the difference between the sheep possessed by these two gentlemen is so great that they have the appearance of being quite different varieties."

If there exist savages so barbarous as never to think of the inherited character of the offspring of their domestic animals, yet any one animal particularly useful to them, for any special purpose, would be carefully preserved during famines and other accidents, to which savages are so liable, and such choice animals would thus generally leave more offspring than the inferior ones; so that in this case there would be a kind of unconscious selection going on. We see the value set on animals even by the barbarians of Tierra del Fuego, by their killing and devouring their old women, in times of dearth, as of less value than their dogs.

In plants the same gradual process of improvement, through the occasional preservation of the best individuals, whether or not sufficiently distinct to be ranked at their first appearance as distinct varieties, and whether or not two or more species or races have become blended together by crossing, may plainly be recognised in the increased size and beauty which we now see in the varieties of the heartsease, rose, pelargonium, dahlia, and other plants, when compared with the older varieties or with their parent-stocks. No one would ever expect to get a first-rate heartsease or dahlia from the seed of a wild plant. No one would expect to raise a first-rate melting pear from the seed of a wild pear, though he might succeed from a poor seedling growing wild, if it had come from a garden-stock. The pear, though cultivated in classical times, appears, from Pliny's description, to have been a fruit of very inferior quality. I have seen great surprise expressed in horticultural works at the wonderful skill of gardeners, in having produced such splendid results from such poor materials; but the art, I cannot doubt, has been simple, and, as far as the final result is concerned, has been followed almost unconsciously. It has consisted in always cultivating the best known variety, sowing its seeds, and, when a slightly better variety has chanced to appear, selecting it, and so onwards. But the gardeners of the classical period, who cultivated the best pear they could procure, never thought what splendid fruit we should eat; though we owe our excellent fruit, in some small degree, to their having naturally chosen and preserved the best varieties they could anywhere find.

A large amount of change in our cultivated plants, thus slowly and unconsciously accumulated, explains, as I believe, the well-known fact, that in a vast number of cases we cannot recognise, and therefore do not know, the wild parent-stocks of the plants which have been longest cultivated in our flower and kitchen gardens. If it has taken centuries or thousands of years to improve or modify most of our plants up to their present standard of usefulness to man, we can understand how it is that neither Australia, the Cape of Good Hope, nor any other region inhabited by quite uncivilised man, has afforded us a single plant worth culture. It is not that these countries, so rich in species, do not by a strange chance possess the aboriginal stocks of any useful plants, but that the native plants have not been improved by continued selection up to a standard of perfection comparable with that given to the plants in countries anciently civilised.

In regard to the domestic animals kept by uncivilised man, it should not be overlooked that they almost always have to struggle for their own food, at least during certain seasons. And in two countries very differently circumstanced, individuals of the same species, having slightly different constitutions or structure, would often succeed better in the one country than in the other, and thus by a process of "natural selection," as will hereafter be more fully explained, two sub-breeds might be formed. This, perhaps, partly explains what has been remarked by some authors, namely, that the varieties kept by savages have more of the character of species than the varieties kept in civilised countries.

On the view here given of the all-important part which selection by man has played, it becomes at once obvious, how it is that our domestic races show adaptation in their structure or in their habits to man's wants or fancies. We can, I think, further understand the frequently abnormal character of our domestic races, and likewise their differences being so great in external characters and relatively so slight in internal parts or organs. Man can hardly select, or only with much difficulty, any deviation of structure excepting such as is externally visible; and indeed he rarely cares for what is internal. He can never act by selection, excepting on variations which are first given to him in some slight degree by nature. No man would ever try to make a fantail, till he saw a pigeon with a tail developed in some slight degree in an unusual manner, or a pouter till he saw a pigeon with a crop of somewhat unusual size; and the more abnormal or unusual any character was when it first appeared, the more likely it would be to catch his attention. But to use such an expression as trying to make a fantail, is, I have no doubt, in most cases, utterly incorrect. The man who first selected a pigeon with a slightly larger tail, never dreamed what the descendants of that pigeon would become through long-continued, partly unconscious and partly methodical selection. Perhaps the parent bird of all fantails had only fourteen tail-feathers somewhat expanded, like the present Java fantail, or like individuals of other and distinct breeds, in which as many as seventeen tail-feathers have been counted. Perhaps the first pouter-pigeon did not inflate its crop much more than the turbit now does the upper part of its œsophagus,—a habit which is disregarded by all fanciers, as it is not one of the points of the breed.

Nor let it be thought that some great deviation of structure would be necessary to catch the fancier's eye: he perceives extremely small differences, and it is in human nature to value any novelty, however slight, in one's own possession. Nor must the value which would formerly be set on any slight differences in the individuals of the same species, be judged of by the value which would now be set on them, after several breeds have once fairly been established. Many slight differences might, and indeed do now, arise amongst pigeons, which are rejected as faults or deviations from the standard of perfection of each breed. The common goose has not given rise to any marked varieties; hence the Thoulouse and the common breed, which differ only in colour, that most fleeting of characters, have lately been exhibited as distinct at our poultry-shows.

I think these views further explain what has sometimes been noticed—namely, that we know nothing about the origin or history of any of our domestic breeds. But, in fact, a breed, like a dialect of a language, can hardly be said to have had a definite origin. A man preserves and breeds from an individual with some slight deviation of structure, or takes more care than usual in matching his best animals and thus improves them, and the improved individuals slowly spread in the immediate neighbourhood. But as yet they will hardly have a distinct name, and from being only slightly valued, their history will be disregarded. When further improved by the same slow and gradual process, they will spread more widely, and will get recognised as something distinct and valuable, and will then probably first receive a provincial name. In semi-civilised countries, with little free communication, the spreading and knowledge of any new sub-breed will be a slow process. As soon as the points of value of the new sub-breed are once fully acknowledged, the principle, as I have called it, of unconscious selection will always tend,—perhaps more at one period than at another, as the breed rises or falls in fashion,—perhaps more in one district than in another, according to the state of civilisation of the inhabitants,—slowly to add to the characteristic features of the breed, whatever they may be. But the chance will be infinitely small of any record having been preserved of such slow, varying, and insensible changes.

[*New domestic varieties are most likely to appear in large populations kept under favourable conditions, particularly when there is some possibility of preventing unwanted crosses.*]

To sum up on the origin of our Domestic Races of animals and plants. I believe that the conditions of life, from their action on the reproductive system, are so far of the highest importance as causing variability. I do not believe that variability is an inherent and necessary contingency, under all circumstances, with all organic beings, as some authors have thought. The effects of variability are modified by various degrees of inheritance and of reversion. Variability is governed by many unknown laws, more especially by that of correlation of growth. Something may be attributed to the direct action of the conditions of life. Something must be attributed to use and disuse. The final result is thus rendered infinitely complex. In some cases, I do not doubt that the intercrossing of species, aboriginally distinct, has played an important part in the origin of our domestic productions. When in any country several domestic breeds have once been established, their occasional intercrossing, with the aid of selection, has, no doubt, largely aided in the formation of new sub-breeds; but the importance of the crossing of varieties has, I believe, been greatly exaggerated, both in regard to animals and to those plants which are propagated by seed. In plants which are temporarily propagated by cuttings, buds, &c., the importance of the crossing both of distinct species and of varieties is immense; for the cultivator here quite disregards the extreme variability both of hybrids and mongrels, and the frequent sterility of hybrids; but the cases of plants not propagated by seed are of little importance to us, for their endurance is only temporary. Over all these causes of Change I am convinced that the accumulative action of Selection, whether applied methodically and more quickly, or unconsciously and more slowly, but more efficiently, is by far the predominant Power.

CHAPTER II

Variation under Nature

BEFORE applying the principles arrived at in the last chapter to organic beings in a state of nature, we must briefly discuss whether these latter are subject to any variation. To treat this subject at all properly, a long catalogue of dry facts should be given; but these I shall reserve for my future work. Nor shall I here discuss the various

definitions which have been given of the term species. No one definition has as yet satisfied all naturalists; yet every naturalist knows
vaguely what he means when he speaks of a species. Generally the
term includes the unknown element of a distinct act of creation. The
term "variety" is almost equally difficult to define; but here community of descent is almost universally implied, though it can rarely be
proved. We have also what are called monstrosities; but they graduate into varieties. By a monstrosity I presume is meant some considerable deviation of structure in one part, either injurious to or not
useful to the species, and not generally propagated. Some authors
use the term "variation" in a technical sense, as implying a modification directly due to the physical conditions of life; and "variations"
in this sense are supposed not to be inherited: but who can say that
the dwarfed condition of shells in the brackish waters of the Baltic,
or dwarfed plants on Alpine summits, or the thicker fur of an animal
from far northwards, would not in some cases be inherited for at least
some few generations? and in this case I presume that the form
would be called a variety.

Again, we have many slight differences which may be called individual differences, such as are known frequently to appear in the
offspring from the same parents, or which may be presumed to have
thus arisen, from being frequently observed in the individuals of the
same species inhabiting the same confined locality. No one supposes
that all the individuals of the same species are cast in the very same
mould. These individual differences are highly important for us, as
they afford materials for natural selection to accumulate, in the same
manner as man can accumulate in any given direction individual
differences in his domesticated productions. These individual differences generally affect what naturalists consider unimportant parts;
but I could show by a long catalogue of facts, that parts which must
be called important, whether viewed under a physiological or classificatory point of view, sometimes vary in the individuals of the same
species. I am convinced that the most experienced naturalist would
be surprised at the number of the cases of variability, even in important parts of structure, which he could collect on good authority, as
I have collected, during a course of years. It should be remembered
that systematists are far from pleased at finding variability in important characters, and that there are not many men who will laboriously
examine internal and important organs, and compare them in many

specimens of the same species. I should never have expected that the branching of the main nerves close to the great central ganglion of an insect would have been variable in the same species; I should have expected that changes of this nature could have been effected only by slow degrees: yet quite recently Mr. Lubbock has shown a degree of variability in these main nerves in Coccus, which may almost be compared to the irregular branching of the stem of a tree. This philosophical naturalist, I may add, has also quite recently shown that the muscles in the larvæ of certain insects are very far from uniform. Authors sometimes argue in a circle when they state that important organs never vary; for these same authors practically rank that character as important (as some few naturalists have honestly confessed) which does not vary; and, under this point of view, no instance of an important part varying will ever be found: but under any other point of view many instances assuredly can be given.

There is one point connected with individual differences, which seems to me extremely perplexing: I refer to those genera which have sometimes been called "protean" or "polymorphic," in which the species present an inordinate amount of variation; and hardly two naturalists can agree which forms to rank as species and which as varieties. We may instance Rubus, Rosa, and Hieracium amongst plants,* several genera of insects, and several genera of Brachiopod shells.* In most polymorphic genera some of the species have fixed and definite characters. Genera which are polymorphic in one country seem to be, with some few exceptions, polymorphic in other countries, and likewise, judging from Brachiopod shells, at former periods of time. These facts seem to be very perplexing, for they seem to show that this kind of variability is independent of the conditions of life. I am inclined to suspect that we see in these polymorphic genera variations in points of structure which are of no service or disservice to the species, and which consequently have not been seized on and rendered definite by natural selection, as hereafter will be explained.

Those forms which possess in some considerable degree the character of species, but which are so closely similar to some other forms, or are so closely linked to them by intermediate gradations, that naturalists do not like to rank them as distinct species, are in several respects the most important for us. We have every reason to believe that many of these doubtful and closely-allied forms have permanently retained their characters in their own country for a long time; for as long, as far as we know, as have good and true species. Practically, when a naturalist can unite two forms together by others having intermediate characters, he treats the one as a variety of the other, ranking the most common, but sometimes the one first described, as the species, and the other as the variety. But cases of great difficulty, which I will not here enumerate, sometimes occur in deciding whether or not to rank one form as a variety of another, even when they are closely connected by intermediate links; nor will the commonly-assumed hybrid nature of the intermediate links always remove the difficulty. In very many cases, however, one form is ranked as a variety of another, not because the intermediate links have actually been found, but because analogy leads the observer to suppose either that they do now somewhere exist, or may formerly have existed; and here a wide door for the entry of doubt and conjecture is opened.

[*The fluidity of the boundary between varieties and species is illustrated by examples, including those from Darwin's own experience.*]

When a young naturalist commences the study of a group of organisms quite unknown to him, he is at first much perplexed to determine what differences to consider as specific, and what as varieties; for he knows nothing of the amount and kind of variation to which the group is subject; and this shows, at least, how very generally there is some variation. But if he confine his attention to one class within one country, he will soon make up his mind how to rank most of the doubtful forms. His general tendency will be to make many species, for he will become impressed, just like the pigeon or poultry-fancier before alluded to, with the amount of difference in the forms which he is continually studying; and he has little general knowledge of analogical variation in other groups and in other countries, by which to correct his first impressions. As he extends the range of his observations, he will meet with more cases of difficulty; for he will encounter a greater number of closely-allied forms. But if his observations be widely extended, he will in the end generally be enabled to make up his own mind which to call varieties and which species; but he will succeed in this at the expense of admitting much variation,—and the truth of this admission will often be disputed by other naturalists. When, moreover, he comes to study allied forms brought from countries not now continuous, in which case he can hardly hope to find the intermediate links between his doubtful forms, he will have to trust almost entirely to analogy, and his difficulties will rise to a climax.

Certainly no clear line of demarcation has as yet been drawn between species and sub-species—that is, the forms which in the opinion of some naturalists come very near to, but do not quite arrive at the rank of species; or, again, between sub-species and well-marked varieties, or between lesser varieties and individual differences. These differences blend into each other in an insensible series; and a series impresses the mind with the idea of an actual passage.

Hence I look at individual differences, though of small interest to the systematist, as of high importance for us, as being the first step towards such slight varieties as are barely thought worth recording in works on natural history. And I look at varieties which are in any degree more distinct and permanent, as steps leading to more strongly marked and more permanent varieties; and at these latter, as leading to sub-species, and to species. The passage from one stage of difference to another and higher stage may be, in some cases, due merely to the long-continued action of different physical conditions in two different regions; but I have not much faith in this view; and I attribute the passage of a variety, from a state in which it differs very slightly from its parent to one in which it differs more, to the action of natural selection in accumulating (as will hereafter be more fully explained) differences of structure in certain definite directions. Hence I believe a well-marked variety may be justly called an incipient species; but whether this belief be justifiable must be judged of by the general weight of the several facts and views given throughout this work.

It need not be supposed that all varieties or incipient species necessarily attain the rank of species. They may whilst in this incipient state become extinct, or they may endure as varieties for very long periods, as has been shown to be the case by Mr. Wollaston with the varieties of certain fossil land-shells in Madeira. If a variety were to flourish so as to exceed in numbers the parent species, it would then rank as the species, and the species as the variety; or it might come to supplant and exterminate the parent species; or both might co-exist, and both rank as independent species. But we shall hereafter have to return to this subject.

From these remarks it will be seen that I look at the term species, as one arbitrarily given for the sake of convenience to a set of individuals closely resembling each other, and that it does not essentially differ from the term variety, which is given to less distinct and more fluctuating forms. The term variety, again, in comparison with mere individual differences, is also applied arbitrarily, and for mere convenience sake.

[*A detailed quantitative investigation indicates that dominant species of plants tend to produce the most varieties or 'incipient species'. The chapter then draws to a close.*]

Finally, then, varieties have the same general characters as species, for they cannot be distinguished from species,—except, firstly, by the discovery of intermediate linking forms, and the occurrence of such links cannot affect the actual characters of the forms which they connect; and except, secondly, by a certain amount of difference, for two forms, if differing very little, are generally ranked as varieties, notwithstanding that intermediate linking forms have not been discovered; but the amount of difference considered necessary to give to two forms the rank of species is quite indefinite. In genera having more than the average number of species in any country, the species of these genera have more than the average number of varieties. In large genera the species are apt to be closely, but unequally, allied together, forming little clusters round certain species. Species very closely allied to other species apparently have restricted ranges. In all these several respects the species of large genera present a strong analogy with varieties. And we can clearly understand these analogies, if species have once existed as varieties, and have thus originated: whereas, these analogies are utterly inexplicable if each species has been independently created.

We have, also, seen that it is the most flourishing and dominant species of the larger genera which on an average vary most; and varieties, as we shall hereafter see, tend to become converted into new and distinct species. The larger genera thus tend to become larger; and throughout nature the forms of life which are now dominant tend to become still more dominant by leaving many modified and dominant descendants. But by steps hereafter to be explained, the larger genera also tend to break up into smaller genera. And thus, the forms of life throughout the universe become divided into groups subordinate to groups.

CHAPTER III

Struggle for Existence

BEFORE entering on the subject of this chapter, I must make a few preliminary remarks, to show how the struggle for existence bears on Natural Selection. It has been seen in the last chapter that amongst organic beings in a state of nature there is some individual variability; indeed I am not aware that this has ever been disputed. It is immaterial for us whether a multitude of doubtful forms be called species or sub-species or varieties; what rank, for instance, the two or three hundred doubtful forms of British plants are entitled to hold, if the existence of any well-marked varieties be admitted. But the mere existence of individual variability and of some few well-marked varieties, though necessary as the foundation for the work, helps us but little in understanding how species arise in nature. How have all those exquisite adaptations of one part of the organisation to another part, and to the conditions of life, and of one distinct organic being to another being, been perfected? We see these beautiful co-adaptations most plainly in the woodpecker and missletoe; and only a little less plainly in the humblest parasite which clings to the hairs of a quadruped or feathers of a bird; in the structure of the beetle which dives through the water; in the plumed seed which is wafted by the gentlest breeze; in short, we see beautiful adaptations everywhere and in every part of the organic world.

Again, it may be asked, how is it that varieties, which I have called incipient species, become ultimately converted into good and distinct species, which in most cases obviously differ from each other far more than do the varieties of the same species? How do those groups of species, which constitute what are called distinct genera, and which differ from each other more than do the species of the same genus, arise? All these results, as we shall more fully see in the next chapter, follow inevitably from the struggle for life. Owing to this struggle for life, any variation, however slight and from whatever cause proceeding, if it be in any degree profitable to an individual of any species, in its infinitely complex relations to other organic beings and to external nature, will tend to the preservation of that individual, and will generally be inherited by its offspring. The offspring, also, will thus have a better chance of surviving, for, of the many individuals of any species which are periodically born, but a small number can survive. I have called this principle, by which each slight variation, if useful, is preserved, by the term of Natural Selection, in order to mark its relation to man's power of selection. We have seen that man by selection can certainly produce great results, and can adapt organic beings to his own uses, through the accumulation of slight but useful variations, given to him by the hand of Nature. But Natural Selection, as we shall hereafter see, is a power incessantly ready for action, and is as immeasurably superior to man's feeble efforts, as the works of Nature are to those of Art.

We will now discuss in a little more detail the struggle for existence. In my future work this subject shall be treated, as it well deserves, at much greater length. The elder De Candolle and Lyell have largely and philosophically shown that all organic beings are exposed to severe competition. In regard to plants, no one has treated this subject with more spirit and ability than W. Herbert, Dean of Manchester, evidently the result of his great horticultural knowledge. Nothing is easier than to admit in words the truth of the universal struggle for life, or more difficult—at least I have found it so—than constantly to bear this conclusion in mind. Yet unless it be thoroughly engrained in the mind, I am convinced that the whole economy of nature, with every fact on distribution, rarity, abundance, extinction, and variation, will be dimly seen or quite misunderstood. We behold the face of nature bright with gladness, we often see superabundance of food; we do not see, or we forget, that the birds which are idly singing round us mostly live on insects or seeds, and are thus constantly destroying life; or we forget how largely these songsters, or their eggs, or their nestlings, are destroyed by birds and beasts of prey; we do not always bear in mind, that though food may be now superabundant, it is not so at all seasons of each recurring year.

I should premise that I use the term Struggle for Existence in a large and metaphorical sense, including dependence of one being on another, and including (which is more important) not only the life of the individual, but success in leaving progeny. Two canine animals in a time of dearth, may be truly said to struggle with each other which shall get food and live. But a plant on the edge of a desert is said to struggle for life against the drought, though more properly it should be said to be dependent on the moisture. A plant which annually produces a thousand seeds, of which on an average only one comes to maturity, may be more truly said to struggle with the plants of the same and other kinds which already clothe the ground. The missletoe is dependent on the apple and a few other trees, but can only in a far-fetched sense be said to struggle with these trees, for if too many of these parasites grow on the same tree, it will languish and die. But several seedling missletoes, growing close together on the same branch, may more truly be said to struggle with each other. As the missletoe is disseminated by birds, its existence depends on birds; and it may metaphorically be said to struggle with other fruit-bearing plants, in order to tempt birds to devour and thus disseminate its seeds rather than those of other plants. In these several senses, which pass into each other, I use for convenience sake the general term of struggle for existence.

A struggle for existence inevitably follows from the high rate at which all organic beings tend to increase. Every being, which during its natural lifetime produces several eggs or seeds, must suffer destruction during some period of its life, and during some season or occasional year, otherwise, on the principle of geometrical increase, its numbers would quickly become so inordinately great that no country could support the product. Hence, as more individuals are produced than can possibly survive, there must in every case be a struggle for existence, either one individual with another of the same species, or with the individuals of distinct species, or with the physical conditions of life. It is the doctrine of Malthus applied with manifold force to the whole animal and vegetable kingdoms; for in this case there can be no artificial increase of food, and no prudential restraint from marriage. Although some species may be now increasing, more or less rapidly, in numbers, all cannot do so, for the world would not hold them.

There is no exception to the rule that every organic being naturally increases at so high a rate, that if not destroyed, the earth would soon be covered by the progeny of a single pair. Even slow-breeding man has doubled in twenty-five years, and at this rate, in a few thousand years, there would literally not be standing room for his progeny. Linnæus has calculated that if an annual plant produced only two seeds—and there is no plant so unproductive as this—and their seedlings next year produced two, and so on, then in twenty years there would be a million plants. The elephant is reckoned to be the slowest breeder of all known animals, and I have taken some pains to estimate its probable minimum rate of natural increase: it will be under the mark to assume that it breeds when thirty years old, and goes on breeding till ninety years old, bringing forth three pair of young in this interval; if this be so, at the end of the fifth century there would be alive fifteen million elephants, descended from the first pair.

But we have better evidence on this subject than mere theoretical calculations, namely, the numerous recorded cases of the astonishingly rapid increase of various animals in a state of nature, when circumstances have been favourable to them during two or three following seasons. Still more striking is the evidence from our domestic animals of many kinds which have run wild in several parts of the world: if the statements of the rate of increase of slow-breeding cattle and horses in South-America, and latterly in Australia, had not been well authenticated, they would have been quite incredible. So it is with plants: cases could be given of introduced plants which have become common throughout whole islands in a period of less than ten years. Several of the plants now most numerous over the wide plains of La Plata, clothing square leagues of surface almost to the exclusion of all other plants, have been introduced from Europe; and there are plants which now range in India, as I hear from Dr. Falconer, from Cape Comorin to the Himalaya, which have been imported from America since its discovery. In such cases, and endless instances could be given, no one supposes that the fertility of these animals or plants has been suddenly and temporarily increased in any sensible degree. The obvious explanation is that the conditions of life have been very favourable, and that there has consequently been less destruction of the old and young, and that nearly all the young have been enabled to breed. In such cases the geometrical ratio of increase, the result of which never fails to be surprising, simply explains the extraordinarily rapid increase and wide diffusion of naturalised productions in their new homes.

In a state of nature almost every plant produces seed, and amongst animals there are very few which do not annually pair. Hence we may confidently assert, that all plants and animals are tending to increase at a geometrical ratio, that all would most rapidly stock every station in which they could any how exist, and that the geometrical tendency to increase must be checked by destruction at some period of life. Our familiarity with the larger domestic animals tends, I think, to mislead us: we see no great destruction falling on them, and we forget that thousands are annually slaughtered for food, and that in a state of nature an equal number would have somehow to be disposed of.

The only difference between organisms which annually produce eggs or seeds by the thousand, and those which produce extremely few, is, that the slow-breeders would require a few more years to people, under favourable conditions, a whole district, let it be ever so large. The condor lays a couple of eggs and the ostrich a score, and yet in the same country the condor may be the more numerous of the two: the Fulmar petrel lays but one egg, yet it is believed to be the most numerous bird in the world. One fly deposits hundreds of eggs, and another, like the hippobosca, a single one; but this difference does not determine how many individuals of the two species can be supported in a district. A large number of eggs is of some importance to those species, which depend on a rapidly fluctuating amount of food, for it allows them rapidly to increase in number. But the real importance of a large number of eggs or seeds is to make up for much destruction at some period of life; and this period in the great majority of cases is an early one. If an animal can in any way protect its own eggs or young, a small number may be produced, and yet the average stock be fully kept up; but if many eggs or young are destroyed, many must be produced, or the species will become extinct. It would suffice to keep up the full number of a tree, which lived on an average for a thousand years, if a single seed were produced once in a thousand years, supposing that this seed were never destroyed, and could be ensured to germinate in a fitting place. So that in all cases, the average number of any animal or plant depends only indirectly on the number of its eggs or seeds.

In looking at Nature, it is most necessary to keep the foregoing considerations always in mind—never to forget that every single organic being around us may be said to be striving to the utmost to increase in numbers; that each lives by a struggle at some period of its life; that heavy destruction inevitably falls either on the young or old, during each generation or at recurrent intervals. Lighten any check, mitigate the destruction ever so little, and the number of the species will almost instantaneously increase to any amount. The face of Nature may be compared to a yielding surface, with ten thousand sharp wedges packed close together and driven inwards by incessant blows, sometimes one wedge being struck, and then another with greater force.

[*Lack of food, destruction of eggs, disease, the vicissitudes of climate: all these keep each species from multiplying indefinitely.*]

In the case of every species, many different checks, acting at different periods of life, and during different seasons or years, probably come into play; some one check or some few being generally the most potent, but all concurring in determining the average number or even the existence of the species. In some cases it can be shown that widely-different checks act on the same species in different districts. When we look at the plants and bushes clothing an entangled bank, we are tempted to attribute their proportional numbers and kinds to what we call chance. But how false a view is this! Every one has heard that when an American forest is cut down, a very different vegetation springs up; but it has been observed that the trees now growing on the ancient Indian mounds, in the Southern United States, display the same beautiful diversity and proportion of kinds as in the surrounding virgin forests. What a struggle between the several kinds of trees must here have gone on during long centuries, each annually scattering its seeds by the thousand; what war between insect and insect — between insects, snails, and other animals with birds and beasts of prey — all striving to increase, and all feeding on each other or on the trees or their seeds and seedlings, or on the other plants which first clothed the ground and thus checked the growth of the trees! Throw up a handful of feathers, and all must fall to the ground according to definite laws; but how simple is this problem compared to the action and reaction of the innumerable plants and animals which have determined, in the course of centuries, the proportional numbers and kinds of trees now growing on the old Indian ruins!

The dependency of one organic being on another, as of a parasite on its prey, lies generally between beings remote in the scale of nature. This is often the case with those which may strictly be said to struggle with each other for existence, as in the case of locusts and grass-feeding quadrupeds. But the struggle almost invariably will be most severe between the individuals of the same species, for they frequent the same districts, require the same food, and are exposed to the same dangers. In the case of varieties of the same species, the struggle will generally be almost equally severe, and we sometimes see the contest soon decided: for instance, if several varieties of wheat be sown together, and the mixed seed be resown, some of the varieties which best suit the soil or climate, or are naturally the most fertile, will beat the others and so yield more seed, and will consequently in a few years quite supplant the other varieties. To keep up a mixed stock of even such extremely close varieties as the variously coloured sweet-peas, they must be each year harvested separately, and the seed then mixed in due proportion, otherwise the weaker kinds will steadily decrease in numbers and disappear. So again with the varieties of sheep: it has been asserted that certain mountain-varieties will starve out other mountain-varieties, so that they cannot be kept together. The same result has followed from keeping together different varieties of the medicinal leech. It may even be doubted whether the varieties of any one of our domestic plants or animals have so exactly the same strength, habits, and constitution, that the original proportions of a mixed stock could be kept up for half a dozen generations, if they were allowed to struggle together, like beings in a state of nature, and if the seed or young were not annually sorted.

As species of the same genus have usually, though by no means invariably, some similarity in habits and constitution, and always in structure, the struggle will generally be more severe between species of the same genus, when they come into competition with each other, than between species of distinct genera. We see this in the recent extension over parts of the United States of one species of swallow having caused the decrease of another species. The recent increase of the missel-thrush in parts of Scotland has caused the decrease of the song-thrush. How frequently we hear of one species of rat taking the place of another species under the most different climates! In Russia the small Asiatic cockroach has everywhere driven before it its great congener. One species of charlock* will supplant another, and so in other cases. We can dimly see why the competition should be most severe between allied forms, which fill nearly the same place in the economy of nature; but probably in no one case could we precisely say why one species has been victorious over another in the great battle of life.

A corollary of the highest importance may be deduced from the foregoing remarks, namely, that the structure of every organic being is related, in the most essential yet often hidden manner, to that of all other organic beings, with which it comes into competition for food or residence, or from which it has to escape, or on which it preys. This is obvious in the structure of the teeth and talons of the tiger; and in that of the legs and claws of the parasite which clings to the hair on the tiger's body. But in the beautifully plumed seed of the dandelion, and in the flattened and fringed legs of the water-beetle, the relation seems at first confined to the elements of air and water. Yet the advantage of plumed seeds no doubt stands in the closest relation to the land being already thickly clothed by other plants; so that the seeds may be widely distributed and fall on unoccupied ground. In the water-beetle, the structure of its legs, so well adapted for diving, allows it to compete with other aquatic insects, to hunt for its own prey, and to escape serving as prey to other animals.

The store of nutriment laid up within the seeds of many plants seems at first sight to have no sort of relation to other plants. But from the strong growth of young plants produced from such seeds (as peas and beans), when sown in the midst of long grass, I suspect that the chief use of the nutriment in the seed is to favour the growth of the young seedling, whilst struggling with other plants growing vigorously all around.

Look at a plant in the midst of its range, why does it not double or quadruple its numbers? We know that it can perfectly well withstand a little more heat or cold, dampness or dryness, for elsewhere it ranges into slightly hotter or colder, damper or drier districts. In this case we can clearly see that if we wished in imagination to give the plant the power of increasing in number, we should have to give it some advantage over its competitors, or over the animals which preyed on it. On the confines of its geographical range, a change of constitution with respect to climate would clearly be an advantage to our plant; but we have reason to believe that only a few plants or animals range so far, that they are destroyed by the rigour of the climate alone. Not until we reach the extreme confines of life, in the arctic regions or on the borders of an utter desert, will competition cease. The land may be extremely cold or dry, yet there will be competition between some few species, or between the individuals of the same species, for the warmest or dampest spots.

Hence, also, we can see that when a plant or animal is placed in a new country amongst new competitors, though the climate may be exactly the same as in its former home, yet the conditions of its life will generally be changed in an essential manner. If we wished to increase its average numbers in its new home, we should have to modify it in a different way to what we should have done in its native country; for we should have to give it some advantage over a different set of competitors or enemies.

It is good thus to try in our imagination to give any form some advantage over another. Probably in no single instance should we know what to do, so as to succeed. It will convince us of our ignorance on the mutual relations of all organic beings; a conviction as necessary, as it seems to be difficult to acquire. All that we can do, is to keep steadily in mind that each organic being is striving to increase at a geometrical ratio; that each at some period of its life, during some season of the year, during each generation or at intervals, has to struggle for life, and to suffer great destruction. When we reflect on this struggle, we may console ourselves with the full belief, that the war of nature is not incessant, that no fear is felt, that death is generally prompt, and that the vigorous, the healthy, and the happy survive and multiply.

CHAPTER IV

Natural Selection

How will the struggle for existence, discussed too briefly in the last chapter, act in regard to variation? Can the principle of selection, which we have seen is so potent in the hands of man, apply in nature? I think we shall see that it can act most effectually. Let it be borne in mind in what an endless number of strange peculiarities our domestic productions, and, in a lesser degree, those under nature, vary; and how strong the hereditary tendency is. Under domestication, it may be truly said that the whole organisation becomes in some degree plastic. Let it be borne in mind how infinitely complex and close-fitting are the mutual relations of all organic beings to each other and to their physical conditions of life. Can it, then, be thought improbable, seeing that variations useful to man have undoubtedly occurred, that other variations useful in some way to each being in the great and complex battle of life, should sometimes occur in the course of thousands of generations? If such do occur, can we doubt (remembering that many more individuals are born than can possibly survive) that individuals having any advantage, however slight, over others, would have the best chance of surviving and of procreating their kind? On the other hand, we may feel sure that any variation in the least degree injurious would be rigidly destroyed. This preservation of favourable variations and the rejection of injurious variations, I call Natural Selection. Variations neither useful nor injurious would not be affected by natural selection, and would be left a fluctuating element, as perhaps we see in the species called polymorphic.

We shall best understand the probable course of natural selection by taking the case of a country undergoing some physical change, for instance, of climate. The proportional numbers of its inhabitants would almost immediately undergo a change, and some species might become extinct. We may conclude, from what we have seen of the intimate and complex manner in which the inhabitants of each country are bound together, that any change in the numerical proportions of some of the inhabitants, independently of the change of climate itself, would most seriously affect many of the others. If the country were open on its borders, new forms would certainly immigrate, and this also would seriously disturb the relations of some of the former inhabitants. Let it be remembered how powerful the influence of a single introduced tree or mammal has been shown to be. But in the case of an island, or of a country partly surrounded by barriers, into which new and better adapted forms could not freely enter, we should then have places in the economy of nature which would assuredly be better filled up, if some of the original inhabitants were in some manner modified; for, had the area been open to immigration, these same places would have been seized on by intruders. In such case, every slight modification, which in the course of ages chanced to arise, and which in any way favoured the individuals of any of the species, by better adapting them to their altered conditions, would tend to be preserved; and natural selection would thus have free scope for the work of improvement.

We have reason to believe, as stated in the first chapter, that a change in the conditions of life, by specially acting on the reproductive system, causes or increases variability; and in the foregoing case the conditions of life are supposed to have undergone a change, and this would manifestly be favourable to natural selection, by giving a better chance of profitable variations occurring; and unless profitable variations do occur, natural selection can do nothing. Not that, as I believe, any extreme amount of variability is necessary; as man can certainly produce great results by adding up in any given direction mere individual differences, so could Nature, but far more easily, from having incomparably longer time at her disposal. Nor do I believe that any great physical change, as of climate, or any unusual degree of isolation to check immigration, is actually necessary to produce new and unoccupied places for natural selection to fill up by modifying and improving some of the varying inhabitants. For as all the inhabitants of each country are struggling together with nicely balanced forces, extremely slight modifications in the structure or habits of one inhabitant would often give it an advantage over others; and still further modifications of the same kind would often still further increase the advantage. No country can be named in which all the native inhabitants are now so perfectly adapted to each other and to the physical conditions under which they live, that none of them could anyhow be improved; for in all countries, the natives have been so far conquered by naturalised productions, that they have allowed foreigners to take firm possession of the land. And as foreigners have thus everywhere beaten some of the natives, we may safely conclude that the natives might have been modified with advantage, so as to have better resisted such intruders.

As man can produce and certainly has produced a great result by his methodical and unconscious means of selection, what may not nature effect? Man can act only on external and visible characters: nature cares nothing for appearances, except in so far as they may be useful to any being. She can act on every internal organ, on every shade of constitutional difference, on the whole machinery of life. Man selects only for his own good; Nature only for that of the being which she tends. Every selected character is fully exercised by her; and the being is placed under well-suited conditions of life. Man keeps the natives of many climates in the same country; he seldom exercises each selected character in some peculiar and fitting manner; he feeds a long and a short beaked pigeon on the same food; he does not exercise a long-backed or long-legged quadruped in any peculiar manner; he exposes sheep with long and short wool to the same climate. He does not allow the most vigorous males to struggle for the females. He does not rigidly destroy all inferior animals, but protects during each varying season, as far as lies in his power, all his productions. He often begins his selection by some half-monstrous form; or at least by some modification prominent enough to catch his eye, or to be plainly useful to him. Under nature, the slightest difference of structure or constitution may well turn the nicely-balanced scale in the struggle for life, and so be preserved. How fleeting are the wishes and efforts of man! how short his time! and consequently how poor will his products be, compared with those accumulated by nature during whole geological periods. Can we wonder, then, that nature's productions should be far "truer" in character than man's productions; that they should be infinitely better adapted to the most complex conditions of life, and should plainly bear the stamp of far higher workmanship?

It may be said that natural selection is daily and hourly scrutinising, throughout the world, every variation, even the slightest; rejecting that which is bad, preserving and adding up all that is good; silently and insensibly working, whenever and wherever opportunity offers, at the improvement of each organic being in relation to its organic and inorganic conditions of life. We see nothing of these slow changes in progress, until the hand of time has marked the long lapse of ages, and then so imperfect is our view into long past geological ages, that we only see that the forms of life are now different from what they formerly were.

Although natural selection can act only through and for the good of each being, yet characters and structures, which we are apt to consider as of very trifling importance, may thus be acted on. When we see leaf-eating insects green, and bark-feeders mottled-grey; the alpine ptarmigan white in winter, the red-grouse the colour of heather, and the black-grouse that of peaty earth, we must believe that these tints are of service to these birds and insects in preserving them from danger. Grouse, if not destroyed at some period of their lives, would increase in countless numbers; they are known to suffer largely from birds of prey; and hawks are guided by eyesight to their prey,—so much so, that on parts of the Continent persons are warned not to keep white pigeons, as being the most liable to destruction. Hence I can see no reason to doubt that natural selection might be most effective in giving the proper colour to each kind of grouse, and in keeping that colour, when once acquired, true and constant. Nor ought we to think that the occasional destruction of an animal of any particular colour would produce little effect: we should remember how essential it is in a flock of white sheep to destroy every lamb with the faintest trace of black. In plants the down on the fruit and the colour of the flesh are considered by botanists as characters of the most trifling importance: yet we hear from an excellent horticulturist, Downing, that in the United States smooth-skinned fruits suffer far more from a beetle, a curculio, than those with down; that purple plums suffer far more from a certain disease than yellow plums; whereas another disease attacks yellow-fleshed peaches far more than those with other coloured flesh. If, with all the aids of art, these slight differences make a great difference in cultivating the several varieties, assuredly, in a state of nature, where the trees would have to struggle with other trees and with a host of enemies, such differences would effectually settle which variety, whether a smooth or downy, a yellow or purple fleshed fruit, should succeed.

In looking at many small points of difference between species, which, as far as our ignorance permits us to judge, seem to be quite unimportant, we must not forget that climate, food, &c., probably produce some slight and direct effect. It is, however, far more necessary to bear in mind that there are many unknown laws of correlation of growth, which, when one part of the organisation is modified through variation, and the modifications are accumulated by natural selection for the good of the being, will cause other modifications, often of the most unexpected nature.

As we see that those variations which under domestication appear at any particular period of life, tend to reappear in the offspring at the same period;—for instance, in the seeds of the many varieties of our culinary and agricultural plants; in the caterpillar and cocoon stages of the varieties of the silkworm; in the eggs of poultry, and in the colour of the down of their chickens; in the horns of our sheep and cattle when nearly adult;—so in a state of nature, natural selection will be enabled to act on and modify organic beings at any age, by the accumulation of profitable variations at that age, and by their inheritance at a corresponding age. If it profit a plant to have its seeds more and more widely disseminated by the wind, I can see no greater difficulty in this being effected through natural selection, than in the cotton-planter increasing and improving by selection the down in the pods on his cotton-trees. Natural selection may modify and adapt the larva of an insect to a score of contingencies, wholly different from those which concern the mature insect. These modifications will no doubt affect, through the laws of correlation, the structure of the adult; and probably in the case of those insects which live only for a few hours, and which never feed, a large part of their structure is merely the correlated result of successive changes in the structure of their larvæ. So, conversely, modifications in the adult will probably often affect the structure of the larva; but in all cases natural selection will ensure that modifications consequent on other modifications at a different period of life, shall not be in the least degree injurious: for if they became so, they would cause the extinction of the species.

Natural selection will modify the structure of the young in relation to the parent, and of the parent in relation to the young. In social animals it will adapt the structure of each individual for the benefit of the community; if each in consequence profits by the selected change. What natural selection cannot do, is to modify the structure of one species, without giving it any advantage, for the good of another species; and though statements to this effect may be found in works of natural history, I cannot find one case which will bear investigation. A structure used only once in an animal's whole life, if of high importance to it, might be modified to any extent by natural selection; for instance, the great jaws possessed by certain insects, and used exclusively for opening the cocoon—or the hard tip to the beak of nestling birds, used for breaking the egg. It has been asserted, that of the best short-beaked tumbler-pigeons more perish in the egg than are able to get out of it; so that fanciers assist in the act of hatching. Now, if nature had to make the beak of a full-grown pigeon very short for the bird's own advantage, the process of modification would be very slow, and there would be simultaneously the most rigorous selection of the young birds within the egg, which had the most powerful and hardest beaks, for all with weak beaks would inevitably perish: or, more delicate and more easily broken shells might be selected, the thickness of the shell being known to vary like every other structure.

Sexual Selection.—Inasmuch as peculiarities often appear under domestication in one sex and become hereditarily attached to that sex, the same fact probably occurs under nature, and if so, natural selection will be able to modify one sex in its functional relations to the other sex, or in relation to wholly different habits of life in the two sexes, as is sometimes the case with insects. And this leads me to say a few words on what I call Sexual Selection. This depends, not on a struggle for existence, but on a struggle between the males for possession of the females; the result is not death to the unsuccessful competitor, but few or no offspring. Sexual selection is, therefore, less rigorous than natural selection. Generally, the most vigorous males, those which are best fitted for their places in nature, will leave most progeny. But in many cases, victory will depend not on general vigour, but on having special weapons, confined to the male sex. A hornless stag or spurless cock would have a poor chance of leaving offspring. Sexual selection by always allowing the victor to breed might surely give indomitable courage, length to the spur, and strength to the wing to strike in the spurred leg, as well as the brutal cock-fighter, who knows well that he can improve his breed by careful selection of the best cocks. How low in the scale of nature this law of battle descends, I know not; male alligators have been described as fighting, bellowing, and whirling round, like Indians in a war-dance, for the possession of the females; male salmons have been seen fighting all day long; male stag-beetles often bear wounds from the huge mandibles of other males. The war is, perhaps, severest between the males of polygamous animals, and these seem oftenest provided with special weapons. The males of carnivorous animals are already well armed; though to them and to others, special means of defence may be given through means of sexual selection, as the mane to the lion, the shoulder-pad to the boar, and the hooked jaw to the male salmon; for the shield may be as important for victory, as the sword or spear.

Amongst birds, the contest is often of a more peaceful character. All those who have attended to the subject, believe that there is the severest rivalry between the males of many species to attract by singing the females. The rock-thrush of Guiana, birds of Paradise, and some others, congregate; and successive males display their gorgeous plumage and perform strange antics before the females, which standing by as spectators, at last choose the most attractive partner. Those who have closely attended to birds in confinement well know that they often take individual preferences and dislikes: thus Sir R. Heron has described how one pied peacock was eminently attractive to all his hen birds. It may appear childish to attribute any effect to such apparently weak means: I cannot here enter on the details necessary to support this view; but if man can in a short time give elegant carriage and beauty to his bantams, according to his standard of beauty, I can see no good reason to doubt that female birds, by selecting, during thousands of generations, the most melodious or beautiful males, according to their standard of beauty, might produce a marked effect. I strongly suspect that some well-known laws with respect to the plumage of male and female birds, in comparison with the plumage of the young, can be explained on the view of plumage having been chiefly modified by sexual selection, acting when the birds have come to the breeding age or during the breeding season; the modifications thus produced being inherited at corresponding ages or seasons, either by the males alone, or by the males and females; but I have not space here to enter on this subject.

Thus it is, as I believe, that when the males and females of any animal have the same general habits of life, but differ in structure, colour, or ornament, such differences have been mainly caused by sexual selection; that is, individual males have had, in successive generations, some slight advantage over other males, in their weapons, means of defence, or charms; and have transmitted these advantages to their male offspring. Yet, I would not wish to attribute all such sexual differences to this agency: for we see peculiarities arising and becoming attached to the male sex in our domestic animals (as the wattle in male carriers, horn-like protuberances in the cocks of certain fowls, &c.), which we cannot believe to be either useful to the males in battle, or attractive to the females. We see analogous cases under nature, for instance, the tuft of hair on the breast of the turkey-cock, which can hardly be either useful or ornamental to this bird; — indeed, had the tuft appeared under domestication, it would have been called a monstrosity.

Illustrations of the action of Natural Selection. — In order to make it clear how, as I believe, natural selection acts, I must beg permission to give one or two imaginary illustrations. Let us take the case of a wolf, which preys on various animals, securing some by craft, some by strength, and some by fleetness; and let us suppose that the fleetest prey, a deer for instance, had from any change in the country increased in numbers, or that other prey had decreased in numbers, during that season of the year when the wolf is hardest pressed for food. I can under such circumstances see no reason to doubt that the swiftest and slimmest wolves would have the best chance of surviving, and so be preserved or selected,—provided always that they retained strength to master their prey at this or at some other period of the year, when they might be compelled to prey on other animals. I can see no more reason to doubt this, than that man can improve the fleetness of his greyhounds by careful and methodical selection, or by that unconscious selection which results from each man trying to keep the best dogs without any thought of modifying the breed.

Even without any change in the proportional numbers of the animals on which our wolf preyed, a cub might be born with an innate tendency to pursue certain kinds of prey. Nor can this be thought very improbable; for we often observe great differences in the natural tendencies of our domestic animals; one cat, for instance, taking to catch rats, another mice; one cat, according to Mr. St. John, bringing home winged game, another hares or rabbits, and another hunting on marshy ground and almost nightly catching woodcocks or snipes. The tendency to catch rats rather than mice is known to be inherited. Now, if any slight innate change of habit or of structure benefited an individual wolf, it would have the best chance of surviving and of leaving offspring. Some of its young would probably inherit the same habits or structure, and by the repetition of this process, a new variety might be formed which would either supplant or coexist with the parent-form of wolf. Or, again, the wolves inhabiting a mountainous district, and those frequenting the lowlands, would naturally be forced to hunt different prey; and from the continued preservation of the individuals best fitted for the two sites, two varieties might slowly be formed. These varieties would cross and blend where they met; but to this subject of intercrossing we shall soon have to return. I may add, that, according to Mr. Pierce, there are two varieties of the wolf inhabiting the Catskill Mountains in the United States, one with a light greyhound-like form, which pursues deer, and the other more bulky, with shorter legs, which more frequently attacks the shepherd's flocks.

Let us now take a more complex case. Certain plants excrete a sweet juice, apparently for the sake of eliminating something injurious from their sap: this is effected by glands at the base of the stipules in some Leguminosæ,* and at the back of the leaf of the common laurel. This juice, though small in quantity, is greedily sought by insects. Let us now suppose a little sweet juice or nectar to be excreted by the inner bases of the petals of a flower. In this case insects in seeking the nectar would get dusted with pollen, and would certainly often transport the pollen from one flower to the stigma of another flower. The flowers of two distinct individuals of the same species would thus get crossed; and the act of crossing, we have good reason to believe (as will hereafter be more fully alluded to), would produce very vigorous seedlings, which consequently would have the best chance of flourishing and surviving. Some of these seedlings would probably inherit the nectar-excreting power. Those individual flowers which had the largest glands or nectaries, and which excreted most nectar, would be oftenest visited by insects, and would be oftenest crossed; and so in the long-run would gain the upper hand. Those flowers, also, which had their stamens and pistils placed, in relation to the size and habits of the particular insects which visited them, so as to favour in any degree the transportal of their pollen from flower to flower, would likewise be favoured or selected. We might have taken the case of insects visiting flowers for the sake of collecting pollen instead of nectar; and as pollen is formed for the sole object of fertilisation, its destruction appears a simple loss to the plant; yet if a little pollen were carried, at first occasionally and then habitually, by the pollen-devouring insects from flower to flower, and a cross thus effected, although nine-tenths of the pollen were destroyed, it might still be a great gain to the plant; and those individuals which produced more and more pollen, and had larger and larger anthers, would be selected.*

When our plant, by this process of the continued preservation or natural selection of more and more attractive flowers, had been rendered highly attractive to insects, they would, unintentionally on their part, regularly carry pollen from flower to flower; and that they can most effectually do this, I could easily show by many striking instances. I will give only one — not as a very striking case, but as likewise illustrating one step in the separation of the sexes of plants, presently to be alluded to. Some holly-trees bear only male flowers, which have four stamens producing rather a small quantity of pollen, and a rudimentary pistil; other holly-trees bear only female flowers; these have a full-sized pistil, and four stamens with shrivelled anthers, in which not a grain of pollen can be detected. Having found a female tree exactly sixty yards from a male tree, I put the stigmas of twenty flowers, taken from different branches, under the microscope, and on all, without exception, there were pollen-grains, and on some a profusion of pollen. As the wind had set for several days from the female to the male tree, the pollen could not thus have been carried. The weather had been cold and boisterous, and therefore not favourable to bees, nevertheless every female flower which I examined had been effectually fertilised by the bees, accidentally dusted with pollen, having flown from tree to tree in search of nectar. But to return to our imaginary case: as soon as the plant had been rendered so highly attractive to insects that pollen was regularly carried from flower to flower, another process might commence. No naturalist doubts the advantage of what has been called the "physiological division of labour;"* hence we may believe that it would be advantageous to a plant to produce stamens alone in one flower or on one whole plant, and pistils alone in another flower or on another plant. In plants under culture and placed under new conditions of life, sometimes the male organs and sometimes the female organs become more or less impotent; now if we suppose this to occur in ever so slight a degree under nature, then as pollen is already carried regularly from flower to flower, and as a more complete separation of the sexes of our plant would be advantageous on the principle of the division of labour, individuals with this tendency more and more increased, would be continually favoured or selected, until at last a complete separation of the sexes would be effected.

Let us now turn to the nectar-feeding insects in our imaginary case: we may suppose the plant of which we have been slowly increasing the nectar by continued selection, to be a common plant; and that certain insects depended in main part on its nectar for food. I could give many facts, showing how anxious bees are to save time; for instance, their habit of cutting holes and sucking the nectar at the bases of certain flowers, which they can, with a very little more trouble, enter by the mouth. Bearing such facts in mind, I can see no reason to doubt that an accidental deviation in the size and form of the body, or in the curvature and length of the proboscis, &c., far too slight to be appreciated by us, might profit a bee or other insect, so that an individual so characterised would be able to obtain its food more quickly, and so have a better chance of living and leaving descendants. Its descendants would probably inherit a tendency to a similar slight deviation of structure. The tubes of the corollas* of the common red and incarnate clovers (Trifolium pratense and incarnatum) do not on a hasty glance appear to differ in length;

yet the hive-bee can easily suck the nectar out of the incarnate clover, but not out of the common red clover, which is visited by humble-bees alone; so that whole fields of the red clover offer in vain an abundant supply of precious nectar to the hive-bee. Thus it might be a great advantage to the hive-bee to have a slightly longer or differently constructed proboscis. On the other hand, I have found by experiment that the fertility of clover greatly depends on bees visiting and moving parts of the corolla, so as to push the pollen on to the stigmatic surface. Hence, again, if humble-bees were to become rare in any country, it might be a great advantage to the red clover to have a shorter or more deeply divided tube to its corolla, so that the hive-bee could visit its flowers. Thus I can understand how a flower and a bee might slowly become, either simultaneously or one after the other, modified and adapted in the most perfect manner to each other, by the continued preservation of individuals presenting mutual and slightly favourable deviations of structure.

I am well aware that this doctrine of natural selection, exemplified in the above imaginary instances, is open to the same objections which were at first urged against Sir Charles Lyell's noble views on "the modern changes of the earth, as illustrative of geology;" but we now very seldom hear the action, for instance, of the coast-waves, called a trifling and insignificant cause, when applied to the excavation of gigantic valleys or to the formation of the longest lines of inland cliffs. Natural selection can act only by the preservation and accumulation of infinitesimally small inherited modifications, each profitable to the preserved being; and as modern geology has almost banished such views as the excavation of a great valley by a single diluvial wave, so will natural selection, if it be a true principle, banish the belief of the continued creation of new organic beings, or of any great and sudden modification in their structure.

[*In a brief digression, Darwin argues that in both the animal and vegetable kingdoms, self-fertilization does not go on for ever: in all species, the intercrossing of two individuals occasionally does occur.*]

Circumstances favourable to Natural Selection. — This is an extremely intricate subject. A large amount of inheritable and diversified variability is favourable, but I believe mere individual differences suffice for the work. A large number of individuals, by giving a better chance for the appearance within any given period of profitable variations, will compensate for a lesser amount of variability in each individual, and is, I believe, an extremely important element of success. Though nature grants vast periods of time for the work of natural selection, she does not grant an indefinite period; for as all organic beings are striving, it may be said, to seize on each place in the economy of nature, if any one species does not become modified and improved in a corresponding degree with its competitors, it will soon be exterminated.

In man's methodical selection, a breeder selects for some definite object, and free intercrossing will wholly stop his work. But when many men, without intending to alter the breed, have a nearly common standard of perfection, and all try to get and breed from the best animals, much improvement and modification surely but slowly follow from this unconscious process of selection, notwithstanding a large amount of crossing with inferior animals. Thus it will be in nature; for within a confined area, with some place in its polity not so perfectly occupied as might be, natural selection will always tend to preserve all the individuals varying in the right direction, though in different degrees, so as better to fill up the unoccupied place. But if the area be large, its several districts will almost certainly present different conditions of life; and then if natural selection be modifying and improving a species in the several districts, there will be inter-crossing with the other individuals of the same species on the confines of each. And in this case the effects of intercrossing can hardly be counterbalanced by natural selection always tending to modify all the individuals in each district in exactly the same manner to the conditions of each; for in a continuous area, the conditions will generally graduate away insensibly from one district to another. The intercrossing will most affect those animals which unite for each birth, which wander much, and which do not breed at a very quick rate. Hence in animals of this nature, for instance in birds, varieties will generally be confined to separated countries; and this I believe to be the case. In hermaphrodite organisms which cross only occasionally, and likewise in animals which unite for each birth, but which wander little and which can increase at a very rapid rate, a new and improved variety might be quickly formed on any one spot, and might there maintain itself in a body, so that whatever intercrossing took place would be chiefly between the individuals of the same new variety. A local variety when once thus formed might subsequently slowly spread to other districts. On the above principle, nurserymen always prefer getting seed from a large body of plants of the same variety, as the chance of intercrossing with other varieties is thus lessened.

Even in the case of slow-breeding animals, which unite for each birth, we must not overrate the effects of intercrosses in retarding natural selection; for I can bring a considerable catalogue of facts, showing that within the same area, varieties of the same animal can long remain distinct, from haunting different stations, from breeding at slightly different seasons, or from varieties of the same kind preferring to pair together.

Intercrossing plays a very important part in nature in keeping the individuals of the same species, or of the same variety, true and uniform in character. It will obviously thus act far more efficiently with those animals which unite for each birth; but I have already attempted to show that we have reason to believe that occasional intercrosses take place with all animals and with all plants. Even if these take place only at long intervals, I am convinced that the young thus produced will gain so much in vigour and fertility over the offspring from long-continued self-fertilisation, that they will have a better chance of surviving and propagating their kind; and thus, in the long run, the influence of intercrosses, even at rare intervals, will be great. If there exist organic beings which never intercross, uniformity of character can be retained amongst them, as long as their conditions of life remain the same, only through the principle of inheritance, and through natural selection destroying any which depart from the proper type; but if their conditions of life change and they undergo modification, uniformity of character can be given to their modified offspring, solely by natural selection preserving the same favourable variations.

Isolation, also, is an important element in the process of natural selection. In a confined or isolated area, if not very large, the organic and inorganic conditions of life will generally be in a great degree uniform; so that natural selection will tend to modify all the individuals of a varying species throughout the area in the same manner in relation to the same conditions. Intercrosses, also, with the individuals of the same species, which otherwise would have inhabited the surrounding and differently circumstanced districts, will be prevented. But isolation probably acts more efficiently in checking the immigration of better adapted organisms, after any physical change, such as of climate or elevation of the land, &c.; and thus new places in the natural economy of the country are left open for the old inhabitants to struggle for, and become adapted to, through modifications in their structure and constitution. Lastly, isolation, by checking immigration and consequently competition, will give time for any new variety to be slowly improved; and this may sometimes be of importance in the production of new species. If, however, an isolated area be very small, either from being surrounded by barriers, or from having very peculiar physical conditions, the total number of the individuals supported on it will necessarily be very small; and fewness of individuals will greatly retard the production of new species through natural selection, by decreasing the chance of the appearance of favourable variations.

If we turn to nature to test the truth of these remarks, and look at any small isolated area, such as an oceanic island, although the total number of the species inhabiting it, will be found to be small, as we shall see in our chapter on geographical distribution; yet of these species a very large proportion are endemic,—that is, have been produced there, and nowhere else. Hence an oceanic island at first sight seems to have been highly favourable for the production of new species. But we may thus greatly deceive ourselves, for to ascertain whether a small isolated area, or a large open area like a continent, has been most favourable for the production of new organic forms, we ought to make the comparison within equal times; and this we are incapable of doing.

Although I do not doubt that isolation is of considerable importance in the production of new species, on the whole I am inclined to believe that largeness of area is of more importance, more especially in the production of species, which will prove capable of enduring for a long period, and of spreading widely. Throughout a great and open area, not only will there be a better chance of favourable variations arising from the large number of individuals of the same species there supported, but the conditions of life are infinitely complex from the large number of already existing species; and if some of these many species become modified and improved, others will have to be improved in a corresponding degree or they will be exterminated. Each new form, also, as soon as it has been much improved, will be able to spread over the open and continuous area, and will thus come into competition with many others. Hence more new places will be formed, and the competition to fill them will be more severe, on a large than on a small and isolated area. Moreover, great areas, though now continuous, owing to oscillations of level, will often have recently existed in a broken condition, so that the good effects of isolation will generally, to a certain extent, have concurred. Finally, I conclude that, although small isolated areas probably have been in some respects highly favourable for the production of new species, yet that the course of modification will generally have been more rapid on large areas; and what is more important, that the new forms produced on large areas, which already have been victorious over many competitors, will be those that will spread most widely, will give rise to most new varieties and species, and will thus play an important part in the changing history of the organic world.

We can, perhaps, on these views, understand some facts which will be again alluded to in our chapter on geographical distribution; for instance, that the productions of the smaller continent of Australia have formerly yielded, and apparently are now yielding, before those of the larger Europæo-Asiatic area. Thus, also, it is that continental productions have everywhere become so largely natural-ised on islands. On a small island, the race for life will have been less severe, and there will have been less modification and less extermin-ation. Hence, perhaps, it comes that the flora of Madeira, according to Oswald Heer, resembles the extinct tertiary* flora of Europe. All fresh-water basins, taken together, make a small area compared with that of the sea or of the land; and, consequently, the competition between fresh-water productions will have been less severe than elsewhere; new forms will have been more slowly formed, and old forms more slowly exterminated. And it is in fresh water that we find seven genera of Ganoid fishes,* remnants of a once preponderant order: and in fresh water we find some of the most anomalous forms now known in the world, as the Ornithorhynchus and Lepidosiren,* which, like fossils, connect to a certain extent orders now widely separated in the natural scale. These anomalous forms may almost be called living fossils; they have endured to the present day, from hav-ing inhabited a confined area, and from having thus been exposed to less severe competition.

To sum up the circumstances favourable and unfavourable to nat-ural selection, as far as the extreme intricacy of the subject permits.

I conclude, looking to the future, that for terrestrial productions a large continental area, which will probably undergo many oscillations of level, and which consequently will exist for long periods in a broken condition, will be the most favourable for the production of many new forms of life, likely to endure long and to spread widely. For the area will first have existed as a continent, and the inhabitants, at this period numerous in individuals and kinds, will have been subjected to very severe competition. When converted by subsidence into large separate islands, there will still exist many individuals of the same species on each island: intercrossing on the confines of the range of each species will thus be checked: after physical changes of any kind, immigration will be prevented, so that new places in the polity of each island will have to be filled up by modifications of the old inhabitants; and time will be allowed for the varieties in each to become well modified and perfected. When, by renewed elevation, the islands shall be re-converted into a continental area, there will again be severe competition: the most favoured or improved varieties will be enabled to spread: there will be much extinction of the less improved forms, and the relative proportional numbers of the various inhabitants of the renewed continent will again be changed; and again there will be a fair field for natural selection to improve still further the inhabitants, and thus produce new species.

That natural selection will always act with extreme slowness, I fully admit. Its action depends on there being places in the polity of nature, which can be better occupied by some of the inhabitants of the country undergoing modification of some kind. The existence of such places will often depend on physical changes, which are generally very slow, and on the immigration of better adapted forms having been checked. But the action of natural selection will probably still oftener depend on some of the inhabitants becoming slowly modified; the mutual relations of many of the other inhabitants being thus disturbed. Nothing can be effected, unless favourable variations occur, and variation itself is apparently always a very slow process. The process will often be greatly retarded by free intercrossing. Many will exclaim that these several causes are amply sufficient wholly to stop the action of natural selection. I do not believe so. On the other hand, I do believe that natural selection will always act very slowly, often only at long intervals of time, and generally on only a very few of the inhabitants of the same region at the same time.

I further believe, that this very slow, intermittent action of natural selection accords perfectly well with what geology tells us of the rate and manner at which the inhabitants of this world have changed.

Slow though the process of selection may be, if feeble man can do much by his powers of artificial selection, I can see no limit to the amount of change, to the beauty and infinite complexity of the co-adaptations between all organic beings, one with another and with their physical conditions of life, which may be effected in the long course of time by nature's power of selection.

Extinction. — This subject will be more fully discussed in our chapter on Geology; but it must be here alluded to from being intimately connected with natural selection. Natural selection acts solely through the preservation of variations in some way advantageous, which consequently endure. But as from the high geometrical powers of increase of all organic beings, each area is already fully stocked with inhabitants, it follows that as each selected and favoured form increases in number, so will the less favoured forms decrease and become rare. Rarity, as geology tells us, is the precursor to extinction. We can, also, see that any form represented by few individuals will, during fluctuations in the seasons or in the number of its enemies, run a good chance of utter extinction. But we may go further than this; for as new forms are continually and slowly being produced, unless we believe that the number of specific forms goes on perpetually and almost indefinitely increasing, numbers inevitably must become extinct. That the number of specific forms has not indefinitely increased, geology shows us plainly; and indeed we can see reason why they should not have thus increased, for the number of places in the polity of nature is not indefinitely great, — not that we have any means of knowing that any one region has as yet got its maximum of species. Probably no region is as yet fully stocked, for at the Cape of Good Hope, where more species of plants are crowded together than in any other quarter of the world, some foreign plants have become naturalised, without causing, as far as we know, the extinction of any natives.

Furthermore, the species which are most numerous in individuals will have the best chance of producing within any given period favourable variations. We have evidence of this, in the facts given in the second chapter, showing that it is the common species which afford the greatest number of recorded varieties, or incipient species. Hence, rare species will be less quickly modified or improved within any given period, and they will consequently be beaten in the race for life by the modified descendants of the commoner species.

From these several considerations I think it inevitably follows, that as new species in the course of time are formed through natural selection, others will become rarer and rarer, and finally extinct. The forms which stand in closest competition with those undergoing modification and improvement, will naturally suffer most. And we have seen in the chapter on the Struggle for Existence that it is the most closely-allied forms,—varieties of the same species, and species of the same genus or of related genera,—which, from having nearly the same structure, constitution, and habits, generally come into the severest competition with each other. Consequently, each new variety or species, during the progress of its formation, will generally press hardest on its nearest kindred, and tend to exterminate them. We see the same process of extermination amongst our domesticated productions, through the selection of improved forms by man. Many curious instances could be given showing how quickly new breeds of cattle, sheep, and other animals, and varieties of flowers, take the place of older and inferior kinds. In Yorkshire, it is historically known that the ancient black cattle were displaced by the long-horns, and that these "were swept away by the short-horns" (I quote the words of an agricultural writer) "as if by some murderous pestilence."*

Divergence of Character.—The principle, which I have designated by this term, is of high importance on my theory, and explains, as I believe, several important facts. In the first place, varieties, even strongly-marked ones, though having somewhat of the character of species—as is shown by the hopeless doubts in many cases how to rank them—yet certainly differ from each other far less than do good and distinct species. Nevertheless, according to my view, varieties are species in the process of formation, or are, as I have called them, incipient species. How, then, does the lesser difference between varieties become augmented into the greater difference between species? That this does habitually happen, we must infer from most of the innumerable species throughout nature presenting well-marked differences; whereas varieties, the supposed prototypes and parents of future well-marked species, present slight and ill-defined differences. Mere chance, as we may call it, might cause one variety to differ in some character from its parents, and the offspring of this variety again to differ from its parent in the very same character and in a greater degree; but this alone would never account for so habitual and large an amount of difference as that between varieties of the same species and species of the same genus.

As has always been my practice, let us seek light on this head from our domestic productions. We shall here find something analogous. A fancier is struck by a pigeon having a slightly shorter beak; another fancier is struck by a pigeon having a rather longer beak; and on the acknowledged principle that "fanciers do not and will not admire a medium standard, but like extremes,"* they both go on (as has actually occurred with tumbler-pigeons) choosing and breeding from birds with longer and longer beaks, or with shorter and shorter beaks. Again, we may suppose that at an early period one man preferred swifter horses; another stronger and more bulky horses. The early differences would be very slight; in the course of time, from the continued selection of swifter horses by some breeders, and of stronger ones by others, the differences would become greater, and would be noted as forming two sub-breeds; finally, after the lapse of centuries, the sub-breeds would become converted into two well-established and distinct breeds. As the differences slowly become greater, the inferior animals with intermediate characters, being neither very swift nor very strong, will have been neglected, and will have tended to disappear. Here, then, we see in man's productions the action of what may be called the principle of divergence, causing differences, at first barely appreciable, steadily to increase, and the breeds to diverge in character both from each other and from their common parent.

But how, it may be asked, can any analogous principle apply in nature? I believe it can and does apply most efficiently, from the simple circumstance that the more diversified the descendants from any one species become in structure, constitution, and habits, by so much will they be better enabled to seize on many and widely diversified places in the polity of nature, and so be enabled to increase in numbers.

We can clearly see this in the case of animals with simple habits. Take the case of a carnivorous quadruped, of which the number that can be supported in any country has long ago arrived at its full average.

If its natural powers of increase be allowed to act, it can succeed in increasing (the country not undergoing any change in its conditions) only by its varying descendants seizing on places at present occupied by other animals: some of them, for instance, being enabled to feed on new kinds of prey, either dead or alive; some inhabiting new stations, climbing trees, frequenting water, and some perhaps becoming less carnivorous. The more diversified in habits and structure the descendants of our carnivorous animal became, the more places they would be enabled to occupy. What applies to one animal will apply throughout all time to all animals—that is, if they vary—for otherwise natural selection can do nothing. So it will be with plants. It has been experimentally proved, that if a plot of ground be sown with one species of grass, and a similar plot be sown with several distinct genera of grasses, a greater number of plants and a greater weight of dry herbage can thus be raised. The same has been found to hold good when first one variety and then several mixed varieties of wheat have been sown on equal spaces of ground. Hence, if any one species of grass were to go on varying, and those varieties were continually selected which differed from each other in at all the same manner as distinct species and genera of grasses differ from each other, a greater number of individual plants of this species of grass, including its modified descendants, would succeed in living on the same piece of ground. And we well know that each species and each variety of grass is annually sowing almost countless seeds; and thus, as it may be said, is striving its utmost to increase its numbers. Consequently, I cannot doubt that in the course of many thousands of generations, the most distinct varieties of any one species of grass would always have the best chance of succeeding and of increasing in numbers, and thus of supplanting the less distinct varieties; and varieties, when rendered very distinct from each other, take the rank of species.

The truth of the principle, that the greatest amount of life can be supported by great diversification of structure, is seen under many natural circumstances. In an extremely small area, especially if freely open to immigration, and where the contest between individual and individual must be severe, we always find great diversity in its inhabitants. For instance, I found that a piece of turf, three feet by four in size, which had been exposed for many years to exactly the same conditions, supported twenty species of plants, and these belonged to eighteen genera and to eight orders, which shows how much these plants differed from each other. So it is with the plants and insects on small and uniform islets; and so in small ponds of fresh water. Farmers find that they can raise most food by a rotation of plants belonging to the most different orders: nature follows what may be called a simultaneous rotation. Most of the animals and plants which live close round any small piece of ground, could live on it (supposing it not to be in any way peculiar in its nature), and may be said to be striving to the utmost to live there; but, it is seen, that where they come into the closest competition with each other, the advantages of diversification of structure, with the accompanying differences of habit and constitution, determine that the inhabitants, which thus jostle each other most closely, shall, as a general rule, belong to what we call different genera and orders.

The same principle is seen in the naturalisation of plants through man's agency in foreign lands. It might have been expected that the plants which have succeeded in becoming naturalised in any land would generally have been closely allied to the indigenes; for these are commonly looked at as specially created and adapted for their own country. It might, also, perhaps have been expected that naturalised plants would have belonged to a few groups more especially adapted to certain stations in their new homes. But the case is very different; and Alph. De Candolle has well remarked in his great and admirable work, that floras gain by naturalisation, proportionally with the number of the native genera and species, far more in new genera than in new species. To give a single instance: in the last edition of Dr. Asa Gray's 'Manual of the Flora of the Northern United States,' 260 naturalised plants are enumerated, and these belong to 162 genera. We thus see that these naturalised plants are of a highly diversified nature. They differ, moreover, to a large extent from the indigenes, for out of the 162 genera, no less than 100 genera are not there indigenous, and thus a large proportional addition is made to the genera of these States.

By considering the nature of the plants or animals which have struggled successfully with the indigenes of any country, and have there become naturalised, we can gain some crude idea in what manner some of the natives would have had to be modified, in order to have gained an advantage over the other natives; and we may, I think, at least safely infer that diversification of structure, amounting to new generic differences, would have been profitable to them.

The advantage of diversification in the inhabitants of the same region is, in fact, the same as that of the physiological division of labour in the organs of the same individual body—a subject so well elucidated by Milne Edwards. No physiologist doubts that a stomach by being adapted to digest vegetable matter alone, or flesh alone, draws most nutriment from these substances. So in the general economy of any land, the more widely and perfectly the animals and plants are diversified for different habits of life, so will a greater number of individuals be capable of there supporting themselves. A set of animals, with their organisation but little diversified, could hardly compete with a set more perfectly diversified in structure. It may be doubted, for instance, whether the Australian marsupials, which are divided into groups differing but little from each other, and feebly representing, as Mr. Waterhouse and others have remarked, our carnivorous, ruminant, and rodent mammals, could successfully compete with these well-pronounced orders. In the Australian mammals, we see the process of diversification in an early and incomplete stage of development. After the foregoing discussion, which ought to have been much amplified, we may, I think, assume that the modified descendants of any one species will succeed by so much the better as they become more diversified in structure, and are thus enabled to encroach on places occupied by other beings. Now let us see how this principle of great benefit being derived from divergence of character, combined with the principles of natural selection and of extinction, will tend to act.

The accompanying diagram will aid us in understanding this rather perplexing subject. Let A to L represent the species of a genus large in its own country; these species are supposed to resemble each other in unequal degrees, as is so generally the case in nature, and as is represented in the diagram by the letters standing at unequal distances. I have said a large genus, because we have seen in the second chapter, that on an average more of the species of large genera vary than of small genera; and the varying species of the large genera present a greater number of varieties. We have, also, seen that the species, which are the commonest and the most widely-diffused, vary more than rare species with restricted ranges. Let (A) be a common, widely-diffused, and varying species, belonging to a genus large in its own country. The little fan of diverging dotted lines of unequal lengths proceeding from (A), may represent its varying offspring.

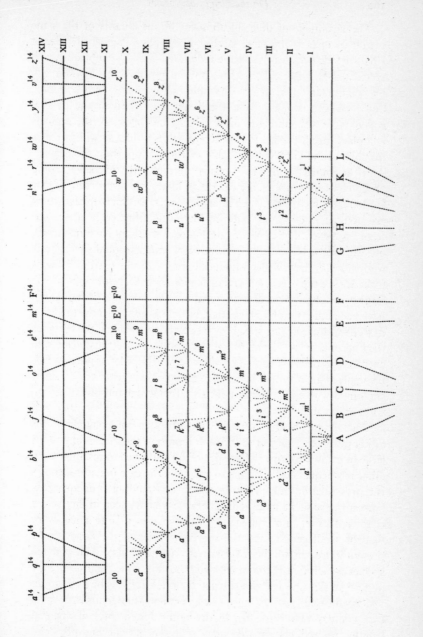

The variations are supposed to be extremely slight, but of the most diversified nature; they are not supposed all to appear simultaneously, but often after long intervals of time; nor are they all supposed to endure for equal periods. Only those variations which are in some way profitable will be preserved or naturally selected. And here the importance of the principle of benefit being derived from divergence of character comes in; for this will generally lead to the most different or divergent variations (represented by the outer dotted lines) being preserved and accumulated by natural selection. When a dotted line reaches one of the horizontal lines, and is there marked by a small numbered letter, a sufficient amount of variation is supposed to have been accumulated to have formed a fairly well-marked variety, such as would be thought worthy of record in a systematic work.

The intervals between the horizontal lines in the diagram, may represent each a thousand generations; but it would have been better if each had represented ten thousand generations. After a thousand generations, species (A) is supposed to have produced two fairly well-marked varieties, namely a^1 and m^1. These two varieties will generally continue to be exposed to the same conditions which made their parents variable, and the tendency to variability is in itself hereditary, consequently they will tend to vary, and generally to vary in nearly the same manner as their parents varied. Moreover, these two varieties, being only slightly modified forms, will tend to inherit those advantages which made their common parent (A) more numerous than most of the other inhabitants of the same country; they will likewise partake of those more general advantages which made the genus to which the parent-species belonged, a large genus in its own country. And these circumstances we know to be favourable to the production of new varieties.

If, then, these two varieties be variable, the most divergent of their variations will generally be preserved during the next thousand generations. And after this interval, variety a^1 is supposed in the diagram to have produced variety a^2, which will, owing to the principle of divergence, differ more from (A) than did variety a^1. Variety m^1 is supposed to have produced two varieties, namely m^2 and s^2, differing from each other, and more considerably from their common parent (A). We may continue the process by similar steps for any length of time; some of the varieties, after each thousand generations, producing only a single variety, but in a more and more modified condition, some producing two or three varieties, and some failing to produce any. Thus the varieties or modified descendants, proceeding from the common parent (A), will generally go on increasing in number and diverging in character. In the diagram the process is represented up to the ten-thousandth generation, and under a condensed and simplified form up to the fourteen-thousandth generation.

But I must here remark that I do not suppose that the process ever goes on so regularly as is represented in the diagram, though in itself made somewhat irregular. I am far from thinking that the most divergent varieties will invariably prevail and multiply: a medium form may often long endure, and may or may not produce more than one modified descendant; for natural selection will always act according to the nature of the places which are either unoccupied or not perfectly occupied by other beings; and this will depend on infinitely complex relations. But as a general rule, the more diversified in structure the descendants from any one species can be rendered, the more places they will be enabled to seize on, and the more their modified progeny will be increased. In our diagram the line of succession is broken at regular intervals by small numbered letters marking the successive forms which have become sufficiently distinct to be recorded as varieties. But these breaks are imaginary, and might have been inserted anywhere, after intervals long enough to have allowed the accumulation of a considerable amount of divergent variation.

As all the modified descendants from a common and widely-diffused species, belonging to a large genus, will tend to partake of the same advantages which made their parent successful in life, they will generally go on multiplying in number as well as diverging in character: this is represented in the diagram by the several divergent branches proceeding from (A). The modified offspring from the later and more highly improved branches in the lines of descent, will, it is probable, often take the place of, and so destroy, the earlier and less improved branches: this is represented in the diagram by some of the lower branches not reaching to the upper horizontal lines. In some cases I do not doubt that the process of modification will be confined to a single line of descent, and the number of the descendants will not be increased; although the amount of divergent modification may have been increased in the successive generations. This case would be represented in the diagram, if all the lines proceeding from (A) were removed, excepting that from a^1 to a^{10}. In the same way, for instance, the English race-horse and English pointer have apparently both gone on slowly diverging in character from their original stocks, without either having given off any fresh branches or races.

After ten thousand generations, species (A) is supposed to have produced three forms, a^{10}, f^{10}, and m^{10}, which, from having diverged in character during the successive generations, will have come to differ largely, but perhaps unequally, from each other and from their common parent. If we suppose the amount of change between each horizontal line in our diagram to be excessively small, these three forms may still be only well-marked varieties; or they may have arrived at the doubtful category of sub-species; but we have only to suppose the steps in the process of modification to be more numerous or greater in amount, to convert these three forms into well-defined species: thus the diagram illustrates the steps by which the small differences distinguishing varieties are increased into the larger differences distinguishing species. By continuing the same process for a greater number of generations (as shown in the diagram in a condensed and simplified manner), we get eight species, marked by the letters between a^{14} and m^{14}, all descended from (A). Thus, as I believe, species are multiplied and genera are formed.

In a large genus it is probable that more than one species would vary. In the diagram I have assumed that a second species (I) has produced, by analogous steps, after ten thousand generations, either two well-marked varieties (w^{10} and z^{10}) or two species, according to the amount of change supposed to be represented between the horizontal lines. After fourteen thousand generations, six new species, marked by the letters n^{14} to z^{14}, are supposed to have been produced. In each genus, the species, which are already extremely different in character, will generally tend to produce the greatest number of modified descendants; for these will have the best chance of filling new and widely different places in the polity of nature: hence in the diagram I have chosen the extreme species (A), and the nearly extreme species (I), as those which have largely varied, and have given rise to new varieties and species. The other nine species (marked by capital letters) of our original genus, may for a long period continue transmitting unaltered descendants; and this is shown in the diagram by the dotted lines not prolonged far upwards from want of space.

But during the process of modification, represented in the diagram, another of our principles, namely that of extinction, will have played an important part. As in each fully stocked country natural selection necessarily acts by the selected form having some advantage in the struggle for life over other forms, there will be a constant tendency in the improved descendants of any one species to supplant and exterminate in each stage of descent their predecessors and their original parent. For it should be remembered that the competition will generally be most severe between those forms which are most nearly related to each other in habits, constitution, and structure. Hence all the intermediate forms between the earlier and later states, that is between the less and more improved state of a species, as well as the original parent-species itself, will generally tend to become extinct. So it probably will be with many whole collateral lines of descent, which will be conquered by later and improved lines of descent. If, however, the modified offspring of a species get into some distinct country, or become quickly adapted to some quite new station, in which child and parent do not come into competition, both may continue to exist.

If then our diagram be assumed to represent a considerable amount of modification, species (A) and all the earlier varieties will have become extinct, having been replaced by eight new species (a^{14} to m^{14}); and (I) will have been replaced by six (n^{14} to z^{14}) new species.

But we may go further than this. The original species of our genus were supposed to resemble each other in unequal degrees, as is so generally the case in nature; species (A) being more nearly related to B, C, and D, than to the other species; and species (I) more to G, H, K, L, than to the others. These two species (A) and (I), were also supposed to be very common and widely diffused species, so that they must originally have had some advantage over most of the other species of the genus. Their modified descendants, fourteen in number at the fourteen-thousandth generation, will probably have inherited some of the same advantages: they have also been modified and improved in a diversified manner at each stage of descent, so as to have become adapted to many related places in the natural economy of their country. It seems, therefore, to me extremely probable that they will have taken the places of, and thus exterminated, not only their parents (A) and (I), but likewise some of the original species which were most nearly related to their parents. Hence very few of the original species will have transmitted offspring to the fourteen-thousandth generation. We may suppose that only one (F), of the two species which were least closely related to the other nine original species, has transmitted descendants to this late stage of descent.

The new species in our diagram descended from the original eleven species, will now be fifteen in number. Owing to the divergent tendency of natural selection, the extreme amount of difference in character between species a^{14} and z^{14} will be much greater than that between the most different of the original eleven species. The new species, moreover, will be allied to each other in a widely different manner. Of the eight descendants from (A) the three marked a^{14}, q^{14}, p^{14}, will be nearly related from having recently branched off from a^{10}; b^{14} and f^{14}, from having diverged at an earlier period from a^5, will be in some degree distinct from the three first-named species; and lastly, o^{14}, e^{14}, and m^{14}, will be nearly related one to the other, but from having diverged at the first commencement of the process of modification, will be widely different from the other five species, and may constitute a sub-genus or even a distinct genus.

The six descendants from (I) will form two sub-genera or even genera. But as the original species (I) differed largely from (A), standing nearly at the extreme points of the original genus, the six descendants from (I) will, owing to inheritance, differ considerably from the eight descendants from (A); the two groups, moreover, are supposed to have gone on diverging in different directions. The intermediate species, also (and this is a very important consideration), which connected the original species (A) and (I), have all become, excepting (F), extinct, and have left no descendants. Hence the six new species descended from (I), and the eight descended from (A), will have to be ranked as very distinct genera, or even as distinct sub-families.

Thus it is, as I believe, that two or more genera are produced by descent, with modification, from two or more species of the same genus. And the two or more parent-species are supposed to have descended from some one species of an earlier genus. In our diagram, this is indicated by the broken lines, beneath the capital letters, converging in sub-branches downwards towards a single point; this point representing a single species, the supposed single parent of our several new sub-genera and genera.

It is worth while to reflect for a moment on the character of the new species F^{14}, which is supposed not to have diverged much in character, but to have retained the form of (F), either unaltered or altered only in a slight degree. In this case, its affinities to the other fourteen new species will be of a curious and circuitous nature. Having descended from a form which stood between the two parent-species (A) and (I), now supposed to be extinct and unknown, it will be in some degree intermediate in character between the two groups descended from these species. But as these two groups have gone on diverging in character from the type of their parents, the new species (F^{14}) will not be directly intermediate between them, but rather between types of the two groups; and every naturalist will be able to bring some such case before his mind.

In the diagram, each horizontal line has hitherto been supposed to represent a thousand generations, but each may represent a million or hundred million generations, and likewise a section of the successive strata of the earth's crust including extinct remains. We shall, when we come to our chapter on Geology, have to refer again to this subject, and I think we shall then see that the diagram throws light on the affinities of extinct beings, which, though generally belonging to the same orders, or families, or genera, with those now living, yet are often, in some degree, intermediate in character between existing groups; and we can understand this fact, for the extinct species lived at very ancient epochs when the branching lines of descent had diverged less.

I see no reason to limit the process of modification, as now explained, to the formation of genera alone. If, in our diagram, we suppose the amount of change represented by each successive group of diverging dotted lines to be very great, the forms marked a^{14} to p^{14}, those marked b^{14} and f^{14}, and those marked o^{14} to m^{14}, will form three very distinct genera. We shall also have two very distinct genera descended from (I) and as these latter two genera, both from continued divergence of character and from inheritance from a different parent, will differ widely from the three genera descended from (A), the two little groups of genera will form two distinct families, or even orders, according to the amount of divergent modification supposed to be represented in the diagram. And the two new families, or orders, will have descended from two species of the original genus; and these two species are supposed to have descended from one species of a still more ancient and unknown genus.

We have seen that in each country it is the species of the larger genera which oftenest present varieties or incipient species. This, indeed, might have been expected; for as natural selection acts through one form having some advantage over other forms in the struggle for existence, it will chiefly act on those which already have some advantage; and the largeness of any group shows that its species have inherited from a common ancestor some advantage in common. Hence, the struggle for the production of new and modified descendants, will mainly lie between the larger groups, which are all trying to increase in number. One large group will slowly conquer another large group, reduce its numbers, and thus lessen its chance of further variation and improvement. Within the same large group, the later and more highly perfected sub-groups, from branching out and seizing on many new places in the polity of Nature, will constantly tend to supplant and destroy the earlier and less improved sub-groups. Small and broken groups and sub-groups will finally tend to disappear. Looking to the future, we can predict that the groups of organic beings which are now large and triumphant, and which are least broken up, that is, which as yet have suffered least extinction, will for a long period continue to increase. But which groups will ultimately prevail, no man can predict; for we well know that many groups, formerly most extensively developed, have now become extinct. Looking still more remotely to the future, we may predict that, owing to the continued and steady increase of the larger groups, a multitude of smaller groups will become utterly extinct, and leave no modified descendants; and consequently that of the species living at any one period, extremely few will transmit descendants to a remote futurity. I shall have to return to this subject in the chapter on Classification, but I may add that on this view of extremely few of the more ancient species having transmitted descendants, and on the view of all the descendants of the same species making a class, we can understand how it is that there exist but very few classes in each main division of the animal and vegetable kingdoms. Although extremely few of the most ancient species may now have living and modified descendants, yet at the most remote geological period, the earth may have been as well peopled with many species of many genera, families, orders, and classes, as at the present day.

Summary of Chapter.—If during the long course of ages and under varying conditions of life, organic beings vary at all in the several parts of their organisation, and I think this cannot be disputed; if there be, owing to the high geometrical powers of increase of each species, at some age, season, or year, a severe struggle for life, and this certainly cannot be disputed; then, considering the infinite complexity of the relations of all organic beings to each other and to their conditions of existence, causing an infinite diversity in structure, constitution, and habits, to be advantageous to them, I think it would be a most extraordinary fact if no variation ever had occurred useful to each being's own welfare, in the same way as so many variations have occurred useful to man. But if variations useful to any organic being do occur, assuredly individuals thus characterised will have the best chance of being preserved in the struggle for life; and from the strong principle of inheritance they will tend to produce offspring similarly characterised. This principle of preservation, I have called, for the sake of brevity, Natural Selection. Natural selection, on the principle of qualities being inherited at corresponding ages, can modify the egg, seed, or young, as easily as the adult. Amongst many animals, sexual selection will give its aid to ordinary selection, by assuring to the most vigorous and best adapted males the greatest number of offspring. Sexual selection will also give characters useful to the males alone, in their struggles with other males.

Whether natural selection has really thus acted in nature, in modifying and adapting the various forms of life to their several conditions and stations, must be judged of by the general tenour and balance of evidence given in the following chapters. But we already see how it entails extinction; and how largely extinction has acted in the world's history, geology plainly declares. Natural selection, also, leads to divergence of character; for more living beings can be supported on the same area the more they diverge in structure, habits, and constitution, of which we see proof by looking at the inhabitants of any small spot or at naturalised productions. Therefore during the modification of the descendants of any one species, and during the incessant struggle of all species to increase in numbers, the more diversified these descendants become, the better will be their chance of succeeding in the battle of life. Thus the small differences distinguishing varieties of the same species, will steadily tend to increase till they come to equal the greater differences between species of the same genus, or even of distinct genera.

We have seen that it is the common, the widely-diffused, and widely-ranging species, belonging to the larger genera, which vary most;

and these will tend to transmit to their modified offspring that superiority which now makes them dominant in their own countries. Natural selection, as has just been remarked, leads to divergence of character and to much extinction of the less improved and intermediate forms of life. On these principles, I believe, the nature of the affinities of all organic beings may be explained. It is a truly wonderful fact—the wonder of which we are apt to overlook from familiarity—that all animals and all plants throughout all time and space should be related to each other in group subordinate to group, in the manner which we everywhere behold—namely, varieties of the same species most closely related together, species of the same genus less closely and unequally related together, forming sections and sub-genera, species of distinct genera much less closely related, and genera related in different degrees, forming sub-families, families, orders, sub-classes, and classes. The several subordinate groups in any class cannot be ranked in a single file, but seem rather to be clustered round points, and these round other points, and so on in almost endless cycles. On the view that each species has been independently created, I can see no explanation of this great fact in the classification of all organic beings; but, to the best of my judgment, it is explained through inheritance and the complex action of natural selection, entailing extinction and divergence of character, as we have seen illustrated in the diagram.

The affinities of all the beings of the same class have sometimes been represented by a great tree. I believe this simile largely speaks the truth. The green and budding twigs may represent existing species; and those produced during each former year may represent the long succession of extinct species. At each period of growth all the growing twigs have tried to branch out on all sides, and to overtop and kill the surrounding twigs and branches, in the same manner as species and groups of species have tried to overmaster other species in the great battle for life. The limbs divided into great branches, and these into lesser and lesser branches, were themselves once, when the tree was small, budding twigs; and this connexion of the former and present buds by ramifying branches may well represent the classification of all extinct and living species in groups subordinate to groups. Of the many twigs which flourished when the tree was a mere bush, only two or three, now grown into great branches, yet survive and bear all the other branches; so with the species which lived during long-past geological periods, very few now have living and modified descendants. From the first growth of the tree, many a limb and branch has decayed and dropped off; and these lost branches of various sizes may represent those whole orders, families, and genera which have now no living representatives, and which are known to us only from having been found in a fossil state. As we here and there see a thin straggling branch springing from a fork low down in a tree, and which by some chance has been favoured and is still alive on its summit, so we occasionally see an animal like the Ornithorhynchus or Lepidosiren, which in some small degree connects by its affinities two large branches of life, and which has apparently been saved from fatal competition by having inhabited a protected station. As buds give rise by growth to fresh buds, and these, if vigorous, branch out and overtop on all sides many a feebler branch, so by generation I believe it has been with the great Tree of Life, which fills with its dead and broken branches the crust of the earth, and covers the surface with its ever branching and beautiful ramifications.

[*Chapter V discusses the source of the variations on which natural selection will act. Darwin attributes variation not to chance, but to shifting environmental conditions and especially to habit, particularly as these affect the reproductive system. In addition, certain characters change because they are correlated with others, while new or unusual ones tend to vary more than those long-established.*]

CHAPTER VI

Difficulties on Theory

LONG before having arrived at this part of my work, a crowd of difficulties will have occurred to the reader. Some of them are so grave that to this day I can never reflect on them without being staggered; but, to the best of my judgment, the greater number are only apparent, and those that are real are not, I think, fatal to my theory.

These difficulties and objections may be classed under the following heads:—Firstly, why, if species have descended from other species by insensibly fine gradations, do we not everywhere see innumerable transitional forms? Why is not all nature in confusion instead of the species being, as we see them, well defined?

Secondly, is it possible that an animal having, for instance, the structure and habits of a bat, could have been formed by the modification of some animal with wholly different habits? Can we believe that natural selection could produce, on the one hand, organs of trifling importance, such as the tail of a giraffe, which serves as a fly-flapper, and, on the other hand, organs of such wonderful structure, as the eye, of which we hardly as yet fully understand the inimitable perfection?

Thirdly, can instincts be acquired and modified through natural selection? What shall we say to so marvellous an instinct as that which leads the bee to make cells, which have practically anticipated the discoveries of profound mathematicians?

Fourthly, how can we account for species, when crossed, being sterile and producing sterile offspring, whereas, when varieties are crossed, their fertility is unimpaired?

The two first heads shall be here discussed—Instinct and Hybridism in separate chapters.

On the absence or rarity of transitional varieties.—As natural selection acts solely by the preservation of profitable modifications, each new form will tend in a fully-stocked country to take the place of, and finally to exterminate, its own less improved parent or other less-favoured forms with which it comes into competition. Thus extinction and natural selection will, as we have seen, go hand in hand. Hence, if we look at each species as descended from some other unknown form, both the parent and all the transitional varieties will generally have been exterminated by the very process of formation and perfection of the new form.

But, as by this theory innumerable transitional forms must have existed, why do we not find them embedded in countless numbers in the crust of the earth? It will be much more convenient to discuss this question in the chapter on the Imperfection of the geological record; and I will here only state that I believe the answer mainly lies in the record being incomparably less perfect than is generally supposed; the imperfection of the record being chiefly due to organic beings not inhabiting profound depths of the sea, and to their remains being embedded and preserved to a future age only in masses of sediment sufficiently thick and extensive to withstand an enormous amount of future degradation; and such fossiliferous masses can be accumulated only where much sediment is deposited on the shallow bed of the sea, whilst it slowly subsides. These contingencies will concur only rarely, and after enormously long intervals. Whilst the bed of the sea is stationary or is rising, or when very little sediment is being deposited, there will be blanks in our geological history. The crust of the earth is a vast museum; but the natural collections have been made only at intervals of time immensely remote.

But it may be urged that when several closely-allied species inhabit the same territory we surely ought to find at the present time many transitional forms. Let us take a simple case: in travelling from north to south over a continent, we generally meet at successive intervals with closely allied or representative species, evidently filling nearly the same place in the natural economy of the land. These representative species often meet and interlock; and as the one becomes rarer and rarer, the other becomes more and more frequent, till the one replaces the other. But if we compare these species where they intermingle, they are generally as absolutely distinct from each other in every detail of structure as are specimens taken from the metropolis inhabited by each. By my theory these allied species have descended from a common parent; and during the process of modification, each has become adapted to the conditions of life of its own region, and has supplanted and exterminated its original parent and all the transitional varieties between its past and present states. Hence we ought not to expect at the present time to meet with numerous transitional varieties in each region, though they must have existed there, and may be embedded there in a fossil condition. But in the intermediate region, having intermediate conditions of life, why do we not now find closely-linking intermediate varieties? This difficulty for a long time quite confounded me. But I think it can be in large part explained.

In the first place we should be extremely cautious in inferring, because an area is now continuous, that it has been continuous during a long period. Geology would lead us to believe that almost every continent has been broken up into islands even during the later tertiary periods; and in such islands distinct species might have been separately formed without the possibility of intermediate varieties existing in the intermediate zones. By changes in the form of the land and of climate, marine areas now continuous must often have existed within recent times in a far less continuous and uniform condition than at present. But I will pass over this way of escaping from the difficulty; for I believe that many perfectly defined species have been formed on strictly continuous areas; though I do not doubt that the formerly broken condition of areas now continuous has played an important part in the formation of new species, more especially with freely-crossing and wandering animals.

In looking at species as they are now distributed over a wide area, we generally find them tolerably numerous over a large territory, then becoming somewhat abruptly rarer and rarer on the confines, and finally disappearing. Hence the neutral territory between two representative species is generally narrow in comparison with the territory proper to each. We see the same fact in ascending mountains, and sometimes it is quite remarkable how abruptly, as Alph. De Candolle has observed, a common alpine species disappears. The same fact has been noticed by Forbes in sounding the depths of the sea with the dredge. To those who look at climate and the physical conditions of life as the all-important elements of distribution, these facts ought to cause surprise, as climate and height or depth graduate away insensibly. But when we bear in mind that almost every species, even in its metropolis, would increase immensely in numbers, were it not for other competing species; that nearly all either prey on or serve as prey for others; in short, that each organic being is either directly or indirectly related in the most important manner to other organic beings, we must see that the range of the inhabitants of any country by no means exclusively depends on insensibly changing physical conditions, but in large part on the presence of other species, on which it depends, or by which it is destroyed, or with which it comes into competition; and as these species are already defined objects (however they may have become so), not blending one into another by insensible gradations, the range of any one species, depending as it does on the range of others, will tend to be sharply defined. Moreover, each species on the confines of its range, where it exists in lessened numbers, will, during fluctuations in the number of its enemies or of its prey, or in the seasons, be extremely liable to utter extermination; and thus its geographical range will come to be still more sharply defined.

If I am right in believing that allied or representative species, when inhabiting a continuous area, are generally so distributed that each has a wide range, with a comparatively narrow neutral territory between them, in which they become rather suddenly rarer and rarer; then, as varieties do not essentially differ from species, the same rule will probably apply to both; and if we in imagination adapt a varying species to a very large area, we shall have to adapt two varieties to two large areas, and a third variety to a narrow intermediate zone. The intermediate variety, consequently, will exist in lesser numbers from inhabiting a narrow and lesser area; and practically, as far as I can make out, this rule holds good with varieties in a state of nature. I have met with striking instances of the rule in the case of varieties intermediate between well-marked varieties in the genus Balanus.* And it would appear from information given me by Mr. Watson, Dr. Asa Gray, and Mr. Wollaston, that generally when varieties intermediate between two other forms occur, they are much rarer numerically than the forms which they connect. Now, if we may trust these facts and inferences, and therefore conclude that varieties linking two other varieties together have generally existed in lesser numbers than the forms which they connect, then, I think, we can understand why intermediate varieties should not endure for very long periods;—why as a general rule they should be exterminated and disappear, sooner than the forms which they originally linked together.

For any form existing in lesser numbers would, as already remarked, run a greater chance of being exterminated than one existing in large numbers; and in this particular case the intermediate form would be eminently liable to the inroads of closely allied forms existing on both sides of it. But a far more important consideration, as I believe, is that, during the process of further modification, by which two varieties are supposed on my theory to be converted and perfected into two distinct species, the two which exist in larger numbers from inhabiting larger areas, will have a great advantage over the intermediate variety, which exists in smaller numbers in a narrow and intermediate zone. For forms existing in larger numbers will always have a better chance, within any given period, of presenting further favourable variations for natural selection to seize on, than will the rarer forms which exist in lesser numbers. Hence, the more common forms, in the race for life, will tend to beat and supplant the less common forms, for these will be more slowly modified and improved. It is the same principle which, as I believe, accounts for the common species in each country, as shown in the second chapter, presenting on an average a greater number of well-marked varieties than do the rarer species. I may illustrate what I mean by supposing three varieties of sheep to be kept, one adapted to an extensive mountainous region; a second to a comparatively narrow, hilly tract; and a third to wide plains at the base; and that the inhabitants are all trying with equal steadiness and skill to improve their stocks by selection; the chances in this case will be strongly in favour of the great holders on the mountains or on the plains improving their breeds more quickly than the small holders on the intermediate narrow, hilly tract; and consequently the improved mountain or plain breed will soon take the place of the less improved hill breed; and thus the two breeds, which originally existed in greater numbers, will come into close contact with each other, without the interposition of the supplanted, intermediate hill-variety.

To sum up, I believe that species come to be tolerably well-defined objects, and do not at any one period present an inextricable chaos of varying and intermediate links: firstly, because new varieties are very slowly formed, for variation is a very slow process, and natural selection can do nothing until favourable variations chance to occur, and until a place in the natural polity of the country can be better filled by some modification of some one or more of its inhabitants. And such new places will depend on slow changes of climate, or on the occasional immigration of new inhabitants, and, probably, in a still more important degree, on some of the old inhabitants becoming slowly modified, with the new forms thus produced and the old ones acting and reacting on each other. So that, in any one region and at any one time, we ought only to see a few species presenting slight modifications of structure in some degree permanent; and this assuredly we do see.

Secondly, areas now continuous must often have existed within the recent period in isolated portions, in which many forms, more especially amongst the classes which unite for each birth and wander much, may have separately been rendered sufficiently distinct to rank as representative species. In this case, intermediate varieties between the several representative species and their common parent, must formerly have existed in each broken portion of the land, but these links will have been supplanted and exterminated during the process of natural selection, so that they will no longer exist in a living state.

Thirdly, when two or more varieties have been formed in different portions of a strictly continuous area, intermediate varieties will, it is probable, at first have been formed in the intermediate zones, but they will generally have had a short duration. For these intermediate varieties will, from reasons already assigned (namely from what we know of the actual distribution of closely allied or representative species, and likewise of acknowledged varieties), exist in the intermediate zones in lesser numbers than the varieties which they tend to connect. From this cause alone the intermediate varieties will be liable to accidental extermination; and during the process of further modification through natural selection, they will almost certainly be beaten and supplanted by the forms which they connect; for these from existing in greater numbers will, in the aggregate, present more variation, and thus be further improved through natural selection and gain further advantages.

Lastly, looking not to any one time, but to all time, if my theory be true, numberless intermediate varieties, linking most closely all the species of the same group together, must assuredly have existed; but the very process of natural selection constantly tends, as has been so often remarked, to exterminate the parent forms and the intermediate links. Consequently evidence of their former existence could be found only amongst fossil remains, which are preserved, as we shall in a future chapter attempt to show, in an extremely imperfect and intermittent record.

On the origin and transitions of organic beings with peculiar habits and structure.—It has been asked by the opponents of such views as I hold, how, for instance, a land carnivorous animal could have been converted into one with aquatic habits; for how could the animal in its transitional state have subsisted? It would be easy to show that within the same group carnivorous animals exist having every intermediate grade between truly aquatic and strictly terrestrial habits; and as each exists by a struggle for life, it is clear that each is well adapted in its habits to its place in nature. Look at the Mustela vison* of North America, which has webbed feet and which resembles an otter in its fur, short legs, and form of tail; during summer this animal dives for and preys on fish, but during the long winter it leaves the frozen waters, and preys like other polecats on mice and land animals. If a different case had been taken, and it had been asked how an insectivorous quadruped could possibly have been converted into a flying bat, the question would have been far more difficult, and I could have given no answer. Yet I think such difficulties have very little weight.

Here, as on other occasions, I lie under a heavy disadvantage, for out of the many striking cases which I have collected, I can give only one or two instances of transitional habits and structures in closely allied species of the same genus; and of diversified habits, either constant or occasional, in the same species. And it seems to me that nothing less than a long list of such cases is sufficient to lessen the difficulty in any particular case like that of the bat.

Look at the family of squirrels; here we have the finest gradation from animals with their tails only slightly flattened, and from others, as Sir J. Richardson has remarked, with the posterior part of their bodies rather wide and with the skin on their flanks rather full, to the so-called flying squirrels; and flying squirrels have their limbs and even the base of the tail united by a broad expanse of skin, which serves as a parachute and allows them to glide through the air to an astonishing distance from tree to tree. We cannot doubt that each structure is of use to each kind of squirrel in its own country, by enabling it to escape birds or beasts of prey, or to collect food more quickly, or, as there is reason to believe, by lessening the danger from occasional falls. But it does not follow from this fact that the structure of each squirrel is the best that it is possible to conceive under all natural conditions. Let the climate and vegetation change, let other competing rodents or new beasts of prey immigrate, or old ones become modified, and all analogy would lead us to believe that some at least of the squirrels would decrease in numbers or become exterminated, unless they also became modified and improved in structure in a corresponding manner. Therefore, I can see no difficulty, more especially under changing conditions of life, in the continued preservation of individuals with fuller and fuller flank-membranes, each modification being useful, each being propagated, until by the accumulated effects of this process of natural selection, a perfect so-called flying squirrel was produced.

[*The development of flight is illustrated by examples ranging from flying lemurs and flying fish to various kinds of birds.*]

I will now give two or three instances of diversified and of changed habits in the individuals of the same species. When either case occurs, it would be easy for natural selection to fit the animal, by some modification of its structure, for its changed habits, or exclusively for one of its several different habits. But it is difficult to tell, and immaterial for us, whether habits generally change first and structure afterwards; or whether slight modifications of structure lead to changed habits; both probably often change almost simultaneously. Of cases of changed habits it will suffice merely to allude to that of the many British insects which now feed on exotic plants, or exclusively on artificial substances. Of diversified habits innumerable instances could be given: I have often watched a tyrant flycatcher (Saurophagus sulphuratus) in South America, hovering over one spot and then proceeding to another, like a kestrel, and at other times standing stationary on the margin of water, and then dashing like a kingfisher at a fish. In our own country the larger titmouse (Parus major) may be seen climbing branches, almost like a creeper; it often, like a shrike, kills small birds by blows on the head; and I have many times seen and heard it hammering the seeds of the yew on a branch, and thus breaking them like a nuthatch. In North America the black bear was seen by Hearne swimming for hours with widely open mouth, thus catching, like a whale, insects in the water. Even in so extreme a case as this, if the supply of insects were constant, and if better adapted competitors did not already exist in the country, I can see no difficulty in a race of bears being rendered, by natural selection, more and more aquatic in their structure and habits, with larger and larger mouths, till a creature was produced as monstrous as a whale.*

As we sometimes see individuals of a species following habits widely different from those both of their own species and of the other species of the same genus, we might expect, on my theory, that such individuals would occasionally have given rise to new species, having anomalous habits, and with their structure either slightly or considerably modified from that of their proper type. And such instances do occur in nature. Can a more striking instance of adaptation be given than that of a woodpecker for climbing trees and for seizing insects in the chinks of the bark? Yet in North America there are woodpeckers which feed largely on fruit, and others with elongated wings which chase insects on the wing; and on the plains of La Plata, where not a tree grows, there is a woodpecker, which in every essential part of its organisation, even in its colouring, in the harsh tone of its voice, and undulatory flight, told me plainly of its close blood-relationship to our common species; yet it is a wood-pecker which never climbs a tree!

Petrels are the most aërial and oceanic of birds, yet in the quiet Sounds of Tierra del Fuego, the Puffinuria berardi, in its general habits, in its astonishing power of diving, its manner of swimming, and of flying when unwillingly it takes flight, would be mistaken by any one for an auk or grebe; nevertheless, it is essentially a petrel, but with many parts of its organisation profoundly modified.* On the other hand, the acutest observer by examining the dead body of the water-ouzel would never have suspected its sub-aquatic habits; yet this anomalous member of the strictly terrestrial thrush family wholly subsists by diving,—grasping the stones with its feet and using its wings under water.

He who believes that each being has been created as we now see it, must occasionally have felt surprise when he has met with an animal having habits and structure not at all in agreement. What can be plainer than that the webbed feet of ducks and geese are formed for swimming? yet there are upland geese with webbed feet which rarely or never go near the water; and no one except Audubon has seen the frigate-bird, which has all its four toes webbed, alight on the surface of the sea. On the other hand, grebes and coots are eminently aquatic, although their toes are only bordered by membrane. What seems plainer than that the long toes of grallatores* are formed for walking over swamps and floating plants, yet the water-hen is nearly as aquatic as the coot; and the landrail nearly as terrestrial as the quail or partridge. In such cases, and many others could be given, habits have changed without a corresponding change of structure. The webbed feet of the upland goose may be said to have become rudi-mentary in function, though not in structure. In the frigate-bird, the deeply-scooped membrane between the toes shows that structure has begun to change.

He who believes in separate and innumerable acts of creation will say, that in these cases it has pleased the Creator to cause a being of one type to take the place of one of another type; but this seems to me only restating the fact in dignified language. He who believes in the struggle for existence and in the principle of natural selection, will acknowledge that every organic being is constantly endeavouring to increase in numbers; and that if any one being vary ever so little, either in habits or structure, and thus gain an advantage over some other inhabitant of the country, it will seize on the place of that inhabitant, however different it may be from its own place. Hence it will cause him no surprise that there should be geese and frigate-birds with webbed feet, either living on the dry land or most rarely alighting on the water; that there should be long-toed corncrakes living in meadows instead of in swamps; that there should be wood-peckers where not a tree grows; that there should be diving thrushes, and petrels with the habits of auks.

Organs of extreme perfection and complication. — To suppose that the eye, with all its inimitable contrivances for adjusting the focus to different distances, for admitting different amounts of light, and for the correction of spherical and chromatic aberration,* could have been formed by natural selection, seems, I freely confess, absurd in the highest possible degree. Yet reason tells me, that if numerous gradations from a perfect and complex eye to one very imperfect and simple, each grade being useful to its possessor, can be shown to exist; if further, the eye does vary ever so slightly, and the variations be inherited, which is certainly the case; and if any variation or modification in the organ be ever useful to an animal under changing conditions of life, then the difficulty of believing that a perfect and complex eye could be formed by natural selection, though insuperable by our imagination, can hardly be considered real. How a nerve comes to be sensitive to light, hardly concerns us more than how life itself first originated; but I may remark that several facts make me suspect that any sensitive nerve may be rendered sensitive to light, and likewise to those coarser vibrations of the air which produce sound.

In looking for the gradations by which an organ in any species has been perfected, we ought to look exclusively to its lineal ancestors; but this is scarcely ever possible, and we are forced in each case to look to species of the same group, that is to the collateral descendants from the same original parent-form, in order to see what gradations are possible, and for the chance of some gradations having been transmitted from the earlier stages of descent, in an unaltered or little altered condition. Amongst existing Vertebrata, we find but a small amount of gradation in the structure of the eye, and from fossil species we can learn nothing on this head. In this great class we should probably have to descend far beneath the lowest known fossiliferous stratum to discover the earlier stages, by which the eye has been perfected.

In the Articulata* we can commence a series with an optic nerve merely coated with pigment, and without any other mechanism; and from this low stage, numerous gradations of structure, branching off in two fundamentally different lines, can be shown to exist, until we reach a moderately high stage of perfection. In certain crustaceans, for instance, there is a double cornea, the inner one divided into facets, within each of which there is a lens-shaped swelling. In other crustaceans the transparent cones which are coated by pigment, and which properly act only by excluding lateral pencils of light, are convex at their upper ends and must act by convergence; and at their lower ends there seems to be an imperfect vitreous substance. With these facts, here far too briefly and imperfectly given, which show that there is much graduated diversity in the eyes of living crustaceans, and bearing in mind how small the number of living animals is in proportion to those which have become extinct, I can see no very great difficulty (not more than in the case of many other structures) in believing that natural selection has converted the simple apparatus of an optic nerve merely coated with pigment and invested by transparent membrane, into an optical instrument as perfect as is possessed by any member of the great Articulate class.

He who will go thus far, if he find on finishing this treatise that large bodies of facts, otherwise inexplicable, can be explained by the theory of descent, ought not to hesitate to go further, and to admit that a structure even as perfect as the eye of an eagle might be formed by natural selection, although in this case he does not know any of the transitional grades. His reason ought to conquer his imagination; though I have felt the difficulty far too keenly to be surprised at any degree of hesitation in extending the principle of natural selection to such startling lengths.

It is scarcely possible to avoid comparing the eye to a telescope. We know that this instrument has been perfected by the long-continued efforts of the highest human intellects; and we naturally infer that the eye has been formed by a somewhat analogous process. But may not this inference be presumptuous? Have we any right to assume that the Creator works by intellectual powers like those of man?

If we must compare the eye to an optical instrument, we ought in imagination to take a thick layer of transparent tissue, with a nerve sensitive to light beneath, and then suppose every part of this layer to be continually changing slowly in density, so as to separate into layers of different densities and thicknesses, placed at different distances from each other, and with the surfaces of each layer slowly changing in form. Further we must suppose that there is a power always intently watching each slight accidental alteration in the transparent layers; and carefully selecting each alteration which, under varied circumstances, may in any way, or in any degree, tend to produce a distincter image. We must suppose each new state of the instrument to be multiplied by the million; and each to be preserved till a better be produced, and then the old ones to be destroyed. In living bodies, variation will cause the slight alterations, generation will multiply them almost infinitely, and natural selection will pick out with unerring skill each improvement. Let this process go on for millions on millions of years; and during each year on millions of individuals of many kinds; and may we not believe that a living optical instrument might thus be formed as superior to one of glass, as the works of the Creator are to those of man?

If it could be demonstrated that any complex organ existed, which could not possibly have been formed by numerous, successive, slight modifications, my theory would absolutely break down. But I can find out no such case. No doubt many organs exist of which we do not know the transitional grades, more especially if we look to much-isolated species, round which, according to my theory, there has been much extinction. Or again, if we look to an organ common to all the members of a large class, for in this latter case the organ must have been first formed at an extremely remote period, since which all the many members of the class have been developed; and in order to discover the early transitional grades through which the organ has passed, we should have to look to very ancient ancestral forms, long since become extinct.

[*After considering other highly complicated organs, including the origin of lungs from a simple swimbladder, Darwin examines the reasons for the persistence of variations that seem to be of little or no apparent use.*]

Summary of Chapter. — We have in this chapter discussed some of the difficulties and objections which may be urged against my theory.

Many of them are very grave; but I think that in the discussion light has been thrown on several facts, which on the theory of independent acts of creation are utterly obscure. We have seen that species at any one period are not indefinitely variable, and are not linked together by a multitude of intermediate gradations, partly because the process of natural selection will always be very slow, and will act, at any one time, only on a very few forms; and partly because the very process of natural selection almost implies the continual supplanting and extinction of preceding and intermediate gradations. Closely allied species, now living on a continuous area, must often have been formed when the area was not continuous, and when the conditions of life did not insensibly graduate away from one part to another. When two varieties are formed in two districts of a continuous area, an intermediate variety will often be formed, fitted for an intermediate zone; but from reasons assigned, the intermediate variety will usually exist in lesser numbers than the two forms which it connects; consequently the two latter, during the course of further modification, from existing in greater numbers, will have a great advantage over the less numerous intermediate variety, and will thus generally succeed in supplanting and exterminating it.

We have seen in this chapter how cautious we should be in concluding that the most different habits of life could not graduate into each other; that a bat, for instance, could not have been formed by natural selection from an animal which at first could only glide through the air.

We have seen that a species may under new conditions of life change its habits, or have diversified habits, with some habits very unlike those of its nearest congeners. Hence we can understand, bearing in mind that each organic being is trying to live wherever it can live, how it has arisen that there are upland geese with webbed feet, ground woodpeckers, diving thrushes, and petrels with the habits of auks.

Although the belief that an organ so perfect as the eye could have been formed by natural selection, is more than enough to stagger any one; yet in the case of any organ, if we know of a long series of gradations in complexity, each good for its possessor, then, under changing conditions of life, there is no logical impossibility in the acquirement of any conceivable degree of perfection through natural selection.

In the cases in which we know of no intermediate or transitional states, we should be very cautious in concluding that none could have existed, for the homologies of many organs and their intermediate states show that wonderful metamorphoses in function are at least possible. For instance, a swim-bladder has apparently been converted into an air-breathing lung. The same organ having performed simultaneously very different functions, and then having been specialised for one function; and two very distinct organs having performed at the same time the same function, the one having been perfected whilst aided by the other, must often have largely facilitated transitions.

We are far too ignorant, in almost every case, to be enabled to assert that any part or organ is so unimportant for the welfare of a species, that modifications in its structure could not have been slowly accumulated by means of natural selection. But we may confidently believe that many modifications, wholly due to the laws of growth, and at first in no way advantageous to a species, have been subsequently taken advantage of by the still further modified descendants of this species. We may, also, believe that a part formerly of high importance has often been retained (as the tail of an aquatic animal by its terrestrial descendants), though it has become of such small importance that it could not, in its present state, have been acquired by natural selection,—a power which acts solely by the preservation of profitable variations in the struggle for life.

Natural selection will produce nothing in one species for the exclusive good or injury of another; though it may well produce parts, organs, and excretions highly useful or even indispensable, or highly injurious to another species, but in all cases at the same time useful to the owner. Natural selection in each well-stocked country, must act chiefly through the competition of the inhabitants one with another, and consequently will produce perfection, or strength in the battle for life, only according to the standard of that country. Hence the inhabitants of one country, generally the smaller one, will often yield, as we see they do yield, to the inhabitants of another and generally larger country. For in the larger country there will have existed more individuals, and more diversified forms, and the competition will have been severer, and thus the standard of perfection will have been rendered higher. Natural selection will not necessarily produce absolute perfection; nor, as far as we can judge by our limited faculties, can absolute perfection be everywhere found.

On the theory of natural selection we can clearly understand the full meaning of that old canon in natural history, "Natura non facit saltum."* This canon, if we look only to the present inhabitants of the world, is not strictly correct, but if we include all those of past times, it must by my theory be strictly true.

It is generally acknowledged that all organic beings have been formed on two great laws—Unity of Type, and the Conditions of Existence. By unity of type is meant that fundamental agreement in structure, which we see in organic beings of the same class, and which is quite independent of their habits of life. On my theory, unity of type is explained by unity of descent. The expression of conditions of existence, so often insisted on by the illustrious Cuvier, is fully embraced by the principle of natural selection. For natural selection acts by either now adapting the varying parts of each being to its organic and inorganic conditions of life; or by having adapted them during long-past periods of time: the adaptations being aided in some cases by use and disuse, being slightly affected by the direct action of the external conditions of life, and being in all cases subjected to the several laws of growth. Hence, in fact, the law of the Conditions of Existence is the higher law; as it includes, through the inheritance of former adaptations, that of Unity of Type.

[*The following three chapters (VII–IX) consider other potential objections to the theory of natural selection. Special attention is devoted to instinct—how do hive-bees make geometrically exact hexagonal cells without some sort of design involved?—and the problems posed by the apparent sterility of most hybrid crossings. The most serious objection is posed by the geological record, which fails to exhibit the wealth of intermediate forms predicted by a gradual theory of species change. Darwin responds by suggesting that the fossil evidence is radically imperfect, far more so than thought by most geologists. The geological record does, however, provide important evidence favourable towards the theory, and this is dealt with in the first of a series of chapters (X–XIII) which demonstrate how the origin, distribution, extinction, and embryological development of living beings in time and space can be understood in evolutionary terms. The final chapter opens by summing up the difficulties the theory has faced.*]

CHAPTER XIV

Recapitulation and Conclusion

As this whole volume is one long argument, it may be convenient to the reader to have the leading facts and inferences briefly recapitulated.

That many and grave objections may be advanced against the theory of descent with modification through natural selection, I do not deny. I have endeavoured to give to them their full force. Nothing at first can appear more difficult to believe than that the more complex organs and instincts should have been perfected, not by means superior to, though analogous with, human reason, but by the accumulation of innumerable slight variations, each good for the individual possessor. Nevertheless, this difficulty, though appearing to our imagination insuperably great, cannot be considered real if we admit the following propositions, namely, — that gradations in the perfection of any organ or instinct, which we may consider, either do now exist or could have existed, each good of its kind, — that all organs and instincts are, in ever so slight a degree, variable, — and, lastly, that there is a struggle for existence leading to the preservation of each profitable deviation of structure or instinct. The truth of these propositions cannot, I think, be disputed.

It is, no doubt, extremely difficult even to conjecture by what gradations many structures have been perfected, more especially amongst broken and failing groups of organic beings; but we see so many strange gradations in nature, as is proclaimed by the canon, "Natura non facit saltum," that we ought to be extremely cautious in saying that any organ or instinct, or any whole being, could not have arrived at its present state by many graduated steps. There are, it must be admitted, cases of special difficulty on the theory of natural selection; and one of the most curious of these is the existence of two or three defined castes of workers or sterile females in the same community of ants; but I have attempted to show* how this difficulty can be mastered.

With respect to the almost universal sterility of species when first crossed, which forms so remarkable a contrast with the almost universal fertility of varieties when crossed, I must refer the reader to the recapitulation of the facts given at the end of the eighth chapter, which seem to me conclusively to show that this sterility is no more a special endowment than is the incapacity of two trees to be grafted together; but that it is incidental on constitutional differences in the reproductive systems of the intercrossed species. We see the truth of this conclusion in the vast difference in the result, when the same two species are crossed reciprocally; that is, when one species is first used as the father and then as the mother.

The fertility of varieties when intercrossed and of their mongrel offspring cannot be considered as universal; nor is their very general fertility surprising when we remember that it is not likely that either their constitutions or their reproductive systems should have been profoundly modified. Moreover, most of the varieties which have been experimentised on have been produced under domestication; and as domestication apparently tends to eliminate sterility, we ought not to expect it also to produce sterility.

The sterility of hybrids is a very different case from that of first crosses, for their reproductive organs are more or less functionally impotent; whereas in first crosses the organs on both sides are in a perfect condition. As we continually see that organisms of all kinds are rendered in some degree sterile from their constitutions having been disturbed by slightly different and new conditions of life, we need not feel surprise at hybrids being in some degree sterile, for their constitutions can hardly fail to have been disturbed from being compounded of two distinct organisations. This parallelism is supported by another parallel, but directly opposite, class of facts; namely, that the vigour and fertility of all organic beings are increased by slight changes in their conditions of life, and that the offspring of slightly modified forms or varieties acquire from being crossed increased vigour and fertility. So that, on the one hand, considerable changes in the conditions of life and crosses between greatly modified forms, lessen fertility; and on the other hand, lesser changes in the conditions of life and crosses between less modified forms, increase fertility.

Turning to geographical distribution, the difficulties encountered on the theory of descent with modification are grave enough. All the individuals of the same species, and all the species of the same genus, or even higher group, must have descended from common parents; and therefore, in however distant and isolated parts of the world they are now found, they must in the course of successive generations have passed from some one part to the others. We are often wholly unable even to conjecture how this could have been effected. Yet, as we have reason to believe that some species have retained the same specific form for very long periods, enormously long as measured by years, too much stress ought not to be laid on the occasional wide diffusion of the same species; for during very long periods of time there will always be a good chance for wide migration by many means. A broken or interrupted range may often be accounted for by the extinction of the species in the intermediate regions. It cannot be denied that we are as yet very ignorant of the full extent of the various climatal and geographical changes which have affected the earth during modern periods; and such changes will obviously have greatly facilitated migration. As an example, I have attempted to show how potent has been the influence of the Glacial period on the distribution both of the same and of representative species throughout the world. We are as yet profoundly ignorant of the many occasional means of transport. With respect to distinct species of the same genus inhabiting very distant and isolated regions, as the process of modification has necessarily been slow, all the means of migration will have been possible during a very long period; and consequently the difficulty of the wide diffusion of species of the same genus is in some degree lessened.

As on the theory of natural selection an interminable number of intermediate forms must have existed, linking together all the species in each group by gradations as fine as our present varieties, it may be asked, Why do we not see these linking forms all around us? Why are not all organic beings blended together in an inextricable chaos? With respect to existing forms, we should remember that we have no right to expect (excepting in rare cases) to discover *directly* connecting links between them, but only between each and some extinct and supplanted form. Even on a wide area, which has during a long period remained continuous, and of which the climate and other conditions of life change insensibly in going from a district occupied by one species into another district occupied by a closely allied species, we have no just right to expect often to find intermediate varieties in the intermediate zone. For we have reason to believe that only a few species are undergoing change at any one period; and all changes are slowly effected. I have also shown that the intermediate varieties which will at first probably exist in the intermediate zones, will be liable to be supplanted by the allied forms on either hand; and the latter, from existing in greater numbers, will generally be modified and improved at a quicker rate than the intermediate varieties, which exist in lesser numbers; so that the intermediate varieties will, in the long run, be supplanted and exterminated.

On this doctrine of the extermination of an infinitude of connecting links, between the living and extinct inhabitants of the world, and at each successive period between the extinct and still older species, why is not every geological formation charged with such links? Why does not every collection of fossil remains afford plain evidence of the gradation and mutation of the forms of life? We meet with no such evidence, and this is the most obvious and forcible of the many objections which may be urged against my theory. Why, again, do whole groups of allied species appear, though certainly they often falsely appear, to have come in suddenly on the several geological stages? Why do we not find great piles of strata beneath the Silurian system,* stored with the remains of the progenitors of the Silurian groups of fossils? For certainly on my theory such strata must somewhere have been deposited at these ancient and utterly unknown epochs in the world's history.

I can answer these questions and grave objections only on the supposition that the geological record is far more imperfect than most geologists believe. It cannot be objected that there has not been time sufficient for any amount of organic change; for the lapse of time has been so great as to be utterly inappreciable by the human intellect. The number of specimens in all our museums is absolutely as nothing compared with the countless generations of countless species which certainly have existed. We should not be able to recognise a species as the parent of any one or more species if we were to examine them ever so closely, unless we likewise possessed many of the intermediate links between their past or parent and present states; and these many links we could hardly ever expect to discover, owing to the imperfection of the geological record. Numerous existing doubtful forms could be named which are probably varieties; but who will pretend that in future ages so many fossil links will be discovered, that naturalists will be able to decide, on the common view, whether or not these doubtful forms are varieties? As long as most of the links between any two species are unknown, if any one link or intermediate variety be discovered, it will simply be classed as another and distinct species. Only a small portion of the world has been geologically explored. Only organic beings of certain classes can be preserved in a fossil condition, at least in any great number. Widely ranging species vary most, and varieties are often at first local, — both causes rendering the discovery of intermediate links less likely. Local varieties will not spread into other and distant regions until they are considerably modified and improved; and when they do spread, if discovered in a geological formation, they will appear as if suddenly created there, and will be simply classed as new species. Most formations have been intermittent in their accumulation; and their duration, I am inclined to believe, has been shorter than the average duration of specific forms. Successive formations are separated from each other by enormous blank intervals of time; for fossiliferous formations, thick enough to resist future degradation, can be accumulated only where much sediment is deposited on the subsiding bed of the sea. During the alternate periods of elevation and of stationary level the record will be blank. During these latter periods there will probably be more variability in the forms of life; during periods of subsidence, more extinction.

With respect to the absence of fossiliferous formations beneath the lowest Silurian strata, I can only recur to the hypothesis given in the ninth chapter. That the geological record is imperfect all will admit; but that it is imperfect to the degree which I require, few will be inclined to admit. If we look to long enough intervals of time, geology plainly declares that all species have changed; and they have changed in the manner which my theory requires, for they have changed slowly and in a graduated manner. We clearly see this in the fossil remains from consecutive formations invariably being much more closely related to each other, than are the fossils from formations distant from each other in time.

Such is the sum of the several chief objections and difficulties which may justly be urged against my theory; and I have now briefly recapitulated the answers and explanations which can be given to them. I have felt these difficulties far too heavily during many years to doubt their weight. But it deserves especial notice that the more important objections relate to questions on which we are confessedly ignorant; nor do we know how ignorant we are. We do not know all the possible transitional gradations between the simplest and the most perfect organs; it cannot be pretended that we know all the varied means of Distribution during the long lapse of years, or that we know how imperfect the Geological Record is. Grave as these several difficulties are, in my judgment they do not overthrow the theory of descent with modification.

Now let us turn to the other side of the argument. Under domestication we see much variability. This seems to be mainly due to the reproductive system being eminently susceptible to changes in the conditions of life; so that this system, when not rendered impotent, fails to reproduce offspring exactly like the parent-form. Variability is governed by many complex laws,—by correlation of growth, by use and disuse, and by the direct action of the physical conditions of life. There is much difficulty in ascertaining how much modification our domestic productions have undergone; but we may safely infer that the amount has been large, and that modifications can be inherited for long periods. As long as the conditions of life remain the same, we have reason to believe that a modification, which has already been inherited for many generations, may continue to be inherited for an almost infinite number of generations. On the other hand we have evidence that variability, when it has once come into play, does not wholly cease; for new varieties are still occasionally produced by our most anciently domesticated productions.

Man does not actually produce variability; he only unintentionally exposes organic beings to new conditions of life, and then nature acts on the organisation, and causes variability. But man can and does select the variations given to him by nature, and thus accumulate them in any desired manner. He thus adapts animals and plants for his own benefit or pleasure. He may do this methodically, or he may do it unconsciously by preserving the individuals most useful to him at the time, without any thought of altering the breed. It is certain that he can largely influence the character of a breed by selecting, in each successive generation, individual differences so slight as to be quite inappreciable by an uneducated eye. This process of selection has been the great agency in the production of the most distinct and useful domestic breeds. That many of the breeds produced by man have to a large extent the character of natural species, is shown by the inextricable doubts whether very many of them are varieties or aboriginal species.

There is no obvious reason why the principles which have acted so efficiently under domestication should not have acted under nature.

In the preservation of favoured individuals and races, during the constantly-recurrent Struggle for Existence, we see the most powerful and ever-acting means of selection. The struggle for existence inevitably follows from the high geometrical ratio of increase which is common to all organic beings. This high rate of increase is proved by calculation, by the effects of a succession of peculiar seasons, and by the results of naturalisation, as explained in the third chapter. More individuals are born than can possibly survive. A grain in the balance will determine which individual shall live and which shall die,—which variety or species shall increase in number, and which shall decrease, or finally become extinct. As the individuals of the same species come in all respects into the closest competition with each other, the struggle will generally be most severe between them; it will be almost equally severe between the varieties of the same species, and next in severity between the species of the same genus. But the struggle will often be very severe between beings most remote in the scale of nature. The slightest advantage in one being, at any age or during any season, over those with which it comes into competition, or better adaptation in however slight a degree to the surrounding physical conditions, will turn the balance.

With animals having separated sexes there will in most cases be a struggle between the males for possession of the females. The most vigorous individuals, or those which have most successfully struggled with their conditions of life, will generally leave most progeny. But success will often depend on having special weapons or means of defence, or on the charms of the males; and the slightest advantage will lead to victory.

As geology plainly proclaims that each land has undergone great physical changes, we might have expected that organic beings would have varied under nature, in the same way as they generally have varied under the changed conditions of domestication. And if there be any variability under nature, it would be an unaccountable fact if natural selection had not come into play. It has often been asserted, but the assertion is quite incapable of proof, that the amount of variation under nature is a strictly limited quantity. Man, though acting on external characters alone and often capriciously, can produce within a short period a great result by adding up mere individual differences in his domestic productions; and every one admits that there are at least individual differences in species under nature.

But, besides such differences, all naturalists have admitted the existence of varieties, which they think sufficiently distinct to be worthy of record in systematic works. No one can draw any clear distinction between individual differences and slight varieties; or between more plainly marked varieties and sub-species, and species. Let it be observed how naturalists differ in the rank which they assign to the many representative forms in Europe and North America.

If then we have under nature variability and a powerful agent always ready to act and select, why should we doubt that variations in any way useful to beings, under their excessively complex relations of life, would be preserved, accumulated, and inherited? Why, if man can by patience select variations most useful to himself, should nature fail in selecting variations useful, under changing conditions of life, to her living products? What limit can be put to this power, acting during long ages and rigidly scrutinising the whole constitution, structure, and habits of each creature,—favouring the good and rejecting the bad? I can see no limit to this power, in slowly and beautifully adapting each form to the most complex relations of life. The theory of natural selection, even if we looked no further than this, seems to me to be in itself probable. I have already recapitulated, as fairly as I could, the opposed difficulties and objections: now let us turn to the special facts and arguments in favour of the theory.

On the view that species are only strongly marked and permanent varieties, and that each species first existed as a variety, we can see why it is that no line of demarcation can be drawn between species, commonly supposed to have been produced by special acts of creation, and varieties which are acknowledged to have been produced by secondary laws. On this same view we can understand how it is that in each region where many species of a genus have been produced, and where they now flourish, these same species should present many varieties; for where the manufactory of species has been active, we might expect, as a general rule, to find it still in action; and this is the case if varieties be incipient species. Moreover, the species of the larger genera, which afford the greater number of varieties or incipient species, retain to a certain degree the character of varieties; for they differ from each other by a less amount of difference than do the species of smaller genera. The closely allied species also of the larger genera apparently have restricted ranges, and they are clustered in little groups round other species—in which respects they resemble varieties. These are strange relations on the view of each species having been independently created, but are intelligible if all species first existed as varieties.

As each species tends by its geometrical ratio of reproduction to increase inordinately in number; and as the modified descendants of each species will be enabled to increase by so much the more as they become more diversified in habits and structure, so as to be enabled to seize on many and widely different places in the economy of nature, there will be a constant tendency in natural selection to preserve the most divergent offspring of any one species. Hence during a long-continued course of modification, the slight differences, characteristic of varieties of the same species, tend to be augmented into the greater differences characteristic of species of the same genus. New and improved varieties will inevitably supplant and exterminate the older, less improved and intermediate varieties; and thus species are rendered to a large extent defined and distinct objects. Dominant species belonging to the larger groups tend to give birth to new and dominant forms; so that each large group tends to become still larger, and at the same time more divergent in character. But as all groups cannot thus succeed in increasing in size, for the world would not hold them, the more dominant groups beat the less dominant. This tendency in the large groups to go on increasing in size and diverging in character, together with the almost inevitable contingency of much extinction, explains the arrangement of all the forms of life, in groups subordinate to groups, all within a few great classes, which we now see everywhere around us, and which has prevailed throughout all time. This grand fact of the grouping of all organic beings seems to me utterly inexplicable on the theory of creation.

As natural selection acts solely by accumulating slight, successive, favourable variations, it can produce no great or sudden modification; it can act only by very short and slow steps. Hence the canon of "Natura non facit saltum," which every fresh addition to our knowledge tends to make more strictly correct, is on this theory simply intelligible. We can plainly see why nature is prodigal in variety, though niggard in innovation. But why this should be a law of nature if each species has been independently created, no man can explain.

Many other facts are, as it seems to me, explicable on this theory. How strange it is that a bird, under the form of woodpecker, should have been created to prey on insects on the ground; that upland geese, which never or rarely swim, should have been created with webbed feet; that a thrush should have been created to dive and feed on sub-aquatic insects; and that a petrel should have been created with habits and structure fitting it for the life of an auk or grebe! and so on in endless other cases. But on the view of each species constantly trying to increase in number, with natural selection always ready to adapt the slowly varying descendants of each to any unoccupied or ill-occupied place in nature, these facts cease to be strange, or perhaps might even have been anticipated.

As natural selection acts by competition, it adapts the inhabitants of each country only in relation to the degree of perfection of their associates; so that we need feel no surprise at the inhabitants of any one country, although on the ordinary view supposed to have been specially created and adapted for that country, being beaten and supplanted by the naturalised productions from another land. Nor ought we to marvel if all the contrivances in nature be not, as far as we can judge, absolutely perfect; and if some of them be abhorrent to our ideas of fitness. We need not marvel at the sting of the bee causing the bee's own death; at drones being produced in such vast numbers for one single act, and being then slaughtered by their sterile sisters; at the astonishing waste of pollen by our fir-trees; at the instinctive hatred of the queen bee for her own fertile daughters; at ichneumonidæ* feeding within the live bodies of caterpillars; and at other such cases. The wonder indeed is, on the theory of natural selection, that more cases of the want of absolute perfection have not been observed.

The complex and little known laws governing variation are the same, as far as we can see, with the laws which have governed the production of so-called specific forms. In both cases physical conditions seem to have produced but little direct effect; yet when varieties enter any zone, they occasionally assume some of the characters of the species proper to that zone. In both varieties and species, use and disuse seem to have produced some effect; for it is difficult to resist this conclusion when we look, for instance, at the logger-headed duck, which has wings incapable of flight, in nearly the same condition as in the domestic duck; or when we look at the burrowing tucutucu,* which is occasionally blind, and then at certain moles, which are habitually blind and have their eyes covered with skin; or when we look at the blind animals inhabiting the dark caves of America and Europe. In both varieties and species correlation of growth seems to have played a most important part, so that when one part has been modified other parts are necessarily modified. In both varieties and species reversions to long-lost characters occur. How inexplicable on the theory of creation is the occasional appearance of stripes on the shoulder and legs of the several species of the horse-genus and in their hybrids! How simply is this fact explained if we believe that these species have descended from a striped progenitor, in the same manner as the several domestic breeds of pigeon have descended from the blue and barred rock-pigeon!

On the ordinary view of each species having been independently created, why should the specific characters, or those by which the species of the same genus differ from each other, be more variable than the generic characters in which they all agree? Why, for instance, should the colour of a flower be more likely to vary in any one species of a genus, if the other species, supposed to have been created independently, have differently coloured flowers, than if all the species of the genus have the same coloured flowers? If species are only well-marked varieties, of which the characters have become in a high degree permanent, we can understand this fact; for they have already varied since they branched off from a common progenitor in certain characters, by which they have come to be specifically distinct from each other; and therefore these same characters would be more likely still to be variable than the generic characters which have been inherited without change for an enormous period. It is inexplicable on the theory of creation why a part developed in a very unusual manner in any one species of a genus, and therefore, as we may naturally infer, of great importance to the species, should be eminently liable to variation; but, on my view, this part has undergone, since the several species branched off from a common progenitor, an unusual amount of variability and modification, and therefore we might expect this part generally to be still variable. But a part may be developed in the most unusual manner, like the wing of a bat, and yet not be more variable than any other structure, if the part be common to many subordinate forms, that is, if it has been inherited for a very long period; for in this case it will have been rendered constant by long-continued natural selection.

Glancing at instincts, marvellous as some are, they offer no greater difficulty than does corporeal structure on the theory of the natural selection of successive, slight, but profitable modifications. We can thus understand why nature moves by graduated steps in endowing different animals of the same class with their several instincts. I have attempted to show* how much light the principle of gradation throws on the admirable architectural powers of the hive-bee. Habit no doubt sometimes comes into play in modifying instincts; but it certainly is not indispensable, as we see, in the case of neuter insects, which leave no progeny to inherit the effects of long-continued habit. On the view of all the species of the same genus having descended from a common parent, and having inherited much in common, we can understand how it is that allied species, when placed under considerably different conditions of life, yet should follow nearly the same instincts; why the thrush of South America, for instance, lines her nest with mud like our British species. On the view of instincts having been slowly acquired through natural selection we need not marvel at some instincts being apparently not perfect and liable to mistakes, and at many instincts causing other animals to suffer.

If species be only well-marked and permanent varieties, we can at once see why their crossed offspring should follow the same complex laws in their degrees and kinds of resemblance to their parents,—in being absorbed into each other by successive crosses, and in other such points,—as do the crossed offspring of acknowledged varieties. On the other hand, these would be strange facts if species have been independently created, and varieties have been produced by secondary laws.

If we admit that the geological record is imperfect in an extreme degree, then such facts as the record gives, support the theory of descent with modification. New species have come on the stage slowly and at successive intervals; and the amount of change, after equal intervals of time, is widely different in different groups. The extinction of species and of whole groups of species, which has played so conspicuous a part in the history of the organic world, almost inevitably follows on the principle of natural selection; for old forms will be supplanted by new and improved forms. Neither single species nor groups of species reappear when the chain of ordinary generation has once been broken. The gradual diffusion of dominant forms, with the slow modification of their descendants, causes the forms of life, after long intervals of time, to appear as if they had changed simultaneously throughout the world. The fact of the fossil remains of each formation being in some degree intermediate in character between the fossils in the formations above and below, is simply explained by their intermediate position in the chain of descent. The grand fact that all extinct organic beings belong to the same system with recent beings, falling either into the same or into intermediate groups, follows from the living and the extinct being the offspring of common parents. As the groups which have descended from an ancient progenitor have generally diverged in character, the progenitor with its early descendants will often be intermediate in character in comparison with its later descendants; and thus we can see why the more ancient a fossil is, the oftener it stands in some degree intermediate between existing and allied groups. Recent forms are generally looked at as being, in some vague sense, higher than ancient and extinct forms; and they are in so far higher as the later and more improved forms have conquered the older and less improved organic beings in the struggle for life. Lastly, the law of the long endurance of allied forms on the same continent, — of marsupials in Australia, of edentata in America, and other such cases, — is intelligible, for within a confined country, the recent and the extinct will naturally be allied by descent.

Looking to geographical distribution, if we admit that there has been during the long course of ages much migration from one part of the world to another, owing to former climatal and geographical changes and to the many occasional and unknown means of dispersal, then we can understand, on the theory of descent with modification, most of the great leading facts in Distribution. We can see why there should be so striking a parallelism in the distribution of organic beings throughout space, and in their geological succession through-out time; for in both cases the beings have been connected by the bond of ordinary generation, and the means of modification have been the same. We see the full meaning of the wonderful fact, which must have struck every traveller, namely, that on the same continent, under the most diverse conditions, under heat and cold, on mountain and lowland, on deserts and marshes, most of the inhabitants within each great class are plainly related; for they will generally be descendants of the same progenitors and early colonists. On this same principle of former migration, combined in most cases with modification, we can understand, by the aid of the Glacial period, the identity of some few plants, and the close alliance of many others, on the most distant mountains, under the most different climates; and likewise the close alliance of some of the inhabitants of the sea in the northern and southern temperate zones, though separated by the whole inter-tropical ocean. Although two areas may present the same physical conditions of life, we need feel no surprise at their inhabitants being widely different, if they have been for a long period completely separated from each other; for as the relation of organism to organism is the most important of all relations, and as the two areas will have received colonists from some third source or from each other, at various periods and in different proportions, the course of modification in the two areas will inevitably be different.

On this view of migration, with subsequent modification, we can see why oceanic islands should be inhabited by few species, but of these, that many should be peculiar. We can clearly see why those animals which cannot cross wide spaces of ocean, as frogs and terrestrial mammals, should not inhabit oceanic islands; and why, on the other hand, new and peculiar species of bats, which can traverse the ocean, should so often be found on islands far distant from any continent. Such facts as the presence of peculiar species of bats, and the absence of all other mammals, on oceanic islands, are utterly inexplicable on the theory of independent acts of creation.

The existence of closely allied or representative species in any two areas, implies, on the theory of descent with modification, that the same parents formerly inhabited both areas; and we almost invariably find that wherever many closely allied species inhabit two areas, some identical species common to both still exist. Wherever many closely allied yet distinct species occur, many doubtful forms and varieties of the same species likewise occur. It is a rule of high gen-erality that the inhabitants of each area are related to the inhabitants of the nearest source whence immigrants might have been derived. We see this in nearly all the plants and animals of the Galapagos archipelago, of Juan Fernandez, and of the other American islands being related in the most striking manner to the plants and animals of the neighbouring American mainland; and those of the Cape de Verde archipelago and other African islands to the African mainland. It must be admitted that these facts receive no explanation on the theory of creation.

The fact, as we have seen, that all past and present organic beings constitute one grand natural system, with group subordinate to group, and with extinct groups often falling in between recent groups, is intelligible on the theory of natural selection with its contingencies of extinction and divergence of character. On these same principles we see how it is, that the mutual affinities of the species and genera within each class are so complex and circuitous. We see why certain characters are far more serviceable than others for classification; — why adaptive characters, though of paramount importance to the being, are of hardly any importance in classification; why characters derived from rudimentary parts, though of no service to the being, are often of high classificatory value; and why embryological characters are the most valuable of all. The real affinities of all organic beings are due to inheritance or community of descent. The natural system is a genealogical arrangement, in which we have to discover the lines of descent by the most permanent characters, however slight their vital importance may be.

The framework of bones being the same in the hand of a man, wing of a bat, fin of the porpoise, and leg of the horse, — the same number of vertebræ forming the neck of the giraffe and of the elephant, — and innumerable other such facts, at once explain themselves on the theory of descent with slow and slight successive modifications. The similarity of pattern in the wing and leg of a bat, though used for such different purpose, — in the jaws and legs of a crab, — in the petals, stamens, and pistils of a flower, is likewise intelligible on the view of the gradual modification of parts or organs, which were alike in the early progenitor of each class. On the principle of successive variations not always supervening at an early age, and being inherited at a corresponding not early period of life, we can clearly see why the embryos of mammals, birds, reptiles, and fishes should be so closely alike, and should be so unlike the adult forms. We may cease marvelling at the embryo of an air-breathing mammal or bird having branchial slits and arteries running in loops, like those in a fish which has to breathe the air dissolved in water, by the aid of well-developed branchiæ.

Disuse, aided sometimes by natural selection, will often tend to reduce an organ, when it has become useless by changed habits or under changed conditions of life; and we can clearly understand on this view the meaning of rudimentary organs. But disuse and selection will generally act on each creature, when it has come to maturity and has to play its full part in the struggle for existence, and will thus have little power of acting on an organ during early life; hence the organ will not be much reduced or rendered rudimentary at this early age. The calf, for instance, has inherited teeth, which never cut through the gums of the upper jaw, from an early progenitor having well-developed teeth; and we may believe, that the teeth in the mature animal were reduced, during successive generations, by disuse or by the tongue and palate having been fitted by natural selection to browse without their aid; whereas in the calf, the teeth have been left untouched by selection or disuse, and on the principle of inheritance at corresponding ages have been inherited from a remote period to the present day. On the view of each organic being and each separate organ having been specially created, how utterly inexplicable it is that parts, like the teeth in the embryonic calf or like the shrivelled wings under the soldered wing-covers of some beetles, should thus so frequently bear the plain stamp of inutility! Nature may be said to have taken pains to reveal, by rudimentary organs and by homologous structures, her scheme of modification, which it seems that we wilfully will not understand.

I have now recapitulated the chief facts and considerations which have thoroughly convinced me that species have changed, and are still slowly changing by the preservation and accumulation of successive slight favourable variations. Why, it may be asked, have all the most eminent living naturalists and geologists rejected this view of the mutability of species? It cannot be asserted that organic beings in a state of nature are subject to no variation; it cannot be proved that the amount of variation in the course of long ages is a limited quantity; no clear distinction has been, or can be, drawn between species and well-marked varieties. It cannot be maintained that species when intercrossed are invariably sterile, and varieties invariably fertile; or that sterility is a special endowment and sign of creation. The belief that species were immutable productions was almost unavoidable as long as the history of the world was thought to be of short duration; and now that we have acquired some idea of the lapse of time, we are too apt to assume, without proof, that the geological record is so perfect that it would have afforded us plain evidence of the mutation of species, if they had undergone mutation.

But the chief cause of our natural unwillingness to admit that one species has given birth to other and distinct species, is that we are always slow in admitting any great change of which we do not see the intermediate steps. The difficulty is the same as that felt by so many geologists, when Lyell first insisted that long lines of inland cliffs had been formed, and great valleys excavated, by the slow action of the coast-waves. The mind cannot possibly grasp the full meaning of the term of a hundred million years; it cannot add up and perceive the full effects of many slight variations, accumulated during an almost infinite number of generations.

Although I am fully convinced of the truth of the views given in this volume under the form of an abstract, I by no means expect to convince experienced naturalists whose minds are stocked with a multitude of facts all viewed, during a long course of years, from a point of view directly opposite to mine. It is so easy to hide our ignorance under such expressions as the "plan of creation," "unity of design," &c., and to think that we give an explanation when we only restate a fact. Any one whose disposition leads him to attach more weight to unexplained difficulties than to the explanation of a certain number of facts will certainly reject my theory. A few naturalists, endowed with much flexibility of mind, and who have already begun to doubt on the immutability of species, may be influenced by this volume; but I look with confidence to the future, to young and rising naturalists, who will be able to view both sides of the question with impartiality. Whoever is led to believe that species are mutable will do good service by conscientiously expressing his conviction; for only thus can the load of prejudice by which this subject is overwhelmed be removed.

Several eminent naturalists have of late published their belief that a multitude of reputed species in each genus are not real species; but that other species are real, that is, have been independently created. This seems to me a strange conclusion to arrive at. They admit that a multitude of forms, which till lately they themselves thought were special creations, and which are still thus looked at by the majority of naturalists, and which consequently have every external characteristic feature of true species,—they admit that these have been produced by variation, but they refuse to extend the same view to other and very slightly different forms. Nevertheless they do not pretend that they can define, or even conjecture, which are the created forms of life, and which are those produced by secondary laws. They admit variation as a *vera causa** in one case, they arbitrarily reject it in another, without assigning any distinction in the two cases. The day will come when this will be given as a curious illustration of the blindness of preconceived opinion. These authors seem no more startled at a miraculous act of creation than at an ordinary birth. But do they really believe that at innumerable periods in the earth's history certain elemental atoms have been commanded suddenly to flash into living tissues? Do they believe that at each supposed act of creation one individual or many were produced? Were all the infinitely numerous kinds of animals and plants created as eggs or seed, or as full grown? and in the case of mammals, were they created bearing the false marks of nourishment from the mother's womb? Although naturalists very properly demand a full explanation of every difficulty from those who believe in the mutability of species, on their own side they ignore the whole subject of the first appearance of species in what they consider reverent silence.

It may be asked how far I extend the doctrine of the modification of species. The question is difficult to answer, because the more distinct the forms are which we may consider, by so much the arguments fall away in force. But some arguments of the greatest weight extend very far. All the members of whole classes can be connected together by chains of affinities, and all can be classified on the same principle, in groups subordinate to groups. Fossil remains sometimes tend to fill up very wide intervals between existing orders. Organs in a rudimentary condition plainly show that an early progenitor had the organ in a fully developed state; and this in some instances necessarily implies an enormous amount of modification in the descendants. Throughout whole classes various structures are formed on the same pattern, and at an embryonic age the species closely resemble each other. Therefore I cannot doubt that the theory of descent with modification embraces all the members of the same class. I believe that animals have descended from at most only four or five progenitors, and plants from an equal or lesser number.

Analogy would lead me one step further, namely, to the belief that all animals and plants have descended from some one prototype. But analogy may be a deceitful guide. Nevertheless all living things have much in common, in their chemical composition, their germinal vesicles,* their cellular structure, and their laws of growth and reproduction. We see this even in so trifling a circumstance as that the same poison often similarly affects plants and animals; or that the poison secreted by the gall-fly produces monstrous growths on the wild rose or oak-tree. Therefore I should infer from analogy that probably all the organic beings which have ever lived on this earth have descended from some one primordial form, into which life was first breathed.*

When the views entertained in this volume on the origin of species, or when analogous views are generally admitted, we can dimly foresee that there will be a considerable revolution in natural history. Systematists will be able to pursue their labours as at present; but they will not be incessantly haunted by the shadowy doubt whether this or that form be in essence a species. This I feel sure, and I speak after experience, will be no slight relief. The endless disputes whether or not some fifty species of British brambles are true species will cease. Systematists will have only to decide (not that this will be easy) whether any form be sufficiently constant and distinct from other forms, to be capable of definition; and if definable, whether the differences be sufficiently important to deserve a specific name. This latter point will become a far more essential consideration than it is at present; for differences, however slight, between any two forms, if not blended by intermediate gradations, are looked at by most naturalists as sufficient to raise both forms to the rank of species. Hereafter we shall be compelled to acknowledge that the only distinction between species and well-marked varieties is, that the latter are known, or believed, to be connected at the present day by intermediate gradations, whereas species were formerly thus connected. Hence, without quite rejecting the consideration of the present existence of intermediate gradations between any two forms, we shall be led to weigh more carefully and to value higher the actual amount of difference between them. It is quite possible that forms now generally acknowledged to be merely varieties may hereafter be thought worthy of specific names, as with the primrose and cowslip; and in this case scientific and common language will come into accordance. In short, we shall have to treat species in the same manner as those naturalists treat genera, who admit that genera are merely artificial combinations made for convenience. This may not be a cheering prospect; but we shall at least be freed from the vain search for the undiscovered and undiscoverable essence of the term species.

The other and more general departments of natural history will rise greatly in interest. The terms used by naturalists of affinity, relationship, community of type, paternity, morphology, adaptive characters, rudimentary and aborted organs, &c., will cease to be metaphorical, and will have a plain signification. When we no longer look at an organic being as a savage looks at a ship, as at something wholly beyond his comprehension; when we regard every production of nature as one which has had a history; when we contemplate every complex structure and instinct as the summing up of many contrivances, each useful to the possessor, nearly in the same way as when we look at any great mechanical invention as the summing up of the labour, the experience, the reason, and even the blunders of numerous workmen; when we thus view each organic being, how far more interesting, I speak from experience, will the study of natural history become!

A grand and almost untrodden field of inquiry will be opened, on the causes and laws of variation, on correlation of growth, on the effects of use and disuse, on the direct action of external conditions, and so forth. The study of domestic productions will rise immensely in value. A new variety raised by man will be a far more important and interesting subject for study than one more species added to the infinitude of already recorded species. Our classifications will come to be, as far as they can be so made, genealogies; and will then truly give what may be called the plan of creation. The rules for classifying will no doubt become simpler when we have a definite object in view. We possess no pedigrees or armorial bearings; and we have to discover and trace the many diverging lines of descent in our natural genealogies, by characters of any kind which have long been inherited. Rudimentary organs will speak infallibly with respect to the nature of long-lost structures. Species and groups of species, which are called aberrant, and which may fancifully be called living fossils, will aid us in forming a picture of the ancient forms of life. Embryology will reveal to us the structure, in some degree obscured, of the prototypes of each great class.

When we can feel assured that all the individuals of the same species, and all the closely allied species of most genera, have within a not very remote period descended from one parent, and have migrated from some one birthplace; and when we better know the many means of migration, then, by the light which geology now throws, and will continue to throw, on former changes of climate and of the level of the land, we shall surely be enabled to trace in an admirable manner the former migrations of the inhabitants of the whole world. Even at present, by comparing the differences of the inhabitants of the sea on the opposite sides of a continent, and the nature of the various inhabitants of that continent in relation to their apparent means of immigration, some light can be thrown on ancient geography.

The noble science of Geology loses glory from the extreme imperfection of the record. The crust of the earth with its embedded remains must not be looked at as a well-filled museum, but as a poor collection made at hazard and at rare intervals. The accumulation of each great fossiliferous formation will be recognised as having depended on an unusual concurrence of circumstances, and the blank intervals between the successive stages as having been of vast duration. But we shall be able to gauge with some security the duration of these intervals by a comparison of the preceding and succeeding organic forms. We must be cautious in attempting to correlate as strictly contemporaneous two formations, which include few identical species, by the general succession of their forms of life. As species are produced and exterminated by slowly acting and still existing causes, and not by miraculous acts of creation and by catastrophes; and as the most important of all causes of organic change is one which is almost independent of altered and perhaps suddenly altered physical conditions, namely, the mutual relation of organism to organism,—the improvement of one being entailing the improvement or the extermination of others; it follows, that the amount of organic change in the fossils of consecutive formations probably serves as a fair measure of the lapse of actual time. A number of species, however, keeping in a body might remain for a long period unchanged, whilst within this same period, several of these species, by migrating into new countries and coming into competition with foreign associates, might become modified; so that we must not overrate the accuracy of organic change as a measure of time. During early periods of the earth's history, when the forms of life were probably fewer and simpler, the rate of change was probably slower; and at the first dawn of life, when very few forms of the simplest structure existed, the rate of change may have been slow in an extreme degree. The whole history of the world, as at present known, although of a length quite incomprehensible by us, will hereafter be recognised as a mere fragment of time, compared with the ages which have elapsed since the first creature, the progenitor of innumerable extinct and living descendants, was created.

In the distant future I see open fields for far more important researches. Psychology will be based on a new foundation, that of the necessary acquirement of each mental power and capacity by gradation. Light will be thrown on the origin of man and his history.

Authors of the highest eminence seem to be fully satisfied with the view that each species has been independently created. To my mind it accords better with what we know of the laws impressed on matter by the Creator, that the production and extinction of the past and present inhabitants of the world should have been due to secondary causes, like those determining the birth and death of the individual. When I view all beings not as special creations, but as the lineal descendants of some few beings which lived long before the first bed of the Silurian system was deposited, they seem to me to become ennobled. Judging from the past, we may safely infer that not one living species will transmit its unaltered likeness to a distant futurity. And of the species now living very few will transmit progeny of any kind to a far distant futurity; for the manner in which all organic beings are grouped, shows that the greater number of species of each genus, and all the species of many genera, have left no descendants, but have become utterly extinct. We can so far take a prophetic glance into futurity as to foretel that it will be the common and widely-spread species, belonging to the larger and dominant groups, which will ultimately prevail and procreate new and dominant species. As all the living forms of life are the lineal descendants of those which lived long before the Silurian epoch, we may feel certain that the ordinary succession by generation has never once been broken, and that no cataclysm has desolated the whole world. Hence we may look with some confidence to a secure future of equally inappreciable length. And as natural selection works solely by and for the good of each being, all corporeal and mental endowments will tend to progress towards perfection.

It is interesting to contemplate an entangled bank, clothed with many plants of many kinds, with birds singing on the bushes, with various insects flitting about, and with worms crawling through the damp earth, and to reflect that these elaborately constructed forms, so different from each other, and dependent on each other in so complex a manner, have all been produced by laws acting around us.

These laws, taken in the largest sense, being Growth with Reproduction; Inheritance which is almost implied by reproduction; Variability from the indirect and direct action of the external conditions of life, and from use and disuse; a Ratio of Increase so high as to lead to a Struggle for Life, and as a consequence to Natural Selection, entailing Divergence of Character and the Extinction of less-improved forms. Thus, from the war of nature, from famine and death, the most exalted object which we are capable of conceiving, namely, the production of the higher animals, directly follows. There is grandeur in this view of life, with its several powers, having been originally breathed* into a few forms or into one; and that, whilst this planet has gone cycling on according to the fixed law of gravity, from so simple a beginning endless forms most beautiful and most wonderful have been, and are being, evolved.

REVIEWS AND RESPONSES

I have gradually learnt to see that it is just as noble a conception of Deity, to believe that he created primal forms capable of self development into all forms needful . . . as to believe that He required a fresh act of intervention to supply the lacunas w^h. he himself had made. I question whether the former be not the loftier thought.

> Novelist and cleric Revd Charles Kingsley in a letter to Darwin of 18 Nov. 1859, *Correspondence*, vii (1992), 379–80

For myself I really think it is the most interesting book I ever read, & can only compare it to the first knowledge of chemistry, getting into a new world or rather behind the scenes. To me the geographical distribution I mean the relation of islands to continents is the most convincing of the proofs, & the relation of the oldest forms to existing species. I dare say I dont feel enough the absence of varieties, but then I dont in the least know if every thing now living were fossilized whether the palæontologists could distinguish them. In fact the a priori reasoning is so entirely satisfactory to me that if the facts wont fit in, why so much the worse for the facts is my feeling.

> Erasmus Darwin in a letter to his brother Charles, 23 Nov. [1859], *Correspondence*, vii (1992), 389–90

I write to thank you for your work on the origin of Species. . . . If I did not think you a good tempered & truth loving man I should not tell you that, (spite of the great knowledge; store of facts; capital views of the corelations of the various parts of organic nature; admirable hints about the diffusions, thro' wide regions, of nearly related organic beings; &c &c) I have read your book with more pain than pleasure. Parts of it I admired greatly; parts I laughed at till my sides were almost sore; other parts I read with absolute sorrow; because I think them utterly false & grievously mischievous— You have *deserted*—after a start in that tram-road of all solid physical truth—the true method of induction—& started up a machinery as wild I think as Bishop Wilkin's locomotive that was to sail with us to the Moon. Many of your wide conclusions are based upon assumptions which can neither be proved nor disproved. Why then express them in the language & arrangements of philosophical induction?. —

As to your grand principle—*natural selection*—what is it but a secondary consequence of supposed, or known, primary facts. Development is a better word because more close to the cause of the fact. For you do not deny causation. I call (in the abstract) causation the will of God: & I can prove that He acts for the good of His creatures. He also acts by laws which we can study & comprehend— Acting by law, & under what is called final cause, comprehends, I think, your whole principle. You write of "natural selection" as if it were done consciously by the selecting agent. 'Tis but a consequence of the pre*supposed* development, & the subsequent battle for life. —

. . . 'Tis the crown & glory of organic science that it *does* thro' *final cause*, link material to moral; & yet *does not* allow us to mingle them in our first conception of laws, & our classification of such laws whether we consider one side of nature or the other— You have ignored this link; &, if I do not mistake your meaning, you have done your best in one or two pregnant cases to break it. Were it possible (which thank God it is not) to break it, humanity in my mind, would suffer a damage that might brutalize it—& sink the human race into a lower grade of degradation than any into which it has fallen since its written records tell us of its history.

Cambridge geologist Revd Prof. Adam Sedgwick writing to Darwin, 24 Nov. 1859, in *Correspondence*, vii (1992), 396–8

We have been reading Darwin's Book on the "Origin of Species" just now: it makes an epoch, as the expression of his thorough adhesion, after long years of study, to the Doctrine of Development—and not the adhesion of an anonym like the author of the "Vestiges," but of a long-celebrated naturalist. The book is ill-written and sadly wanting in illustrative facts—of which he has collected a vast number, but reserves these for a future book of which this smaller one is the avant courier. This will prevent the work from becoming popular, as the "Vestiges" did, but it will have a great effect in the scientific world, causing a thorough and open discussion of a question about which people have hitherto felt timid. So the world gets on step by step towards brave clearness and honesty!

Novelist Marian Evans (George Eliot) in a letter to Barbara Bodichon, 5 Dec. 1859, in *The George Eliot Letters*, ed. G. Haight (New Haven, 1954–78), iii, 227

One great difficulty to my mind in the way of your theory is the fact of the existence of Man. I was beginning to think you had entirely passed over this question, till almost in the last page I find you saying that 'light will be thrown on the origin of man & his history'. By this I suppose is meant that he is to be considered a modified & no doubt *greatly* improved orang! I doubt if this will find acceptance with the generality of readers— . . . Neither can I easily bring myself to the idea that man's reasoning faculties & above all his *moral sense*, c$^{d.}$ ever have been obtained from irrational progenitors, by mere natural selection—acting however gradually & for whatever length of time that may be required. This seems to be doing away altogether with the Divine Image which forms the insurmountable distinction between man & brutes.

Naturalist and cleric Leonard Jenyns, writing to his friend Darwin, 4 Jan. 1860, in *Correspondence*, viii (1993), 14

Prof. [Louis] Agassiz made a verbal communication in opposition to the theory of Mr. Darwin, recently put forth in his work on the origin of species. Mr. Darwin he acknowledged to be one of the best naturalists of England, a laborious and successful writer; his works on the coral reefs, on the cirripeds, and his narrative of the voyage of the Beagle, show him to be a skilful and well-prepared naturalist; but this great knowledge and experience had, in the present instance, been brought to the support, in his opinion, of an ingenious but fanciful theory. . . .

Prof. [William Barton] Rogers stated in reply that the present work is a *résumé* of his conviction on the subject, without the presentation of the facts upon which it rests, which he has not had time to arrange. The problem is admitted to be of transcendent difficulty, and such as no observer or theorist can hope now or perhaps ever positively to resolve. Mr. Darwin makes no pretension to an absolute demonstration, but, after an impartial survey of the facts bearing on the subject and a candid appreciation of the opposing considerations, adopts the view set forth in his book, as offering, in his opinion, a more rational and satisfactory explanation of the history of living nature than the hypothesis of innumerable successive creations.

Report of the first of a series of debates at the Boston Society of Natural History, 15 Feb. 1860, *Proceedings of the Boston Society of Natural History*, 7 (1861), 231–5

You have wrought a considerable modific^n in the views I held— While having the same general conception of the relation of species genera, orders &c as gradually arising by differentiation & divergence like the branches of a tree & while regarding these cumulative modifications as wholly due to the influence of surrounding circums^s. I was under the erroneous impression that the sole cause was adaptation to changing conditions of existence brought about by habit. . . . But you have convinced me that throughout a great proportion of cases, direct adaptation does not explain the facts, but that they are explained only by adaptation through Natural Selection—

Philosopher Herbert Spencer in a letter to Darwin, 22 Feb. 1860, *Correspondence*, viii (1993), 98–9

I rather regret that C. D. went out of his way two or three times (I think not more) to speak of "the Creator" in the popular sense of the First Cause; and also *once* of the "final cause" of certain cuckoo affairs. This latter is sure to be misunderstood, in the face of all the rest of the book: and the other gives occasion for people to ride off from the argument in a way which need not have been granted to them. It is curious to see how those who would otherwise agree with him turn away because his view is "derived from," or "based on," "theology", while he admits critics, by this opening, who would otherwise have no business with the book at all. It seems to me that having carried us up to the earliest group of forms, or to the single primitive one, he and we have nothing to do with how those few forms, or that one, came there. His subject is the "Origin of Species", and not the origin of Organisation; and it seems a needless mischief to have opened the latter speculation at all.—There now! I have delivered my mind.

Journalist and mesmerist Harriet Martineau to Fanny Wedgwood, 13 Mar. 1860, in *Harriet Martineau's Letters to Fanny Wedgwood*, ed. E. S. Arbuckle (Stanford, Calif., 1983), 189

After much consideration, and with assuredly no bias against Mr. Darwin's views, it is our clear conviction that, as the evidence stands, it is not absolutely proven that a group of animals, having all the characters exhibited by species in Nature, has ever been originated by selection, whether artificial or natural. . . . there is no positive evidence, at present, that any group of animals has, by variation and selective breeding, given rise to another group which was, even in the least degree, infertile with the first. Mr. Darwin is perfectly aware of this weak point, and brings forward a multitude of ingenious and important arguments to diminish the force of the objection. We admit the value of these arguments to their fullest extent; nay, we will go so far as to express our belief that experiments, conducted by a skilful physiologist, would very probably obtain the desired production of mutually more or less infertile breeds from a common stock, in a comparatively few years; but still, as the case stands at present, this "little rift within the lute" is not to be disguised nor overlooked.

English man of science Thomas Henry Huxley, writing anonymously on 'The Origin of Species', *Westminster Review*, NS 17 (Apr. 1860), 541–70, at 567–8

Are the transmutationists to monopolise the privilege of conceiving the possibility of the occurrence of unknown phenomena, to be the exclusive propounders of beliefs and surmises, to cry down every kindred barren speculation, and to allow no indulgence in any mere hypothesis save their own? . . .

The natural phenomena already possessed by science are far from being exhausted on which hypotheses, other than transmutative, of the production of species by law might be based, and on a foundation at least as broad as that which Mr. Darwin has exposed in this Essay.

English comparative anatomist Richard Owen, in an unsigned review on 'Darwin on the Origin of Species', *Edinburgh Review*, 111 (Apr. 1860), 487–532, at 502–3

This question is no less 'What am I?' 'What is man?' a created being under the direct government of his Creator, or only an accidental sprout of some primordial type that was the common progenitor of both animals and vegetables. The theologian has no doubt answered those questions, but . . . the naturalist finds himself on the horns of a dilemma. For, either from the facts he observes, he must believe in a special creation of organised species, which creation has been progressive and is now in full operation, or he must adopt some such view as that of Darwin, viz. that the primordial material cell of life has been constantly sprouting forth of itself by 'natural selection' into all the various forms of animals and vegetables. Darwin, indeed, for no reason that I can perceive, except his fear of alarming the clergy, speake of a *Creator* of the original material cell.

. . . Of his three kinds of selection by which he says the world is managed without special interference on the part of a Creator, I only believe in the variation of species by 'human selection,'—i.e. human selection operating within certain limits assigned by the Creator. As for his two other kinds of 'selection' by which he accounts for all the species of animals and vegetables—viz. sexual selection and natural selection, I find them quite impossible to digest. . . .

The theory is almost a materialistic one—nay, even so far atheistic that, if it allows of a deity at all, He has been ever since the institution of the primordial type of life fast asleep. . . . It is far easier for me to believe in the direct and constant Government of the Creation of God, than that He should have created the world and then left it to manage itself, which is Darwin's theory in a few words. Nonetheless, Charles Darwin is an old friend of mine and I feel grateful to him for his work. I hope it will make people attend to such matters, and to be no longer prevented by the first chapter of Genesis from asking for themselves what the Book of Nature says on the subject of Creation.

> London-born Australian naturalist William Sharp Macleay, in a letter to Viscount Sherbrooke, May 1860, in A. P. Martin, *Life and Letters of the Right Honourable Robert Lowe, Viscount Sherbrooke* (London, 1893), ii, 204–7

If Mr. Darwin's theory be true, nothing can prevent its ultimate and general reception, however much it may pain and shock those to whom it is propounded for the first time. If it be merely a clever hypothesis, an ingenious hallucination, to which a very industrious and able man has devoted the greater and the best part of his life, its failure will be nothing new in the history of science.

> 'Natural Selection', in *All the Year Round*, 2 (7 July 1860), 293–9, at 299, a weekly periodical edited by Charles Dickens

The theory of Dr. Darwin, however, on the origin of species by natural selection, gave rise to the hottest of all debates. Here again the men of science were the aggressors, or, if you will, the reformers, while the divines defend the ancient bulwarks. Professor Huxley, it appears, had somewhat facetiously remarked that they had nothing to fear even should it be shown that apes were their ancestors. The Bishop of Oxford on a subsequent day, taking the hint, asked the Professor whether he would prefer a monkey for his grandfather or his grandmother. To this the man of science retorted that he would much rather have a monkey for his grandfather than a man who could indulge in jokes on such a subject. The Bishop then of course had a right to the answer, that the commencement was not made by him. In the debate the Bishop brought all the well-known powers of his eloquence to substantiate the permanence of species. On the other hand, men eminent in science, falling under the direct influence of facts, probably, moreover, in some measure overwhelmed and confounded by the indefinite and all but infinite multiplication of species, gave at least a provisional and partial adhesion to the hypothesis of Dr. Darwin, which may possibly relieve them from an ever-increasing perplexity. In the midst of the discussions on this and other topics, involving at their root questions the most vital, it has been pleasant to mark, for the most part, a spirit of wide and wise toleration,—as if truth and not any narrow party-victory were the earnest search of all. It has been, indeed, of special interest in the present meeting held at Oxford to see, as it were, the ancient city of the University open its arms in friendly embrace to the younger sons of science; to see, as it were, the greeting of ages mediæval with times present; to mark how well it is possible for the Christian, the classic, and the scientific to co-operate in the one grand end,—the advancement of man and the glory of God.

> Report of the Oxford meeting of the British Association for the Advancement of Science in the summer of 1860, in the conservative London weekly *Press* (7 July 1860), 656

The publication by Mr. Darwin of a philosophical work on "the Origin of Species," strongly marks the change which has for some time been silently going on. Its reception equally marks the new temper of the public on this important subject. It is no insignificant fact that the Edinburgh Review, which fifteen years ago gave forth the Rev. Mr. Sedgwick's diatribes against creation under natural law, now proclaims itself has having "no sympathy with the sacerdotal revilers of those who would explain such law". . . .

It seems to the author that Mr. Darwin has only been enabled by his infinitely superior knowledge to point out a principle in what may be called practical animal life, which appears capable of bringing about the modifications theoretically assumed in the earlier work. His book, in no essential respect, contradicts the present: on the contrary, while adding to its explanations of nature, it expresses substantially the same general ideas.

> Scottish author Robert Chambers, writing anonymously in the first post-*Origin* edition of *Vestiges of the Natural History of Creation*, 11th edn. (London, 1860), pp. lxiii–lxiv

By the way Mr Darwins book upsets my ideas somewhat. There does not seem to be any great struggle for existence going on in this wide continent. There is room enough and to spare for both man and beast. The latter seem to live quite jovially and often attain old age. They are subject to various diseases—whole herds are sometimes swept off by epidemics and we meet with diseased animals constantly. Disease does not select or elect to leave the strongest, for it cuts off the ox & horse and leaves the goat & sheep.—A buffalo was caught young at Quihmane and went with the cows—she brought forth two calves by a common bull. The *bos caffer* is certainly stronger than the ox yet the progeny both died. My thoughts however may only shew my ignorance for the book itself has not yet come this length—I speak only from what I see in Reviews. The only case of "natural selection" that occurs in my darkened understanding is this that the genus *Felis* select the neck as the most vulnerable part of their assailants and the Ratel [honey badger] always makes straight for the tendo Achilles and bites it through if he can. How is it that a very light blow on the nose from before backwards kills this animal & the large anteater stone dead and any amount of mauling on the body does not prove fatal—

> Scottish explorer David Livingstone, writing from Senna on the Zambezi in southern Africa, to the anatomist Richard Owen in a letter dated 29 Dec. 1860. Natural History Museum, London, Owen Collection 17/415–17

To insist, therefore, that the new hypothesis of the derivative origin of the actual species is incompatible with final causes and design is to take a position which we must consider philosophically untenable.

We must also regard it as unwise or dangerous, in the present state and present prospects of physical and physiological science. We should expect the philosophical atheist or skeptic to take this ground; also, until better informed, the unlearned and unphilosophical believer; but we should think that the thoughtful theistic philosopher would take the other side. Not to do so seems to concede that only supernatural events can be shown to be designed, which no theist can admit,—seems also to misconceive the scope and meaning of all ordinary arguments for design in Nature. This misconception is shared both by the reviewers and the reviewed. At least, Mr. Darwin uses expressions which seem to imply that the natural forms which surround us, because they have a history or natural sequence, could have been only generally, but not particularly designed,—a view at once superficial and contradictory; whereas his true line should be, that his hypothesis concerns the order and not the cause, the *how* and not the *why* of the phenomena, and so leaves the question of design just where it was before.

> Harvard botanist Asa Gray, writing anonymously in 'Darwin and his Reviewers', *Atlantic Monthly*, 6 (Oct. 1860), 406–25, at 413–14

. . . . in the ev[enin]g at 8 lectured to the working men on succession of life in Time, breaks in succession & the Darwinian doctrines of a struggle for life & change of species by Natural Selection. Some 5 or 600 there, utterly rapt & still for the last 20 minutes. One could have heard a pin head fall. I lectured 1¼ hours, & when done I heard the thunders of applause long after I had left the room.

> British geologist Andrew Ramsay, lecturing on *Origin* at the Royal School of Mines in London, diary entry for 4 Mar. 1861, Imperial College Archives, KGA/Ramsay/1/35 fo. 10ᵛ

Mr. Darwin's remarkable speculation on the Origin of Species is another unimpeachable example of a legitimate hypothesis. What he terms "natural selection" is not only a *vera causa*, but one proved to be capable of producing effects of the same *kind* with those which the hypothesis ascribes to it: the question of possibility is entirely one of degree. It is unreasonable to accuse Mr. Darwin (as has been done) of violating the rules of Induction. The rules of Induction are concerned with the conditions of Proof. Mr. Darwin has never pretended that his doctrine was proved. He was not bound by the rules of Induction, but by those of Hypothesis. And these last have seldom been more completely fulfilled. He has opened a path of inquiry full of promise, the results of which none can foresee. And is it not a wonderful feat of scientific knowledge and ingenuity to have rendered so bold a suggestion, which the first impulse of every one was to reject at once, admissible and discussable, even as a conjecture?

English philosopher and economist John Stuart Mill, *A System of Logic Ratiocinative and Inductive* (London, 1862), ii, 18

The doctrine of M. Darwin is the rational revelation of progress, pitting itself in logical antagonism with the irrational revelation of the fall. These are two principles, two religions in struggle, a thesis and an antithesis of which I defy any German to synthesize. It is a categorical yes and no between which it is necessary to choose, and whoever declares for the one is against the other. For myself, the choice is made.

I believe in progress.

French author Clémence-Auguste Royer in her translator's preface to Darwin, *De l'origine des espèces, ou des lois du progrès chez les êtres organisés* (Paris, 1862), pp. lxiii–lxiv

We have used the words "mechanical life," the "mechanical kingdom," "the mechanical world," and so forth, and we have done so advisedly, for as the vegetable kingdom was slowly developed from the mineral, and as, in like manner, the animal supervened upon the vegetable, so now in these last few ages an entirely new kingdom has sprung up, of which we as yet have only seen what will be one day considered the antediluvian prototypes of the race. . . .

Day by day however the machines are gaining ground upon us; day by day we are becoming more subservient to them; more men are daily bound down as slaves to tend them, more men are daily devoting the energies of their whole lives to the development of mechanical life. The upshot is simply a question of time, but that the time will come when the machines will hold the real supremacy over the world and its inhabitants, is what no person of a truly philosophical mind can for a moment question.

Our opinion is that war to the death should be instantly proclaimed against them. Every machine of every sort should be destroyed by the well-wisher of his species. Let there be no exceptions made, no quarter shown; let us at once go back to the primeval condition of the race. If it be urged that this is impossible under the present condition of human affairs, this at once proves that the mischief is already done, that our servitude has commenced in good earnest, that we have raised a race of beings whom it is beyond our power to destroy, and that we are not only enslaved but are absolutely acquiescent in our bondage.

English-born author Samuel Butler, writing anonymously in 'Darwin Among the Machines' in the Christchurch New Zealand *Press* (13 June 1863)

I'm amused that Darwin, at whom I've been taking another look, should say that he *also* applies the 'Malthusian' theory to plants and animals, as though in Mr Malthus's case the whole thing didn't lie in its *not* being applied to plants and animals, but only — with its geometric progression — to humans as against plants and animals. It is remarkable how Darwin rediscovers, among the beasts and plants, the society of England with its division of labour, competition, opening up of new markets, 'inventions' and Malthusian 'struggle for existence'. It is Hobbes' *bellum omnium contra omnes* and is reminiscent of Hegel's *Phenomenology*, in which civil society figures as an 'intellectual animal kingdom', whereas, in Darwin, the animal kingdom figures as civil society.

Economic writer Karl Marx in a letter to Frederick Engels, 18 June 1862, K. Marx and F. Engels, *Collected Works*, xli (1985), 380

The more I reflect, the more sure I am that America will never settle untill she has the equivalent of an Aristocracy (used in best sense) wherefrom to chuse able Governors & statesmen. There is no more certain fruits of your doctrines than this — that the laws of nature lead infallibly to an aristocracy, as the only security for a *settled condition of improvement* — What has prevented America having one of same sort hitherto?, but the incessant pouring in of democratic elements from the West — which has prevented the sorting of the masses, & frustrated all good effects of Natural Selection.

English botanist Joseph Hooker in a letter to Darwin, 26 Nov. 1862, in *Correspondence*, x (1997), 569

God is the fountain of all sciences. For this cause Catholics have no fear of science, scientifically elaborated and scientifically treated.

They have no fear of any accumulation of facts and phenomena, truly such, nor of any induction or conclusion scientifically established. They fear only science unscientifically handled, superficial observations, hasty generalisations, reckless opposition to revelation, and undissembled readiness to reject revelation rather than doubt of a modern theory about flint instruments and hyena's bones. It is indeed true that Catholics have an intense dislike and hostility to such science as this, and to all its modifications. They hold it to be guilty, not only of *leze majesté* against the Christian revelation, but against the truth and dignity of science itself. They abhor — and I accuse myself of being a ringleader in this abhorrence — the science now in fashion, which I take leave to call 'the brutal philosophy,' to wit, there is no God, and the ape is our Adam.

> English Catholic leader Henry E. Manning, 'The Subjects Proper to the Academia', lecture in 1863, published in his *Essays on Religion and Literature* (London, 1865), 31–66, at 50–1

It still appears to me that in tracing the history of the world backwards . . . all the lines of connexion stop short of a beginning explicable by natural causes; and the absence of any conceivable natural beginning leaves room for, and requires, a supernatural origin. Nor do Mr Darwin's speculations alter this result. For when he has accumulated a vast array of hypotheses, still there is an inexplicable gap at the beginning of his series. To which is to be added, that most of his hypotheses are quite unproven by fact. We can no more adduce an example of a new species, generated in the way which his hypotheses suppose, than Cuvier could. . . . A person who ventures into the controversies which are at present agitated ought to have a great deal of specific knowledge, which I do not possess.

> Revd William Whewell of Cambridge in a letter to the Scottish theologian Revd Dr D. Brown, 26 Oct. 1863, in I. Todhunter, *William Whewell* (London, 1876), ii, 433–4

Many great problems agitate men's minds these days: the unity or multiplicity of human races; creation of man thousands of years ago or several thousand centuries ago; fixity of species or the slow, progressive transformation of species from one to the other; matter reputed to be eternal rather than created; the idea of God being useless. These are some of the questions discussed in our time.

Do not fear that I come here with the pretension of resolving any one of these profound subjects; but on the side, related to these mysteries, there is a question . . . that I can perhaps attempt to discuss, because it is amenable to experiment, and from that point of view I have undertaken severe and conscientious research.

The subject is that kind of generation called spontaneous.

French chemist Louis Pasteur, 'Chimie appliqué à la physiologie: Des générations spontanées', *Revue des cours scientifiques* (28 Apr. 1864), 257–65, at 257

Of all the books I have ever read, not a single one has come even close to making such an overpowering and lasting impression on me, as your theory of the evolution of species. In your book I found all at once the harmonious solution of all the fundamental problems that I had continually tried to solve since I had come to know nature as she really is. Since then your theory—I can say so without exaggerating—has occupied my mind *every day* most pressingly, and whatever I investigate in the life of humans, animals, or plants, your theory of descent always offers me a harmonious solution to all problems, however knotty.

Since it must interest you to hear about the spread of your theory in Germany, I venture to mention the following. Most of the older scientists, and among them many authorities of the first rank, are still your ardent opponents. . . .

Among the younger scientists, by contrast, the number of your sincere and enthusiastic followers is growing daily, and I believe that in a few years their numbers will be greater even that those of your sincere followers in England itself. For overall the Germans seem to me (insofar as I can judge it) less fettered by religious and social prejudice than the English, although the latter surely are politically more advanced and more versatile in their development. . . .

I am only 30 years old, but a stroke of fate has destroyed all prospects of happiness in my life, and this has left me so mature, determined, and immune to praise and blame, that, entirely unswayed by outside influence of any kind, I shall pursue the one goal in my life, namely to disseminate, to support and to perfect your theory of descent.

Naturalist and embryologist Ernst Haeckel to Darwin, 9 [July 1864], trans. from German in *Correspondence*, xii (2001), 482–3

What is the question now placed before society with a glib assurance the most astounding? The question is this—Is man an ape or an angel? My Lord, I am on the side of the angels.

Conservative English politician Benjamin Disraeli, speaking at the Sheldonian Theatre in Oxford on 25 Nov. 1864, in W. F. Monypenny and G. E. Buckle, *The Life of Benjamin Disraeli* (London, 1929), ii, 108

There are priests, who, while defrauding the state of taxes, mount the pulpit and preach: that when materialists and Darwinists do not commit all sorts of crimes, it is not from righteousness but from hypocrisy.

Let them rage! *They* require the fear of punishment, the hope of reward in a dreamt-of beyond, to keep in the right path—for us suffices the consciousness of being men amongst men, and the acknowledgement of their equal rights. We have no other hope than that of receiving the acknowledgements of our fellow-men; no other fear than that of seeing our human dignity violated—a dignity we value the more, since it has been conquered with the greatest labour by us and our ancestors, down to the ape.

German zoologist Carl Vogt in *Lectures on Man* (London, 1864), 469

My own impression has always been,—not only that *your* theory was quite *compatible* with the faith to which I have just tried to give expression,—but that your books afforded me a clue which would guide me in applying that faith to the solution of certain complicated psychological problems which it was of practical importance to me, as a mother, to solve. I felt that you had supplied one of the missing links,—not to say *the* missing link,—between the facts of Science & the promises of religion. Every year's experience tends to deepen in me that impression.

But I have lately read remarks, on the probable bearing of your theory on religious & moral questions, which have perplexed & pained me sorely. I know that the persons who make such remarks must be cleverer & wise than myself. I cannot feel sure that they are mistaken unless you will tell me so.

English mathematician and teacher Mary Boole in a letter to Darwin, 13 Dec. 1866, *Correspondence*, xiv (2004), 424

Any one of the main pleas of our argument, if established, is fatal to Darwin's theory. What then shall we say if we believe that experiment has shown a sharp limit to the variation of every species, that natural selection is powerless to perpetuate new organs even should they appear, that countless ages of a habitable globe are rigidly proven impossible by the physical laws which forbid the assumption of infinite power in a finite mass? What can we believe but that Darwin's theory is an ingenious and plausible speculation, to which future physiologists will look back with the kind of admiration we bestow on the atoms of Lucretius, or the crystal spheres of Eudoxus, containing like these some faint half-truths, marking at once the ignorance of the age and the ability of the philosopher. . . . A plausible theory should not be accepted while unproven; and if the arguments of this essay be admitted, Darwin's theory of the origin of species is not only without sufficient support from evidence, but is proved false by a cumulative proof.

Scottish engineer Fleeming Jenkin, writing anonymously in 'The *Origin of Species*', *North British Review*, 46 (June 1867), 277–318, at 289–90, and 317–18

We are sometimes tempted to ask whether the time will ever arrive, when science shall have obtained such an ascendancy in the education of the millions, that it will be possible to welcome new truths, instead of always looking upon them with fear and disquiet, and to hail every important victory gained over error, instead of resisting the new discovery, long after the evidence in its favour is conclusive. . . . The future now opening before us begins already to reveal new doctrines, if possible more than ever out of harmony with cherished associations of thought. It is therefore desirable, when we contrast ourselves with the rude and superstitious savages who preceded us, to remember, as cultivators of science, that the high comparative place which we have reached in the scale of being has been gained step by step by a conscientious study of natural phenomena, and by fearlessly teaching the doctrines to which they point. It is by faithfully weighing evidence, without regard to preconceived notions, by earnestly and patiently searching for what is true, not what we wish to be true, that we have attained that dignity, which we may in vain hope to claim through the rank of an ideal parentage.

British geologist Charles Lyell, *Principles of Geology*, 10th edn. (London, 1868), ii, 493–4

I am afraid of the subject of Darwin. I am myself so ignorant on it, that I should fear to be in the position of editing a paper on the subject—

Novelist Anthony Trollope, in a letter to the science writer John Ellor Taylor, 25 Sept. 1868, about a potential contribution to *St. Pauls Magazine.* In *The Letters of Anthony Trollope,* ed. N. J. Hall (Stanford, Calif., 1983), i, 447

At the end of your "Origin of Species" you make an appeal to young naturalists, perhaps you meant to include also mere observers and I venture to put myself in this latter class being 24 years of age and having always been fond of natural history especially ornithology and thinking with Gilbert White that labour should be divided and every place should have its own monographer I have made occasional ornithological notes having in view more particularly the making of a list of birds seen in this neighbourhood, then the rest of Portugal and those migrating through. My occupation (being in business as a commission merchant) will not allow me to apportion more than a limited time to my favourite study but as I am fond of shooting curiosities in natural history sometimes come before my notice This country having been hitherto isolated from the rest of Europe in a great measure has never been properly explored and I have little doubt that much might be learnt here—

The merchant William Chester Tait writing from Oporto, Portugal, 26 Jan. 1869, CUL: DAR 178: 43

It would be idle to speak of the delight your letter has given me, but there is no one in the world whose approbation in these matters can have the same weight as yours. Neither is there any one whose approbation I prize more highly, on purely personal grounds, because I always think of you in the same way as converts from barbarism think of the teacher who first relieved them from the intolerable burden of their superstition. I used to be wretched under the weight of the old fashioned 'arguments from design', of which I felt though I was unable to prove to myself, the worthlessness. Consequently the appearance of your 'Origin of Species' formed a real crisis in my life; your book drove away the constraint of my old superstition as if it had been a nightmare and was the first to give me freedom of thought.

Eugenicist and statistician Francis Galton thanking Darwin for his approval of *Hereditary Genius* (1869), letter dated 24 Dec. 1869, CUL: DAR 105: A3–4

It is the same blind mania of imitation and schematization. . . . In this connection nothing has had a more devastating effect than the crude schematization of the Darwinists. It was certainly somewhat surprising, for those of us who were still acquainted with the old nature-philosophy, to see how the genius of a single man restored to its rightful place an idea already given official status as an *a priori* necessity by the nature-philosophers, not only reactivating it, after its long and alas not entirely unjustified banishment, but making it the basis of a general conception of the history of the organic world. But to make an article of faith out of a problem, a principle of synthesis out of a ground for investigation, thereby drugging oneself with assumptions instead of seeking further, was almost worse than the *a priori* approach of the nature-philosophers. For all the valid facts which had been brought out in the meantime were also forced into the new system, and in this context they ran a very definite risk of losing their true meaning under a cloak of hypothesis. To many people the 'struggle for existence' seemed to be something entirely new and unheard of, just as if the doctrine of self-preservation and the instinct of self-preservation had not always been the basis of biology. Even the doctrine of heredity, a phenomenon so commonplace in pathological experience, in its new form dazzled many an otherwise poorly illumined eye.

Pathologist and politician Rudolf Virchow, 'Über die Standpunkte in der wissenschaftlichen Medicin' (On Standpoints in Scientific Medicine), *Virchows Archiv*, 70 (1877), trans. in *Disease, Life, and Man*, ed. L. J. Rather (London, 1959), 145–6

Just a few years ago . . . right at the dawn of our rural science, there was a great debate among cattlemen. Its motive was really interesting. Would there be selection or crossing? In this Darwin, the scientific authority on these matters, was brought to bear, as well as other European and American experts in this science: breeders, *éléveurs*, zootechnicians in a word. The debate began in the Rural Association, first in its journal, then in lectures, reaching the point where it was brought up in general meetings where all our principal ranchers gave their opinions. Two outspoken partisans stood out, one in favour of selection, the other of crossing.

With the exposition of each doctrine, the debate concluded, and the *leaders*, with no clear victory on that occasion, left the meeting with the assembly divided into selectionists and those who favoured crossing. But, the *leaders*? We must applaud them, for once they defended their causes with pen and word, they set out to prove their points with deeds. One practiced selection on the Uruguayan coast, the other crossing on the Rio Negro. Let us display similar patience while we await their Stud-Books.

Description of debate among cattlemen in Uruguay, by 'Un aficionado XY', 'Selección y cruzamiento', *Revista de la Asociación Rural*, 12 (1883), 134; trans. in T. Glick, 'The Reception of Darwinism in Uruguay', in T. Glick, M. A. Puig-Samper, and R. Ruiz (eds.), *The Reception of Darwinism in the Iberian World* (Dordrecht, 2001), 29-52, at 31

According to the rules . . . of the Muhammadan *sharī'ah*, it is incumbent that its followers believe in the apparent or plain meaning of the successive and well-known scriptural texts unless their apparent meaning is contradicted by decisive and rational evidence which necessitates their interpretation [*tā'wīl*]. Therefore the belief of those who follow Muhammad, peace be upon him, should be that God created each species on earth independently from the others, and not according to evolution and by deriving one from another, even though He is capable of either two possibilities. And whether species were created all at once or gradually through natural laws that God laid out is a point which is not addressed in their *sharī'ah* so as to help them determine between the two options. . . . The evidence that you evolutionists mention in your books is speculative and conjectural. . . . Yes, if there was decisive, rational evidence in opposition to the apparent meaning of these passages, they [Muhammadans] would have to interpret them so as to reconcile them with the proofs given in their favour and according to those aforementioned rules (and I believe that it would be easier to strip the Goat's thorn of its thorns by hand). And so you, oh materialists, if it were determined that your proofs for evolution had reached certainty and if you were to be guided by the religion of Muhammad, peace be upon him, whose foundation is that there is no creator but God the Exalted, then you would not be barred from interpreting scripture [in this fashion] or from taking its meaning beyond the apparent one and from applying to it what the decisive evidence indicates regarding evolution, while believing in creation by God the Exalted and not rejecting that. In that case you would be considered among the people of the Islamic religion, and the process of deduction from all creation to the existence of God and the absoluteness of His power, knowledge and wisdom would not escape you. However, I warn you against error and against imagining that [any] presumptive evidence you come upon is absolute. So be thorough and let God be your Guide.

> Lebanese legal scholar Ḥusayn al-Jisr, *A Hamidian Essay on the Truthfulness of Islam and Muhammidian Law* (*Kitāb al-risālah al-Ḥamīdīyah fī ḥaqīqat al-diyānah al-Islāmīyah wa-ḥaqīqat al-sharī'ah al-Muḥammadīyah*) (repr.; Cairo, 1322 A.H./1905 CE; originally published 1887), 224–5; trans. Marwa Elshakry and Ahmed Ragab

Old ideas give way slowly; for they are more than abstract logical forms and categories. They are habits, predispositions, deeply engrained attitudes of aversion and preference. Moreover, the conviction persists—though history shows it to be a hallucination—that all the questions that the human mind has asked are questions that can be answered in terms of the alternatives that the questions themselves present. But in fact intellectual progress usually occurs through sheer abandonment of questions together with both of the alternatives they assume—an abandonment that results from decreasing vitality and interest in their point of view. We do not solve them: we get over them. Old questions are solved by disappearing, evaporating, while new questions corresponding to the changed attitude of endeavor and preference take their place. Doubtless the greatest dissolvent in contemporary thought of old questions, the greatest precipitant of new methods, new intentions, new problems, is the one effected by the scientific revolution that found its climax in the "Origin of Species".

> American philosopher John Dewey, 'Darwin's Influence upon Philosophy', *Popular Science Monthly*, 75 (July 1909), 90–8, at 98

DESCENT OF MAN

THE DESCENT OF MAN,

AND SELECTION IN RELATION TO SEX

Introduction

THE nature of the following work will be best understood by a brief account of how it came to be written. During many years I collected notes on the origin or descent of man, without any intention of publishing on the subject, but rather with the determination not to publish, as I thought that I should thus only add to the prejudices against my views. It seemed to me sufficient to indicate, in the first edition of my 'Origin of Species,' that by this work "light would be thrown on the origin of man and his history;"* and this implies that man must be included with other organic beings in any general conclusion respecting his manner of appearance on this earth. Now the case wears a wholly different aspect. When a naturalist like Carl Vogt ventures to say in his address as President of the National Institution of Geneva (1869), "personne, en Europe au moins, n'ose plus soutenir la création indépendante et de toutes pièces, des espèces,"* it is manifest that at least a large number of naturalists must admit that species are the modified descendants of other species; and this especially holds good with the younger and rising naturalists. The greater number accept the agency of natural selection; though some urge, whether with justice the future must decide, that I have greatly overrated its importance. Of the older and honoured chiefs in natural science, many unfortunately are still opposed to evolution in every form.

In consequence of the views now adopted by most naturalists, and which will ultimately, as in every other case, be followed by other men, I have been led to put together my notes, so as to see how far the general conclusions arrived at in my former works were applicable to man. This seemed all the more desirable as I had never deliberately applied these views to a species taken singly. When we confine our attention to any one form, we are deprived of the weighty arguments derived from the nature of the affinities which connect together whole groups of organisms—their geographical distribution in past and present times, and their geological succession. The homo-logical structure,* embryological development, and rudimentary organs* of a species, whether it be man or any other animal, to which our attention may be directed, remain to be considered; but these great classes of facts afford, as it appears to me, ample and conclusive evidence in favour of the principle of gradual evolution. The strong support derived from the other arguments should, however, always be kept before the mind.

The sole object of this work is to consider, firstly, whether man, like every other species, is descended from some pre-existing form; secondly, the manner of his development; and thirdly, the value of the differences between the so-called races of man. As I shall confine myself to these points, it will not be necessary to describe in detail the differ-ences between the several races—an enormous subject which has been fully discussed in many valuable works. The high antiquity of man has recently been demonstrated by the labours of a host of eminent men, beginning with M. Boucher de Perthes; and this is the indispensable basis for understanding his origin. I shall, therefore, take this conclu-sion for granted, and may refer my readers to the admirable treatises of Sir Charles Lyell, Sir John Lubbock, and others. Nor shall I have occasion to do more than to allude to the amount of difference between man and the anthropomorphous apes;* for Prof. Huxley, in the opinion of most competent judges, has conclusively shewn that in every single visible character man differs less from the higher apes than these do from the lower members of the same order of Primates.

This work contains hardly any original facts in regard to man; but as the conclusions at which I arrived, after drawing up a rough draft, appeared to me interesting, I thought that they might interest others. It has often and confidently been asserted, that man's origin can never be known: but ignorance more frequently begets confidence than does knowledge: it is those who know little, and not those who know much, who so positively assert that this or that problem will never be solved by science. The conclusion that man is the co-descendant with other species of some ancient, lower, and extinct form, is not in any degree new. Lamarck long ago came to this conclusion, which has lately been maintained by several eminent naturalists and philosophers; for instance by Wallace, Huxley, Lyell, Vogt, Lubbock, Büchner, Rolle, &c., and especially by Häckel. This last naturalist, besides his great work, 'Generelle Morphologie' (1866), has recently (1868, with a second edit. in 1870), published his 'Natürliche Schöpfungsgeschichte,'* in which he fully discusses the genealogy of man. If this work had appeared before my essay had been written, I should probably never have completed it. Almost all the conclusions at which I have arrived I find confirmed by this naturalist, whose knowledge on many points is much fuller than mine. Wherever I have added any fact or view from Prof. Häckel's writings, I give his authority in the text, other statements I leave as they originally stood in my manuscript, occasionally giving in the foot-notes references to his works, as a confirmation of the more doubtful or interesting points.

During many years it has seemed to me highly probable that sexual selection has played an important part in differentiating the races of man; but in my 'Origin of Species' (first edition, p. 199)* I contented myself by merely alluding to this belief. When I came to apply this view to man, I found it indispensable to treat the whole subject in full detail.* Consequently the second part of the present work, treating of sexual selection, has extended to an inordinate length, compared with the first part; but this could not be avoided.

I had intended adding to the present volumes an essay on the expression of the various emotions by man and the lower animals. My attention was called to this subject many years ago by Sir Charles Bell's admirable work. This illustrious anatomist maintains that man is endowed with certain muscles solely for the sake of expressing his emotions. As this view is obviously opposed to the belief that man is descended from some other and lower form, it was necessary for me to consider it. I likewise wished to ascertain how far the emotions are expressed in the same manner by the different races of man. But owing to the length of the present work, I have thought it better to reserve my essay, which is partially completed, for separate publication.

PART I.—THE DESCENT OF MAN

[*Chapter I uses evidence from anatomy, embryology, and vestigial organs to show that the human body is derived by descent from other animals. Most of this was well known and relatively uncontroversial.*]

CHAPTER II

Comparison of the Mental Powers of Man and the Lower Animals

WE have seen in the last chapter that man bears in his bodily structure clear traces of his descent from some lower form; but it may be urged that, as man differs so greatly in his mental power from all other animals, there must be some error in this conclusion. No doubt the difference in this respect is enormous, even if we compare the mind of one of the lowest savages, who has no words to express any number higher than four, and who uses no abstract terms for the commonest objects or affections, with that of the most highly organised ape. The difference would, no doubt, still remain immense, even if one of the higher apes had been improved or civilised as much as a dog has been in comparison with its parent-form, the wolf or jackal. The Fuegians rank amongst the lowest barbarians; but I was continually struck with surprise how closely the three natives on board H.M.S. "Beagle," who had lived some years in England and could talk a little English, resembled us in disposition and in most of our mental faculties. If no organic being excepting man had possessed any mental power, or if his powers had been of a wholly different nature from those of the lower animals, then we should never have been able to convince our-selves that our high faculties had been gradually developed. But it can be clearly shewn that there is no fundamental difference of this kind. We must also admit that there is a much wider interval in mental power between one of the lowest fishes, as a lamprey or lancelet,* and one of the higher apes, than between an ape and man; yet this immense interval is filled up by numberless gradations.

Nor is the difference slight in moral disposition between a barbar-ian, such as the man described by the old navigator Byron, who dashed his child on the rocks for dropping a basket of sea-urchins, and a Howard or Clarkson; and in intellect, between a savage who does not use any abstract terms, and a Newton or Shakspeare. Differences of this kind between the highest men of the highest races and the lowest savages, are connected by the finest gradations. Therefore it is possible that they might pass and be developed into each other.

My object in this chapter is solely to shew that there is no funda-mental difference between man and the higher mammals in their mental faculties. Each division of the subject might have been extended into a separate essay, but must here be treated briefly. As no classification of the mental powers has been universally accepted, I shall arrange my remarks in the order most convenient for my purpose; and will select those facts which have most struck me, with the hope that they may produce some effect on the reader.

With respect to animals very low in the scale, I shall have to give some additional facts under Sexual Selection, shewing that their mental powers are higher than might have been expected. The variability of the faculties in the individuals of the same species is an important point for us, and some few illustrations will here be given. But it would be superfluous to enter into many details on this head, for I have found on frequent enquiry, that it is the unanimous opinion of all those who have long attended to animals of many kinds, including birds, that the individuals differ greatly in every mental characteristic. In what manner the mental powers were first developed in the lowest organisms, is as hopeless an enquiry as how life first originated. These are problems for the distant future, if they are ever to be solved by man.

[*Darwin distinguishes the higher mental powers from instinctual behaviours; he then examines complex emotions and mental faculties that seem to distinguish man from other animals, ranging from pleasure and fear to imagination, reason, and the use of tools. The most difficult case is human language.*]

Language.—This faculty has justly been considered as one of the chief distinctions between man and the lower animals. But man, as a highly competent judge, Archbishop Whately remarks, "is not the only animal that can make use of language to express what is passing in his mind, and can understand, more or less, what is so expressed by another." In Paraguay the *Cebus azaræ* when excited utters at least six distinct sounds, which excite in other monkeys similar emotions. The movements of the features and gestures of monkeys are understood by us, and they partly understand ours, as Rengger and others declare. It is a more remarkable fact that the dog, since being domesticated, has learnt to bark in at least four or five distinct tones. Although barking is a new art, no doubt the wild species, the parents of the dog, expressed their feelings by cries of various kinds. With the domesticated dog we have the bark of eagerness, as in the chase; that of anger; the yelping or howling bark of despair, as when shut up; that of joy, as when starting on a walk with his master; and the very distinct one of demand or supplication, as when wishing for a door or window to be opened.

Articulate language is, however, peculiar to man; but he uses in common with the lower animals inarticulate cries to express his meaning, aided by gestures and the movements of the muscles of the face. This especially holds good with the more simple and vivid feelings, which are but little connected with our higher intelligence. Our cries of pain, fear, surprise, anger, together with their appropriate actions, and the murmur of a mother to her beloved child, are more expressive than any words. It is not the mere power of articulation that distinguishes man from other animals, for as every one knows, parrots can talk; but it is his large power of connecting definite sounds with definite ideas; and this obviously depends on the development of the mental faculties.

As Horne Tooke, one of the founders of the noble science of philology,* observes, language is an art, like brewing or baking; but writing would have been a much more appropriate simile. It certainly is not a true instinct, as every language has to be learnt. It differs, however, widely from all ordinary arts, for man has an instinctive tendency to speak, as we see in the babble of our young children; whilst no child has an instinctive tendency to brew, bake, or write. Moreover, no philologist now supposes that any language has been deliberately invented; each has been slowly and unconsciously developed by many steps. The sounds uttered by birds offer in several respects the nearest analogy to language, for all the members of the same species utter the same instinctive cries expressive of their emotions; and all the kinds that have the power of singing exert this power instinctively; but the actual song, and even the call-notes, are learnt from their parents or foster-parents. These sounds, as Daines Barrington has proved, "are no more innate than language is in man." The first attempts to sing "may be compared to the imperfect endeavour in a child to babble." The young males continue practising, or, as the bird-catchers say, recording, for ten or eleven months. Their first essays show hardly a rudiment of the future song; but as they grow older we can perceive what they are aiming at; and at last they are said "to sing their song round." Nestlings which have learnt the song of a distinct species, as with the canary-birds educated in the Tyrol, teach and transmit their new song to their offspring. The slight natural differences of song in the same species inhabiting different districts may be appositely compared, as Barrington remarks, "to provincial dialects;" and the songs of allied, though distinct species may be compared with the languages of distinct races of man. I have given the foregoing details to shew that an instinctive tendency to acquire an art is not a peculiarity confined to man.

With respect to the origin of articulate language, after having read on the one side the highly interesting works of Mr. Hensleigh Wedgwood, the Rev. F. Farrar, and Prof. Schleicher, and the celebrated lectures of Prof. Max Müller on the other side, I cannot doubt that language owes its origin to the imitation and modification, aided by signs and gestures, of various natural sounds, the voices of other animals, and man's own instinctive cries. When we treat of sexual selection we shall see that primeval man, or rather some early progenitor of man, probably used his voice largely, as does one of the gibbon-apes at the present day, in producing true musical cadences, that is in singing; we may conclude from a widely-spread analogy that this power would have been especially exerted during the courtship of the sexes, serving to express various emotions, as love, jealousy, triumph, and serving as a challenge to their rivals. The imitation by articulate sounds of musical cries might have given rise to words expressive of various complex emotions. As bearing on the subject of imitation, the strong tendency in our nearest allies, the monkeys, in microcephalous* idiots, and in the barbarous races of mankind, to imitate whatever they hear deserves notice. As monkeys certainly understand much that is said to them by man, and as in a state of nature they utter signal-cries of danger to their fellows, it does not appear altogether incredible, that some unusually wise ape-like animal should have thought of imitating the growl of a beast of prey, so as to indicate to his fellow monkeys the nature of the expected danger. And this would have been a first step in the formation of a language.

As the voice was used more and more, the vocal organs would have been strengthened and perfected through the principle of the inherited effects of use; and this would have reacted on the power of speech. But the relation between the continued use of language and the development of the brain has no doubt been far more important. The mental powers in some early progenitor of man must have been more highly developed than in any existing ape, before even the most imperfect form of speech could have come into use; but we may confidently believe that the continued use and advancement of this power would have reacted on the mind by enabling and encouraging it to carry on long trains of thought. A long and complex train of thought can no more be carried on without the aid of words, whether spoken or silent, than a long calculation without the use of figures or algebra. It appears, also, that even ordinary trains of thought almost require some form of language, for the dumb, deaf, and blind girl, Laura Bridgman, was observed to use her fingers whilst dreaming. Nevertheless a long succession of vivid and connected ideas, may pass through the mind without the aid of any form of language, as we may infer from the prolonged dreams of dogs. We have, also, seen that retriever-dogs are able to reason to a certain extent; and this they manifestly do without the aid of language. The intimate connection between the brain, as it is now developed in us, and the faculty of speech, is well shewn by those curious cases of brain-disease, in which speech is specially affected, as when the power to remember substantives is lost, whilst other words can be correctly used. There is no more improbability in the effects of the continued use of the vocal and mental organs being inherited, than in the case of hand-writing, which depends partly on the structure of the hand and partly on the disposition of the mind; and hand-writing is certainly inherited.

Why the organs now used for speech should have been originally perfected for this purpose, rather than any other organs, it is not difficult to see. Ants have considerable powers of intercommunication by means of their antennæ, as shewn by Huber, who devotes a whole chapter to their language. We might have used our fingers as efficient instruments, for a person with practice can report to a deaf man every word of a speech rapidly delivered at a public meeting; but the loss of our hands, whilst thus employed, would have been a serious inconvenience. As all the higher mammals possess vocal organs constructed on the same general plan with ours, and which are used as a means of communication, it was obviously probable, if the power of communication had to be improved, that these same organs would have been still further developed; and this has been effected by the aid of adjoining and well-adapted parts, namely the tongue and lips. The fact of the higher apes not using their vocal organs for speech, no doubt depends on their intelligence not having been sufficiently advanced. The possession by them of organs, which with long-continued practice might have been used for speech, although not thus used, is paralleled by the case of many birds which possess organs fitted for singing, though they never sing. Thus, the nightingale and crow have vocal organs similarly constructed, these being used by the former for diversified song, and by the latter merely for croaking.

The formation of different languages and of distinct species, and the proofs that both have been developed through a gradual process, are curiously the same. But we can trace the origin of many words further back than in the case of species, for we can perceive that they have arisen from the imitation of various sounds, as in alliterative poetry. We find in distinct languages striking homologies due to community of descent, and analogies due to a similar process of formation. The manner in which certain letters or sounds change when others change is very like correlated growth. We have in both cases the reduplication of parts, the effects of long-continued use, and so forth. The frequent presence of rudiments, both in languages and in species, is still more remarkable. The letter *m* in the word *am*, means *I*; so that in the expression *I am*, a superfluous and useless rudiment has been retained. In the spelling also of words, letters often remain as the rudiments of ancient forms of pronunciation. Languages, like organic beings, can be classed in groups under groups; and they can be classed either naturally according to descent, or artificially by other characters. Dominant languages and dialects spread widely and lead to the gradual extinction of other tongues. A language, like a species, when once extinct, never, as Sir C. Lyell remarks, reappears. The same language never has two birth-places. Distinct languages may be crossed or blended together. We see variability in every tongue, and new words are continually cropping up; but as there is a limit to the powers of the memory, single words, like whole languages, gradually become extinct. As Max Müller has well remarked:—"A struggle for life is constantly going on amongst the words and grammatical forms in each language. The better, the shorter, the easier forms are constantly gaining the upper hand, and they owe their success to their own inherent virtue." To these more important causes of the survival of certain words, mere novelty may, I think, be added; for there is in the mind of man a strong love for slight changes in all things. The survival or preservation of certain favoured words in the struggle for existence is natural selection.

The perfectly regular and wonderfully complex construction of the languages of many barbarous nations has often been advanced as a proof, either of the divine origin of these languages, or of the high art and former civilisation of their founders. Thus F. von Schlegel writes: "In those languages which appear to be at the lowest grade of intellectual culture, we frequently observe a very high and elaborate degree of art in their grammatical structure. This is especially the case with the Basque and the Lapponian, and many of the American languages." But it is assuredly an error to speak of any language as an art in the sense of its having been elaborately and methodically formed. Philologists now admit that conjugations, declensions, &c., originally existed as distinct words, since joined together; and as such words express the most obvious relations between objects and persons, it is not surprising that they should have been used by the men of most races during the earliest ages. With respect to perfection, the following illustration will best shew how easily we may err: a Crinoid* sometimes consists of no less than 150,000 pieces of shell, all arranged with perfect symmetry in radiating lines; but a naturalist does not consider an animal of this kind as more perfect than a bilateral one with comparatively few parts, and with none of these alike, excepting on the opposite sides of the body. He justly considers the differentiation and specialisation of organs as the test of perfection. So with languages, the most symmetrical and complex ought not to be ranked above irregular, abbreviated, and bastardised languages, which have borrowed expressive words and useful forms of construction from various conquering, or conquered, or immigrant races.

From these few and imperfect remarks I conclude that the extremely complex and regular construction of many barbarous languages, is no proof that they owe their origin to a special act of creation. Nor, as we have seen, does the faculty of articulate speech in itself offer any insuperable objection to the belief that man has been developed from some lower form.

Self-consciousness, Individuality, Abstraction, General Ideas, &c.— It would be useless to attempt discussing these high faculties, which, according to several recent writers, make the sole and complete distinction between man and the brutes, for hardly two authors agree in their definitions. Such faculties could not have been fully developed in man until his mental powers had advanced to a high standard, and this implies the use of a perfect language. No one supposes that one of the lower animals reflects whence he comes or whither he goes,— what is death or what is life, and so forth. But can we feel sure that an old dog with an excellent memory and some power of imagination, as shewn by his dreams, never reflects on his past pleasures in the chase? and this would be a form of self-consciousness. On the other hand, as Büchner has remarked, how little can the hard-worked wife of a degraded Australian savage, who uses hardly any abstract words and cannot count above four, exert her self-consciousness, or reflect on the nature of her own existence.

That animals retain their mental individuality is unquestionable. When my voice awakened a train of old associations in the mind of the above-mentioned dog, he must have retained his mental individuality, although every atom of his brain had probably undergone change more than once during the interval of five years. This dog might have brought forward the argument lately advanced to crush all evolutionists, and said, "I abide amid all mental moods and all material changes. . . . The teaching that atoms leave their impressions as legacies to other atoms falling into the places they have vacated is contradictory of the utterance of consciousness, and is therefore false; but it is the teaching necessitated by evolutionism, consequently the hypothesis is a false one."*

Sense of Beauty.—This sense has been declared to be peculiar to man. But when we behold male birds elaborately displaying their plumes and splendid colours before the females, whilst other birds not thus decorated make no such display, it is impossible to doubt that the females admire the beauty of their male partners. As women everywhere deck themselves with these plumes, the beauty of such ornaments cannot be disputed. The Bower-birds by tastefully ornamenting their playing-passages with gaily-coloured objects, as do certain humming-birds their nests, offer additional evidence that they possess a sense of beauty. So with the song of birds, the sweet strains poured forth by the males during the season of love are certainly admired by the females, of which fact evidence will hereafter be given. If female birds had been incapable of appreciating the beautiful colours, the ornaments, and voices of their male partners, all the labour and anxiety exhibited by them in displaying their charms before the females would have been thrown away; and this it is impossible to admit. Why certain bright colours and certain sounds should excite pleasure, when in harmony, cannot, I presume, be explained any more than why certain flavours and scents are agreeable; but assuredly the same colours and the same sounds are admired by us and by many of the lower animals.

The taste for the beautiful, at least as far as female beauty is concerned, is not of a special nature in the human mind; for it differs widely in the different races of man, as will hereafter be shewn, and is not quite the same even in the different nations of the same race. Judging from the hideous ornaments and the equally hideous music admired by most savages, it might be urged that their æsthetic faculty was not so highly developed as in certain animals, for instance, in birds. Obviously no animal would be capable of admiring such scenes as the heavens at night, a beautiful landscape, or refined music; but such high tastes, depending as they do on culture and complex associations, are not enjoyed by barbarians or by uneducated persons.

Many of the faculties, which have been of inestimable service to man for his progressive advancement, such as the powers of the imagination, wonder, curiosity, an undefined sense of beauty, a tendency to imitation, and the love of excitement or novelty, could not fail to have led to the most capricious changes of customs and fashions. I have alluded to this point, because a recent writer has oddly fixed on Caprice "as one of the most remarkable and typical differences between savages and brutes."* But not only can we perceive how it is that man is capricious, but the lower animals are, as we shall hereafter see, capricious in their affections, aversions, and sense of beauty. There is also good reason to suspect that they love novelty, for its own sake.

Belief in God—Religion.—There is no evidence that man was aboriginally endowed with the ennobling belief in the existence of an Omnipotent God. On the contrary there is ample evidence, derived not from hasty travellers, but from men who have long resided with savages, that numerous races have existed and still exist, who have no idea of one or more gods, and who have no words in their languages to express such an idea. The question is of course wholly distinct from that higher one, whether there exists a Creator and Ruler of the universe; and this has been answered in the affirmative by the highest intellects that have ever lived.

If, however, we include under the term "religion" the belief in unseen or spiritual agencies, the case is wholly different; for this belief seems to be almost universal with the less civilised races. Nor is it difficult to comprehend how it arose. As soon as the important faculties of the imagination, wonder, and curiosity, together with some power of reasoning, had become partially developed, man would naturally have craved to understand what was passing around him, and have vaguely speculated on his own existence. As Mr. M'Lennan has remarked, "Some explanation of the phenomena of life, a man must feign for himself; and to judge from the universality of it, the simplest hypothesis, and the first to occur to men, seems to have been that natural phenomena are ascribable to the presence in animals, plants, and things, and in the forces of nature, of such spirits prompting to action as men are conscious they themselves possess." It is probable, as Mr. Tylor has clearly shewn, that dreams may have first given rise to the notion of spirits; for savages do not readily distinguish between subjective and objective impressions. When a savage dreams, the figures which appear before him are believed to have come from a distance and to stand over him; or "the soul of the dreamer goes out on its travels, and comes home with a remembrance of what it has seen." But until the above-named faculties of imagination, curiosity, reason, &c., had been fairly well developed in the mind of man, his dreams would not have led him to believe in spirits, any more than in the case of a dog.

The tendency in savages to imagine that natural objects and agencies are animated by spiritual or living essences, is perhaps illustrated by a little fact which I once noticed: my dog, a full-grown and very sensible animal, was lying on the lawn during a hot and still day; but at a little distance a slight breeze occasionally moved an open parasol, which would have been wholly disregarded by the dog, had any one stood near it. As it was, every time that the parasol slightly moved, the dog growled fiercely and barked. He must, I think, have reasoned to himself in a rapid and unconscious manner, that movement without any apparent cause indicated the presence of some strange living agent, and no stranger had a right to be on his territory.

The belief in spiritual agencies would easily pass into the belief in the existence of one or more gods. For savages would naturally attribute to spirits the same passions, the same love of vengeance or simplest form of justice, and the same affections which they themselves experienced. The Fuegians appear to be in this respect in an intermediate condition,* for when the surgeon on board the "Beagle" shot some young ducklings as specimens, York Minster declared in the most solemn manner, "Oh! Mr. Bynoe, much rain, much snow, blow much;" and this was evidently a retributive punishment for wasting human food. So again he related how, when his brother killed a "wild man," storms long raged, much rain and snow fell. Yet we could never discover that the Fuegians believed in what we should call a God, or practised any religious rites; and Jemmy Button, with justifiable pride, stoutly maintained that there was no devil in his land. This latter assertion is the more remarkable, as with savages the belief in bad spirits is far more common than the belief in good spirits.

The feeling of religious devotion is a highly complex one, consisting of love, complete submission to an exalted and mysterious superior, a strong sense of dependence, fear, reverence, gratitude, hope for the future, and perhaps other elements. No being could experience so complex an emotion until advanced in his intellectual and moral faculties to at least a moderately high level. Nevertheless we see some distant approach to this state of mind, in the deep love of a dog for his master, associated with complete submission, some fear, and perhaps other feelings. The behaviour of a dog when returning to his master after an absence, and, as I may add, of a monkey to his beloved keeper, is widely different from that towards their fellows. In the latter case the transports of joy appear to be somewhat less, and the sense of equality is shewn in every action. Professor Braubach goes so far as to maintain that a dog looks on his master as on a god.

The same high mental faculties which first led man to believe in unseen spiritual agencies, then in fetishism, polytheism, and ultimately in monotheism, would infallibly lead him, as long as his reasoning powers remained poorly developed, to various strange superstitions and customs. Many of these are terrible to think of—such as the sacrifice of human beings to a blood-loving god; the trial of innocent persons by the ordeal of poison or fire; witchcraft, &c.—yet it is well occasionally to reflect on these superstitions, for they shew us what an infinite debt of gratitude we owe to the improvement of our reason, to science, and our accumulated knowledge. As Sir J. Lubbock has well observed, "it is not too much to say that the horrible dread of unknown evil hangs like a thick cloud over savage life, and embitters every pleasure." These miserable and indirect consequences of our highest faculties may be compared with the incidental and occasional mistakes of the instincts of the lower animals.

CHAPTER III

Comparison of the Mental Powers of Man and the Lower Animals — continued

I FULLY subscribe to the judgment of those writers who maintain that of all the differences between man and the lower animals, the moral sense or conscience is by far the most important. This sense, as Mackintosh remarks, "has a rightful supremacy over every other principle of human action;" it is summed up in that short but imperious word *ought*, so full of high significance. It is the most noble of all the attributes of man, leading him without a moment's hesitation to risk his life for that of a fellow-creature; or after due deliberation, impelled simply by the deep feeling of right or duty, to sacrifice it in some great cause. Immanuel Kant exclaims, "Duty! Wondrous thought, that workest neither by fond insinuation, flattery, nor by any threat, but merely by holding up thy naked law in the soul, and so extorting for thyself always reverence, if not always obedience; before whom all appetites are dumb, however secretly they rebel; whence thy original?"*

This great question has been discussed by many writers of consummate ability; and my sole excuse for touching on it is the impossibility of here passing it over, and because, as far as I know, no one has approached it exclusively from the side of natural history. The investigation possesses, also, some independent interest, as an attempt to see how far the study of the lower animals can throw light on one of the highest psychical faculties of man.

The following proposition seems to me in a high degree probable— namely, that any animal whatever, endowed with well-marked social instincts, would inevitably acquire a moral sense or conscience, as soon as its intellectual powers had become as well developed, or nearly as well developed, as in man. For, *firstly*, the social instincts lead an animal to take pleasure in the society of its fellows, to feel a certain amount of sympathy with them, and to perform various services for them. The services may be of a definite and evidently instinctive nature; or there may be only a wish and readiness, as with most of the higher social animals, to aid their fellows in certain general ways. But these feelings and services are by no means extended to all the individuals of the same species, only to those of the same association. *Secondly*, as soon as the mental faculties had become highly developed, images of all past actions and motives would be incessantly passing through the brain of each individual; and that feeling of dissatisfaction which invariably results, as we shall hereafter see, from any unsatisfied instinct, would arise, as often as it was perceived that the enduring and always present social instinct had yielded to some other instinct, at the time stronger, but neither enduring in its nature, nor leaving behind it a very vivid impression. It is clear that many instinctive desires, such as that of hunger, are in their nature of short duration; and after being satisfied are not readily or vividly recalled. *Thirdly*, after the power of language had been acquired and the wishes of the members of the same community could be distinctly expressed, the common opinion how each member ought to act for the public good, would naturally become to a large extent the guide to action. But the social instincts would still give the impulse to act for the good of the community, this impulse being strengthened, directed, and sometimes even deflected by public opinion, the power of which rests, as we shall presently see, on instinctive sympathy. *Lastly*, habit in the individual would ultimately play a very important part in guiding the conduct of each member; for the social instincts and impulses, like all other instincts, would be greatly strengthened by habit, as would obedience to the wishes and judgment of the community. These several subordinate propositions must now be discussed; and some of them at considerable length.

It may be well first to premise that I do not wish to maintain that any strictly social animal, if its intellectual faculties were to become as active and as highly developed as in man, would acquire exactly the same moral sense as ours. In the same manner as various animals have some sense of beauty, though they admire widely different objects, so they might have a sense of right and wrong, though led by it to follow widely different lines of conduct. If, for instance, to take an extreme case, men were reared under precisely the same conditions as hive-bees, there can hardly be a doubt that our unmarried females would, like the worker-bees, think it a sacred duty to kill their brothers, and mothers would strive to kill their fertile daughters; and no one would think of interfering. Nevertheless the bee, or any other social animal, would in our supposed case gain, as it appears to me, some feeling of right and wrong, or a conscience. For each individual would have an inward sense of possessing certain stronger or more enduring instincts, and other less strong or enduring; so that there would often be a struggle which impulse should be followed; and satisfaction or dissatisfaction would be felt, as past impressions were compared during their incessant passage through the mind. In this case an inward monitor would tell the animal that it would have been better to have followed the one impulse rather than the other. The one course ought to have been followed: the one would have been right and the other wrong; but to these terms I shall have to recur.

[*The main body of the chapter stresses continuities in the sociability of animals and humans. The moral sense originates in the social instincts, which in humans have gradually conquered the less powerful instincts of self-preservation, envy, and lust.*]

Concluding Remarks.—Philosophers of the derivative school of morals formerly assumed that the foundation of morality lay in a form of Selfishness; but more recently in the "Greatest Happiness principle."* According to the view given above, the moral sense is fundamentally identical with the social instincts; and in the case of the lower animals it would be absurd to speak of these instincts as having been developed from selfishness, or for the happiness of the community. They have, however, certainly been developed for the general good of the community. The term, general good, may be defined as the means by which the greatest possible number of individuals can be reared in full vigour and health, with all their faculties perfect, under the conditions to which they are exposed. As the social instincts both of man and the lower animals have no doubt been developed by the same steps, it would be advisable, if found practicable, to use the same definition in both cases, and to take as the test of morality, the general good or welfare of the community, rather than the general happiness; but this definition would perhaps require some limitation on account of political ethics.

When a man risks his life to save that of a fellow-creature, it seems more appropriate to say that he acts for the general good or welfare, rather than for the general happiness of mankind. No doubt the welfare and the happiness of the individual usually coincide; and a contented, happy tribe will flourish better than one that is discontented and unhappy. We have seen that at an early period in the history of man, the expressed wishes of the community will have naturally influenced to a large extent the conduct of each member; and as all wish for happiness, the "greatest happiness principle" will have become a most important secondary guide and object; the social instincts, including sympathy, always serving as the primary impulse and guide. Thus the reproach of laying the foundation of the most noble part of our nature in the base principle of selfishness is removed; unless indeed the satisfaction which every animal feels when it follows its proper instincts, and the dissatisfaction felt when prevented, be called selfish.

The expression of the wishes and judgment of the members of the same community, at first by oral and afterwards by written language, serves, as just remarked, as a most important secondary guide of conduct, in aid of the social instincts, but sometimes in opposition to them. This latter fact is well exemplified by the *Law of Honour*, that is the law of the opinion of our equals, and not of all our countrymen. The breach of this law, even when the breach is known to be strictly accordant with true morality, has caused many a man more agony than a real crime. We recognise the same influence in the burning sense of shame which most of us have felt even after the interval of years, when calling to mind some accidental breach of a trifling though fixed rule of etiquette. The judgment of the community will generally be guided by some rude experience of what is best in the long run for all the members; but this judgment will not rarely err from ignorance and from weak powers of reasoning. Hence the strangest customs and superstitions, in complete opposition to the true welfare and happiness of mankind, have become all-powerful throughout the world. We see this in the horror felt by a Hindoo who breaks his caste, in the shame of a Mahometan woman who exposes her face, and in innumerable other instances. It would be difficult to distinguish between the remorse felt by a Hindoo who has eaten unclean food, from that felt after committing a theft; but the former would probably be the more severe.

How so many absurd rules of conduct, as well as so many absurd religious beliefs, have originated we do not know; nor how it is that they have become, in all quarters of the world, so deeply impressed on the mind of men; but it is worthy of remark that a belief constantly inculcated during the early years of life, whilst the brain is impressible, appears to acquire almost the nature of an instinct; and the very essence of an instinct is that it is followed independently of reason. Neither can we say why certain admirable virtues, such as the love of truth, are much more highly appreciated by some savage tribes than by others; nor, again, why similar differences prevail even amongst civilised nations. Knowing how firmly fixed many strange customs and superstitions have become, we need feel no surprise that the self-regarding virtues should now appear to us so natural, supported as they are by reason, as to be thought innate, although they were not valued by man in his early condition.

Notwithstanding many sources of doubt, man can generally and readily distinguish between the higher and lower moral rules. The higher are founded on the social instincts, and relate to the welfare of others. They are supported by the approbation of our fellowmen and by reason. The lower rules, though some of them when implying self-sacrifice hardly deserve to be called lower, relate chiefly to self, and owe their origin to public opinion, when matured by experience and cultivated; for they are not practised by rude tribes.

As man advances in civilisation, and small tribes are united into larger communities, the simplest reason would tell each individual that he ought to extend his social instincts and sympathies to all the members of the same nation, though personally unknown to him. This point being once reached, there is only an artificial barrier to prevent his sympathies extending to the men of all nations and races. If, indeed, such men are separated from him by great differences in appearance or habits, experience unfortunately shews us how long it is before we look at them as our fellow-creatures. Sympathy beyond the confines of man, that is humanity to the lower animals, seems to be one of the latest moral acquisitions. It is apparently unfelt by savages, except towards their pets. How little the old Romans knew of it is shewn by their abhorrent gladiatorial exhibitions. The very idea of humanity, as far as I could observe, was new to most of the Gauchos of the Pampas. This virtue, one of the noblest with which man is endowed, seems to arise incidentally from our sympathies becoming more tender and more widely diffused, until they are extended to all sentient beings. As soon as this virtue is honoured and practised by some few men, it spreads through instruction and example to the young, and eventually through public opinion.

The highest stage in moral culture at which we can arrive, is when we recognise that we ought to control our thoughts, and "not even in inmost thought to think again the sins that made the past so pleasant to us."* Whatever makes any bad action familiar to the mind, renders its performance by so much the easier. As Marcus Aurelius long ago said, "Such as are thy habitual thoughts, such also will be the character of thy mind; for the soul is dyed by the thoughts."

Our great philosopher, Herbert Spencer, has recently explained his views on the moral sense. He says, "I believe that the experiences of utility organised and consolidated through all past generations of the human race, have been producing corresponding modifications, which, by continued transmission and accumulation, have become in us certain faculties of moral intuition—certain emotions responding to right and wrong conduct, which have no apparent basis in the individual experiences of utility." There is not the least inherent improbability, as it seems to me, in virtuous tendencies being more or less strongly inherited; for, not to mention the various dispositions and habits transmitted by many of our domestic animals, I have heard of cases in which a desire to steal and a tendency to lie appeared to run in families of the upper ranks; and as stealing is so rare a crime in the wealthy classes, we can hardly account by accidental coincidence for the tendency occurring in two or three members of the same family. If bad tendencies are transmitted, it is probable that good ones are likewise transmitted. Excepting through the principle of the transmission of moral tendencies, we cannot understand the differences believed to exist in this respect between the various races of mankind. We have, however, as yet, hardly sufficient evidence on this head.

Even the partial transmission of virtuous tendencies would be an immense assistance to the primary impulse derived directly from the social instincts, and indirectly from the approbation of our fellow-men. Admitting for the moment that virtuous tendencies are inherited, it appears probable, at least in such cases as chastity, temperance, humanity to animals, &c., that they become first impressed on the mental organisation through habit, instruction, and example, continued during several generations in the same family, and in a quite subordinate degree, or not at all, by the individuals possessing such virtues, having succeeded best in the struggle for life. My chief source of doubt with respect to any such inheritance, is that senseless customs, superstitions, and tastes, such as the horror of a Hindoo for unclean food, ought on the same principle to be transmitted. Although this in itself is perhaps not less probable than that animals should acquire inherited tastes for certain kinds of food or fear of certain foes, I have not met with any evidence in support of the transmission of superstitious customs or senseless habits.

Finally, the social instincts which no doubt were acquired by man, as by the lower animals, for the good of the community, will from the first have given to him some wish to aid his fellows, and some feeling of sympathy. Such impulses will have served him at a very early period as a rude rule of right and wrong. But as man gradually advanced in intellectual power and was enabled to trace the more remote consequences of his actions; as he acquired sufficient knowledge to reject baneful customs and superstitions; as he regarded more and more not only the welfare but the happiness of his fellow-men; as from habit, following on beneficial experience, instruction, and example, his sympathies became more tender and widely diffused, so as to extend to the men of all races, to the imbecile, the maimed, and other useless members of society, and finally to the lower animals,—so would the standard of his morality rise higher and higher. And it is admitted by moralists of the derivative school and by some intuitionists,* that the standard of morality has risen since an early period in the history of man.

As a struggle may sometimes be seen going on between the various instincts of the lower animals, it is not surprising that there should be a struggle in man between his social instincts, with their derived virtues, and his lower, though at the moment, stronger impulses or desires. This, as Mr. Galton has remarked, is all the less surprising, as man has emerged from a state of barbarism within a comparatively recent period. After having yielded to some temptation we feel a sense of dissatisfaction, analogous to that felt from other unsatisfied instincts, called in this case conscience; for we cannot prevent past images and impressions continually passing through our minds, and these in their weakened state we compare with the ever-present social instincts, or with habits gained in early youth and strengthened during our whole lives, perhaps inherited, so that they are at last rendered almost as strong as instincts. Looking to future generations, there is no cause to fear that the social instincts will grow weaker, and we may expect that virtuous habits will grow stronger, becoming perhaps fixed by inheritance. In this case the struggle between our higher and lower impulses will be less severe, and virtue will be triumphant.

Summary of the two last Chapters.—There can be no doubt that the difference between the mind of the lowest man and that of the highest animal is immense. An anthropomorphous ape, if he could take a dispassionate view of his own case, would admit that though he could form an artful plan to plunder a garden—though he could use stones for fighting or for breaking open nuts, yet that the thought of fashioning a stone into a tool was quite beyond his scope. Still less, as he would admit, could he follow out a train of metaphysical reasoning, or solve a mathematical problem, or reflect on God, or admire a grand natural scene. Some apes, however, would probably declare that they could and did admire the beauty of the coloured skin and fur of their partners in marriage. They would admit, that though they could make other apes understand by cries some of their perceptions and simpler wants, the notion of expressing definite ideas by definite sounds had never crossed their minds. They might insist that they were ready to aid their fellow-apes of the same troop in many ways, to risk their lives for them, and to take charge of their orphans; but they would be forced to acknowledge that disinterested love for all living creatures, the most noble attribute of man, was quite beyond their comprehension.

Nevertheless the difference in mind between man and the higher animals, great as it is, is certainly one of degree and not of kind. We have seen that the senses and intuitions, the various emotions and faculties, such as love, memory, attention, curiosity, imitation, reason, &c., of which man boasts, may be found in an incipient, or even sometimes in a well-developed condition, in the lower animals. They are also capable of some inherited improvement, as we see in the domestic dog compared with the wolf or jackal. If it be maintained that certain powers, such as self-consciousness, abstraction, &c., are peculiar to man, it may well be that these are the incidental results of other highly-advanced intellectual faculties; and these again are mainly the result of the continued use of a highly developed language. At what age does the new-born infant possess the power of abstraction, or become self-conscious and reflect on its own existence? We cannot answer; nor can we answer in regard to the ascending organic scale. The half-art and half-instinct of language still bears the stamp of its gradual evolution. The ennobling belief in God is not universal with man; and the belief in active spiritual agencies naturally follows from his other mental powers. The moral sense perhaps affords the best and highest distinction between man and the lower animals; but I need not say anything on this head, as I have so lately endeavoured to shew that the social instincts,—the prime principle of man's moral constitution—with the aid of active intellectual powers and the effects of habit, naturally lead to the golden rule, "As ye would that men should do to you, do ye to them likewise;"* and this lies at the foundation of morality.

In a future chapter I shall make some few remarks on the probable steps and means by which the several mental and moral faculties of man have been gradually evolved. That this at least is possible ought not to be denied, when we daily see their development in every infant; and when we may trace a perfect gradation from the mind of an utter idiot, lower than that of the lowest animal, to the mind of a Newton.

[*In Chapter IV, Darwin argues that the humans have evolved according to the same laws of natural selection that apply to the lower animals.*]

CHAPTER V

On the Development of the Intellectual and Moral Faculties during Primeval and Civilised Times

THE subjects to be discussed in this chapter are of the highest interest, but are treated by me in a most imperfect and fragmentary manner. Mr. Wallace, in an admirable paper before referred to,* argues that man after he had partially acquired those intellectual and moral faculties which distinguish him from the lower animals, would have been but little liable to have had his bodily structure modified through natural selection or any other means. For man is enabled through his mental faculties "to keep with an unchanged body in harmony with the changing universe." He has great power of adapting his habits to new conditions of life. He invents weapons, tools and various stratagems, by which he procures food and defends himself. When he migrates into a colder climate he uses clothes, builds sheds, and makes fires; and, by the aid of fire, cooks food otherwise indigestible. He aids his fellow-men in many ways, and anticipates future events. Even at a remote period he practised some subdivision of labour.

The lower animals, on the other hand, must have their bodily structure modified in order to survive under greatly changed conditions. They must be rendered stronger, or acquire more effective teeth or claws, in order to defend themselves from new enemies; or they must be reduced in size so as to escape detection and danger. When they migrate into a colder climate they must become clothed with thicker fur, or have their constitutions altered. If they fail to be thus modified, they will cease to exist.

The case, however, is widely different, as Mr. Wallace has with justice insisted, in relation to the intellectual and moral faculties of man. These faculties are variable; and we have every reason to believe that the variations tend to be inherited. Therefore, if they were formerly of high importance to primeval man and to his ape-like progenitors, they would have been perfected or advanced through natural selection. Of the high importance of the intellectual faculties there can be no doubt, for man mainly owes to them his preeminent position in the world. We can see that, in the rudest state of society, the individuals who were the most sagacious, who invented and used the best weapons or traps, and who were best able to defend themselves, would rear the greatest number of offspring. The tribes which included the largest number of men thus endowed would increase in number and supplant other tribes. Numbers depend primarily on the means of subsistence, and this, partly on the physical nature of the country, but in a much higher degree on the arts which are there practised. As a tribe increases and is victorious, it is often still further increased by the absorption of other tribes. The stature and strength of the men of a tribe are likewise of some importance for its success, and these depend in part on the nature and amount of the food which can be obtained. In Europe the men of the Bronze period* were supplanted by a more powerful and, judging from their sword-handles, larger-handed race; but their success was probably due in a much higher degree to their superiority in the arts.

All that we know about savages, or may infer from their traditions and from old monuments, the history of which is quite forgotten by the present inhabitants, shew that from the remotest times successful tribes have supplanted other tribes. Relics of extinct or forgotten tribes have been discovered throughout the civilised regions of the earth, on the wild plains of America, and on the isolated islands in the Pacific Ocean. At the present day civilised nations are every-where supplanting barbarous nations, excepting where the climate opposes a deadly barrier; and they succeed mainly, though not exclu-sively, through their arts, which are the products of the intellect. It is, therefore, highly probable that with mankind the intellectual faculties have been gradually perfected through natural selection; and this conclusion is sufficient for our purpose. Undoubtedly it would have been very interesting to have traced the development of each separate faculty from the state in which it exists in the lower animals to that in which it exists in man; but neither my ability nor knowledge permit the attempt.

It deserves notice that as soon as the progenitors of man became social (and this probably occurred at a very early period), the advance-ment of the intellectual faculties will have been aided and modified in an important manner, of which we see only traces in the lower animals, namely, through the principle of imitation, together with reason and experience. Apes are much given to imitation, as are the lowest savages; and the simple fact previously referred to, that after a time no animal can be caught in the same place by the same sort of trap, shews that animals learn by experience, and imitate each others' caution. Now, if some one man in a tribe, more sagacious than the others, invented a new snare or weapon, or other means of attack or defence, the plainest self-interest, without the assistance of much reasoning power, would prompt the other members to imitate him; and all would thus profit. The habitual practice of each new art must likewise in some slight degree strengthen the intellect. If the new invention were an important one, the tribe would increase in number, spread, and supplant other tribes. In a tribe thus rendered more numerous there would always be a rather better chance of the birth of other superior and inventive members. If such men left children to inherit their mental superiority, the chance of the birth of still more ingenious members would be somewhat better, and in a very small tribe decidedly better. Even if they left no children, the tribe would still include their blood-relations; and it has been ascertained by agriculturists that by preserving and breeding from the family of an animal, which when slaughtered was found to be valuable, the desired character has been obtained.

Turning now to the social and moral faculties. In order that primeval men, or the ape-like progenitors of man, should have become social, they must have acquired the same instinctive feelings which impel other animals to live in a body; and they no doubt exhibited the same general disposition. They would have felt uneasy when separated from their comrades, for whom they would have felt some degree of love; they would have warned each other of danger, and have given mutual aid in attack or defence. All this implies some degree of sympathy, fidelity, and courage. Such social qualities, the paramount importance of which to the lower animals is disputed by no one, were no doubt acquired by the progenitors of man in a similar manner, namely, through natural selection, aided by inherited habit. When two tribes of primeval man, living in the same country, came into competition, if the one tribe included (other circumstances being equal) a greater number of courageous, sympathetic, and faithful members, who were always ready to warn each other of danger, to aid and defend each other, this tribe would without doubt succeed best and conquer the other. Let it be borne in mind how all-important, in the never-ceasing wars of savages, fidelity and courage must be. The advantage which disciplined soldiers have over undisciplined hordes follows chiefly from the confidence which each man feels in his comrades. Obedience, as Mr. Bagehot has well shewn, is of the highest value, for any form of government is better than none. Selfish and contentious people will not cohere, and without coherence nothing can be effected. A tribe possessing the above qualities in a high degree would spread and be victorious over other tribes; but in the course of time it would, judging from all past history, be in its turn overcome by some other and still more highly endowed tribe. Thus the social and moral qualities would tend slowly to advance and be diffused throughout the world.

But it may be asked, how within the limits of the same tribe did a large number of members first become endowed with these social and moral qualities, and how was the standard of excellence raised? It is extremely doubtful whether the offspring of the more sympathetic and benevolent parents, or of those which were the most faithful to their comrades, would be reared in greater number than the children of selfish and treacherous parents of the same tribe. He who was ready to sacrifice his life, as many a savage has been, rather than betray his comrades, would often leave no offspring to inherit his noble nature. The bravest men, who were always willing to come to the front in war, and who freely risked their lives for others, would on an average perish in larger number than other men. Therefore it seems scarcely possible (bearing in mind that we are not here speaking of one tribe being victorious over another) that the number of men gifted with such virtues, or that the standard of their excellence, could be increased through natural selection, that is, by the survival of the fittest.

Although the circumstances which lead to an increase in the number of men thus endowed within the same tribe are too complex to be clearly followed out, we can trace some of the probable steps. In the first place, as the reasoning powers and foresight of the members became improved, each man would soon learn from experience that if he aided his fellow-men, he would commonly receive aid in return. From this low motive he might acquire the habit of aiding his fellows; and the habit of performing benevolent actions certainly strengthens the feeling of sympathy, which gives the first impulse to benevolent actions. Habits, moreover, followed during many generations probably tend to be inherited.

But there is another and much more powerful stimulus to the development of the social virtues, namely, the praise and the blame of our fellow-men. The love of approbation and the dread of infamy, as well as the bestowal of praise or blame, are primarily due, as we have seen in the third chapter, to the instinct of sympathy; and this instinct no doubt was originally acquired, like all the other social instincts, through natural selection. At how early a period the progenitors of man, in the course of their development, became capable of feeling and being impelled by the praise or blame of their fellow-creatures, we cannot, of course, say. But it appears that even dogs appreciate encouragement, praise, and blame. The rudest savages feel the sentiment of glory, as they clearly show by preserving the trophies of their prowess, by their habit of excessive boasting, and even by the extreme care which they take of their personal appearance and decorations; for unless they regarded the opinion of their comrades, such habits would be senseless.

They certainly feel shame at the breach of some of their lesser rules; but how far they experience remorse is doubtful. I was at first surprised that I could not recollect any recorded instances of this feeling in savages; and Sir J. Lubbock states that he knows of none. But if we banish from our minds all cases given in novels and plays and in death-bed confessions made to priests, I doubt whether many of us have actually witnessed remorse; though we may have often seen shame and contrition for smaller offences. Remorse is a deeply hidden feeling. It is incredible that a savage, who will sacrifice his life rather than betray his tribe, or one who will deliver himself up as a prisoner rather than break his parole, would not feel remorse in his inmost soul, though he might conceal it, if he had failed in a duty which he held sacred.

We may therefore conclude that primeval man, at a very remote period, would have been influenced by the praise and blame of his fellows. It is obvious, that the members of the same tribe would approve of conduct which appeared to them to be for the general good, and would reprobate that which appeared evil. To do good unto others—to do unto others as ye would they should do unto you,—is the foundation-stone of morality. It is, therefore, hardly possible to exaggerate the importance during rude times of the love of praise and the dread of blame. A man who was not impelled by any deep, instinctive feeling, to sacrifice his life for the good of others, yet was roused to such actions by a sense of glory, would by his example excite the same wish for glory in other men, and would strengthen by exercise the noble feeling of admiration. He might thus do far more good to his tribe than by begetting offspring with a tendency to inherit his own high character.

With increased experience and reason, man perceives the more remote consequences of his actions, and the self-regarding virtues, such as temperance, chastity, &c., which during early times are, as we have before seen, utterly disregarded, come to be highly esteemed or even held sacred. I need not, however, repeat what I have said on this head in the third chapter. Ultimately a highly complex sentiment, having its first origin in the social instincts, largely guided by the approbation of our fellow-men, ruled by reason, self-interest, and in later times by deep religious feelings, confirmed by instruction and habit, all combined, constitute our moral sense or conscience.

It must not be forgotten that although a high standard of morality gives but a slight or no advantage to each individual man and his children over the other men of the same tribe, yet that an advancement in the standard of morality and an increase in the number of well-endowed men will certainly give an immense advantage to one tribe over another. There can be no doubt that a tribe including many members who, from possessing in a high degree the spirit of patriotism, fidelity, obedience, courage, and sympathy, were always ready to give aid to each other and to sacrifice themselves for the common good, would be victorious over most other tribes; and this would be natural selection. At all times throughout the world tribes have supplanted other tribes; and as morality is one element in their success, the standard of morality and the number of well-endowed men will thus everywhere tend to rise and increase.

It is, however, very difficult to form any judgment why one particular tribe and not another has been successful and has risen in the scale of civilisation. Many savages are in the same condition as when first discovered several centuries ago. As Mr. Bagehot has remarked, we are apt to look at progress as the normal rule in human society; but history refutes this. The ancients did not even entertain the idea; nor do the oriental nations at the present day. According to another high authority, Mr. Maine, "the greatest part of mankind has never shewn a particle of desire that its civil institutions should be improved." Progress seems to depend on many concurrent favourable conditions, far too complex to be followed out. But it has often been remarked, that a cool climate from leading to industry and the various arts has been highly favourable, or even indispensable for this end. The Esquimaux, pressed by hard necessity, have succeeded in many ingenious inventions, but their climate has been too severe for continued progress. Nomadic habits, whether over wide plains, or through the dense forests of the tropics, or along the shores of the sea, have in every case been highly detrimental. Whilst observing the barbarous inhabitants of Tierra del Fuego, it struck me that the possession of some property, a fixed abode, and the union of many families under a chief, were the indispensable requisites for civilisation.

Such habits almost necessitate the cultivation of the ground; and the first steps in cultivation would probably result, as I have elsewhere shewn, from some such accident as the seeds of a fruit-tree falling on a heap of refuse and producing an unusually fine variety. The problem, however, of the first advance of savages towards civilisation is at present much too difficult to be solved.

Natural Selection as affecting Civilised Nations.—In the last and present chapters I have considered the advancement of man from a former semi-human condition to his present state as a barbarian. But some remarks on the agency of natural selection on civilised nations may be here worth adding. This subject has been ably discussed by Mr. W. R. Greg, and previously by Mr. Wallace and Mr. Galton. Most of my remarks are taken from these three authors. With savages, the weak in body or mind are soon eliminated; and those that survive commonly exhibit a vigorous state of health. We civilised men, on the other hand, do our utmost to check the process of elimination; we build asylums for the imbecile, the maimed, and the sick; we institute poor-laws;* and our medical men exert their utmost skill to save the life of every one to the last moment. There is reason to believe that vaccination has preserved thousands, who from a weak constitution would formerly have succumbed to small-pox. Thus the weak members of civilised societies propagate their kind. No one who has attended to the breeding of domestic animals will doubt that this must be highly injurious to the race of man. It is surprising how soon a want of care, or care wrongly directed, leads to the degeneration of a domestic race; but excepting in the case of man himself, hardly any one is so ignorant as to allow his worst animals to breed.

The aid which we feel impelled to give to the helpless is mainly an incidental result of the instinct of sympathy, which was originally acquired as part of the social instincts, but subsequently rendered, in the manner previously indicated, more tender and more widely diffused. Nor could we check our sympathy, if so urged by hard reason, without deterioration in the noblest part of our nature. The surgeon may harden himself whilst performing an operation, for he knows that he is acting for the good of his patient; but if we were intentionally to neglect the weak and helpless, it could only be for a contingent benefit, with a certain and great present evil. Hence we must bear without complaining the undoubtedly bad effects of the weak surviving and propagating their kind; but there appears to be at least one check in steady action, namely the weaker and inferior members of society not marrying so freely as the sound; and this check might be indefinitely increased, though this is more to be hoped for than expected, by the weak in body or mind refraining from marriage.

In all civilised countries man accumulates property and bequeaths it to his children. So that the children in the same country do not by any means start fair in the race for success. But this is far from an unmixed evil; for without the accumulation of capital the arts could not progress; and it is chiefly through their power that the civilised races have extended, and are now everywhere extending, their range, so as to take the place of the lower races. Nor does the moderate accumulation of wealth interfere with the process of selection. When a poor man becomes rich, his children enter trades or professions in which there is struggle enough, so that the able in body and mind succeed best. The presence of a body of well-instructed men, who have not to labour for their daily bread, is important to a degree which cannot be over-estimated; as all high intellectual work is carried on by them, and on such work material progress of all kinds mainly depends, not to mention other and higher advantages. No doubt wealth when very great tends to convert men into useless drones, but their number is never large; and some degree of elimination here occurs, as we daily see rich men, who happen to be fools or profligate, squandering away all their wealth.

Primogeniture with entailed estates* is a more direct evil, though it may formerly have been a great advantage by the creation of a dominant class, and any government is better than anarchy. The eldest sons, though they may be weak in body or mind, generally marry, whilst the younger sons, however superior in these respects, do not so generally marry. Nor can worthless eldest sons with entailed estates squander their wealth. But here, as elsewhere, the relations of civilised life are so complex that some compensatory checks intervene. The men who are rich through primogeniture are able to select generation after generation the more beautiful and charming women; and these must generally be healthy in body and active in mind. The evil consequences, such as they may be, of the continued preservation of the same line of descent, without any selection, are checked by men of rank always wishing to increase their wealth and power; and this they effect by marrying heiresses. But the daughters of parents who have produced single children, are themselves, as Mr. Galton has shewn, apt to be sterile; and thus noble families are continually cut off in the direct line, and their wealth flows into some side channel; but unfortunately this channel is not determined by superiority of any kind.

Although civilisation thus checks in many ways the action of natural selection, it apparently favours, by means of improved food and the freedom from occasional hardships, the better development of the body. This may be inferred from civilised men having been found, wherever compared, to be physically stronger than savages. They appear also to have equal powers of endurance, as has been proved in many adventurous expeditions. Even the great luxury of the rich can be but little detrimental; for the expectation of life of our aristocracy, at all ages and of both sexes, is very little inferior to that of healthy English lives in the lower classes.

We will now look to the intellectual faculties alone. If in each grade of society the members were divided into two equal bodies, the one including the intellectually superior and the other the inferior, there can be little doubt that the former would succeed best in all occupations and rear a greater number of children. Even in the lowest walks of life, skill and ability must be of some advantage, though in many occupations, owing to the great division of labour, a very small one. Hence in civilised nations there will be some tendency to an increase both in the number and in the standard of the intellectually able. But I do not wish to assert that this tendency may not be more than counterbalanced in other ways, as by the multiplication of the reckless and improvident; but even to such as these, ability must be some advantage.

It has often been objected to views like the foregoing, that the most eminent men who have ever lived have left no offspring to inherit their great intellect. Mr. Galton says, "I regret I am unable to solve the simple question whether, and how far, men and women who are prodigies of genius are infertile. I have, however, shewn that men of eminence are by no means so." Great lawgivers, the founders of beneficent religions, great philosophers and discoverers in science, aid the progress of mankind in a far higher degree by their works than by leaving a numerous progeny. In the case of corporeal structures, it is the selection of the slightly better-endowed and the elimination of the slightly less well-endowed individuals, and not the preservation of strongly-marked and rare anomalies, that leads to the advancement of a species. So it will be with the intellectual faculties, namely from the somewhat more able men in each grade of society succeeding rather better than the less able, and consequently increasing in number, if not otherwise prevented. When in any nation the standard of intellect and the number of intellectual men have increased, we may expect from the law of the deviation from an average, as shewn by Mr. Galton, that prodigies of genius will appear somewhat more frequently than before.

In regard to the moral qualities, some elimination of the worst dispositions is always in progress even in the most civilised nations. Malefactors are executed, or imprisoned for long periods, so that they cannot freely transmit their bad qualities. Melancholic and insane persons are confined, or commit suicide. Violent and quarrelsome men often come to a bloody end. Restless men who will not follow any steady occupation—and this relic of barbarism is a great check to civilisation—emigrate to newly-settled countries, where they prove useful pioneers. Intemperance is so highly destructive, that the expectation of life of the intemperate, at the age, for instance, of thirty, is only 13·8 years; whilst for the rural labourers of England at the same age it is 40·59 years. Profligate women bear few children, and profligate men rarely marry; both suffer from disease. In the breeding of domestic animals, the elimination of those individuals, though few in number, which are in any marked manner inferior, is by no means an unimportant element towards success. This especially holds good with injurious characters which tend to reappear through reversion, such as blackness in sheep; and with mankind some of the worst dispositions, which occasionally without any assignable cause make their appearance in families, may perhaps be reversions to a savage state, from which we are not removed by very many generations. This view seems indeed recognised in the common expression that such men are the black sheep of the family.

With civilised nations, as far as an advanced standard of morality, and an increased number of fairly well-endowed men are concerned, natural selection apparently effects but little; though the fundamental social instincts were originally thus gained. But I have already said enough, whilst treating of the lower races, on the causes which lead to the advance of morality, namely, the approbation of our fellow-men—the strengthening of our sympathies by habit—example and imitation—reason—experience and even self-interest—instruction during youth, and religious feelings.

A most important obstacle in civilised countries to an increase in the number of men of a superior class has been strongly urged by Mr. Greg and Mr. Galton, namely, the fact that the very poor and reckless, who are often degraded by vice, almost invariably marry early, whilst the careful and frugal, who are generally otherwise virtuous, marry late in life, so that they may be able to support themselves and their children in comfort. Those who marry early produce within a given period not only a greater number of generations, but, as shewn by Dr. Duncan, they produce many more children. The children, moreover, that are born by mothers during the prime of life are heavier and larger, and therefore probably more vigorous, than those born at other periods. Thus the reckless, degraded, and often vicious members of society, tend to increase at a quicker rate than the provident and generally virtuous members. Or as Mr. Greg puts the case: "The careless, squalid, unaspiring Irishman multiplies like rabbits: the frugal, foreseeing, self-respecting, ambitious Scot, stern in his morality, spiritual in his faith, sagacious and disciplined in his intelligence, passes his best years in struggle and in celibacy, marries late, and leaves few behind him. Given a land originally peopled by a thousand Saxons and a thousand Celts—and in a dozen generations five-sixths of the population would be Celts, but five-sixths of the property, of the power, of the intellect, would belong to the one-sixth of Saxons that remained. In the eternal 'struggle for existence,' it would be the inferior and *less* favoured race that had prevailed—and prevailed by virtue not of its good qualities but of its faults."

There are, however, some checks to this downward tendency. We have seen that the intemperate suffer from a high rate of mortality, and the extremely profligate leave few offspring. The poorest classes crowd into towns, and it has been proved by Dr. Stark from the statistics of ten years in Scotland, that at all ages the death-rate is higher in towns than in rural districts, "and during the first five years of life the town death-rate is almost exactly double that of the rural districts." As these returns include both the rich and the poor, no doubt more than double the number of births would be requisite to keep up the number of the very poor inhabitants in the towns, relatively to those in the country. With women, marriage at too early an age is highly injurious; for it has been found in France that, "twice as many wives under twenty die in the year, as died out of the same number of the unmarried." The mortality, also, of husbands under twenty is "excessively high,"* but what the cause of this may be seems doubtful. Lastly, if the men who prudently delay marrying until they can bring up their families in comfort, were to select, as they often do, women in the prime of life, the rate of increase in the better class would be only slightly lessened.

It was established from an enormous body of statistics, taken during 1853, that the unmarried men throughout France, between the ages of twenty and eighty, die in a much larger proportion than the married: for instance, out of every 1000 unmarried men, between the ages of twenty and thirty, 11·3 annually died, whilst of the married only 6·5 died. A similar law was proved to hold good, during the years 1863 and 1864, with the entire population above the age of twenty in Scotland: for instance, out of every 1000 unmarried men, between the ages of twenty and thirty, 14·97 annually died, whilst of the married only 7·24 died, that is less than half. Dr. Stark remarks on this, "Bachelorhood is more destructive to life than the most unwholesome trades, or than residence in an unwholesome house or district where there has never been the most distant attempt at sanitary improvement." He considers that the lessened mortality is the direct result of "marriage, and the more regular domestic habits which attend that state." He admits, however, that the intemperate, profligate, and criminal classes, whose duration of life is low, do not commonly marry; and it must likewise be admitted that men with a weak constitution, ill health, or any great infirmity in body or mind, will often not wish to marry, or will be rejected. Dr. Stark seems to have come to the conclusion that marriage in itself is a main cause of prolonged life, from finding that aged married men still have a considerable advantage in this respect over the unmarried of the same advanced age; but every one must have known instances of men, who with weak health during youth did not marry, and yet have survived to old age, though remaining weak and therefore always with a lessened chance of life. There is another remarkable circumstance which seems to support Dr. Stark's conclusion, namely, that widows and widowers in France suffer in comparison with the married a very heavy rate of mortality; but Dr. Farr attributes this to the poverty and evil habits consequent on the disruption of the family, and to grief.

On the whole we may conclude with Dr. Farr that the lesser mortality of married than of unmarried men, which seems to be a general law, "is mainly due to the constant elimination of imperfect types, and to the skilful selection of the finest individuals out of each successive generation;" the selection relating only to the marriage state, and acting on all corporeal, intellectual, and moral qualities. We may, therefore, infer that sound and good men who out of prudence remain for a time unmarried do not suffer a high rate of mortality.

If the various checks specified in the two last paragraphs, and perhaps others as yet unknown, do not prevent the reckless, the vicious and otherwise inferior members of society from increasing at a quicker rate than the better class of men, the nation will retrograde, as has occurred too often in the history of the world. We must remember that progress is no invariable rule. It is most difficult to say why one civilised nation rises, becomes more powerful, and spreads more widely, than another; or why the same nation progresses more at one time than at another. We can only say that it depends on an increase in the actual number of the population, on the number of the men endowed with high intellectual and moral faculties, as well as on their standard of excellence. Corporeal structure, except so far as vigour of body leads to vigour of mind, appears to have little influence.

It has been urged by several writers that as high intellectual powers are advantageous to a nation, the old Greeks, who stood some grades higher in intellect than any race that has ever existed, ought to have risen, if the power of natural selection were real, still higher in the scale, increased in number, and stocked the whole of Europe. Here we have the tacit assumption, so often made with respect to corporeal structures, that there is some innate tendency towards continued development in mind and body. But development of all kinds depends on many concurrent favourable circumstances. Natural selection acts only in a tentative manner. Individuals and races may have acquired certain indisputable advantages, and yet have perished from failing in other characters. The Greeks may have retrograded from a want of coherence between the many small states, from the small size of their whole country, from the practice of slavery, or from extreme sensuality; for they did not succumb until "they were enervated and corrupt to the very core."* The western nations of Europe, who now so immeasurably surpass their former savage progenitors and stand at the summit of civilisation, owe little or none of their superiority to direct inheritance from the old Greeks; though they owe much to the written works of this wonderful people.

Who can positively say why the Spanish nation, so dominant at one time, has been distanced in the race. The awakening of the nations of Europe from the dark ages is a still more perplexing problem. At this early period, as Mr. Galton has remarked, almost all the men of a gentle nature, those given to meditation or culture of the mind, had no refuge except in the bosom of the Church which demanded celibacy; and this could hardly fail to have had a deteriorating influence on each successive generation. During this same period the Holy Inquisition selected with extreme care the freest and boldest men in order to burn or imprison them. In Spain alone some of the best men—those who doubted and questioned, and without doubting there can be no progress—were eliminated during three centuries at the rate of a thousand a year. The evil which the Catholic Church has thus effected, though no doubt counterbalanced to a certain, perhaps large extent in other ways, is incalculable; nevertheless, Europe has progressed at an unparalleled rate.

The remarkable success of the English as colonists over other European nations, which is well illustrated by comparing the progress of the Canadians of English and French extraction, has been ascribed to their "daring and persistent energy;"* but who can say how the English gained their energy. There is apparently much truth in the belief that the wonderful progress of the United States, as well as the character of the people, are the results of natural selection; the more energetic, restless, and courageous men from all parts of Europe having emigrated during the last ten or twelve generations to that great country, and having there succeeded best. Looking to the distant future, I do not think that the Rev. Mr. Zincke takes an exaggerated view when he says: "All other series of events—as that which resulted in the culture of mind in Greece, and that which resulted in the empire of Rome—only appear to have purpose and value when viewed in connection with, or rather as subsidiary to . . . . the great stream of Anglo-Saxon emigration to the west." Obscure as is the problem of the advance of civilisation, we can at least see that a nation which produced during a lengthened period the greatest number of highly intellectual, energetic, brave, patriotic, and benevolent men, would generally prevail over less favoured nations.

Natural selection follows from the struggle for existence; and this from a rapid rate of increase. It is impossible not bitterly to regret, but whether wisely is another question, the rate at which man tends to increase; for this leads in barbarous tribes to infanticide and many other evils, and in civilised nations to abject poverty, celibacy, and to the late marriages of the prudent. But as man suffers from the same physical evils with the lower animals, he has no right to expect an immunity from the evils consequent on the struggle for existence. Had he not been subjected to natural selection, assuredly he would never have attained to the rank of manhood. When we see in many parts of the world enormous areas of the most fertile land peopled by a few wandering savages, but which are capable of supporting numerous happy homes, it might be argued that the struggle for existence had not been sufficiently severe to force man upwards to his highest standard. Judging from all that we know of man and the lower animals, there has always been sufficient variability in the intellectual and moral faculties, for their steady advancement through natural selection. No doubt such advancement demands many favourable concurrent circumstances; but it may well be doubted whether the most favourable would have sufficed, had not the rate of increase been rapid, and the consequent struggle for existence severe to an extreme degree.

On the evidence that all civilised nations were once barbarous.—As we have had to consider the steps by which some semi-human creature has been gradually raised to the rank of man in his most perfect state, the present subject cannot be quite passed over. But it has been treated in so full and admirable a manner by Sir J. Lubbock, Mr. Tylor, Mr. M'Lennan, and others, that I need here give only the briefest summary of their results. The arguments recently advanced by the Duke of Argyll and formerly by Archbishop Whately, in favour of the belief that man came into the world as a civilised being and that all savages have since undergone degradation, seem to me weak in comparison with those advanced on the other side. Many nations, no doubt, have fallen away in civilisation, and some may have lapsed into utter barbarism, though on this latter head I have not met with any evidence. The Fuegians were probably compelled by other conquering hordes to settle in their inhospitable country, and they may have become in consequence somewhat more degraded; but it would

be difficult to prove that they have fallen much below the Botocudos*
who inhabit the finest parts of Brazil.

The evidence that all civilised nations are the descendants of
barbarians, consists, on the one side, of clear traces of their former
low condition in still-existing customs, beliefs, language, &c.; and
on the other side, of proofs that savages are independently able to
raise themselves a few steps in the scale of civilisation, and have actually thus risen. The evidence on the first head is extremely curious,
but cannot be here given: I refer to such cases as that, for instance,
of the art of enumeration, which, as Mr. Tylor clearly shews by the
words still used in some places, originated in counting the fingers,
first of one hand and then of the other, and lastly of the toes. We have
traces of this in our own decimal system, and in the Roman numerals,
which after reaching to the number V., change into VI., &c., when
the other hand no doubt was used. So again, "when we speak of
three-score and ten, we are counting by the vigesimal* system, each
score thus ideally made, standing for 20—for 'one man' as a Mexican
or Carib would put it." According to a large and increasing school of
philologists, every language bears the marks of its slow and gradual
evolution. So it is with the art of writing, as letters are rudiments of
pictorial representations. It is hardly possible to read Mr. M'Lennan's
work and not admit that almost all civilised nations still retain some
traces of such rude habits as the forcible capture of wives. What ancient
nation, as the same author asks, can be named that was originally
monogamous? The primitive idea of justice, as shewn by the law of
battle and other customs of which traces still remain, was likewise
most rude. Many existing superstitions are the remnants of former
false religious beliefs. The highest form of religion—the grand idea
of God hating sin and loving righteousness—was unknown during
primeval times.

Turning to the other kind of evidence: Sir J. Lubbock has shewn
that some savages have recently improved a little in some of their
simpler arts. From the extremely curious account which he gives of
the weapons, tools, and arts, used or practised by savages in various
parts of the world, it cannot be doubted that these have nearly all
been independent discoveries, excepting perhaps the art of making
fire. The Australian boomerang is a good instance of one such
independent discovery. The Tahitians when first visited had advanced
in many respects beyond the inhabitants of most of the other

Polynesian islands. There are no just grounds for the belief that the high culture of the native Peruvians and Mexicans was derived from any foreign source; many native plants were there cultivated, and a few native animals domesticated. We should bear in mind that a wandering crew from some semi-civilised land, if washed to the shores of America, would not, judging from the small influence of most missionaries, have produced any marked effect on the natives, unless they had already become somewhat advanced. Looking to a very remote period in the history of the world, we find, to use Sir J. Lubbock's well-known terms, a paleolithic and neolithic* period; and no one will pretend that the art of grinding rough flint tools was a borrowed one. In all parts of Europe, as far east as Greece, in Palestine, India, Japan, New Zealand, and Africa, including Egypt, flint tools have been discovered in abundance; and of their use the existing inhabitants retain no tradition. There is also indirect evidence of their former use by the Chinese and ancient Jews. Hence there can hardly be a doubt that the inhabitants of these many countries, which include nearly the whole civilised world, were once in a barbarous condition. To believe that man was aboriginally civilised and then suffered utter degradation in so many regions, is to take a pitiably low view of human nature. It is apparently a truer and more cheerful view that progress has been much more general than retrogression; that man has risen, though by slow and interrupted steps, from a lowly condition to the highest standard as yet attained by him in knowledge, morals, and religion.

[*In Chapter VI Darwin examines the place of humans in zoological classification and the history of the earth, and concludes with a speculative genealogy.*]

Thus we have given to man a pedigree of prodigious length, but not, it may be said, of noble quality. The world, it has often been remarked, appears as if it had long been preparing for the advent of man; and this, in one sense is strictly true, for he owes his birth to a long line of progenitors. If any single link in this chain had never existed, man would not have been exactly what he now is. Unless we wilfully close our eyes, we may, with our present knowledge, approximately recognise our parentage; nor need we feel ashamed of it. The most humble organism is something much higher than the inorganic dust under our feet; and no one with an unbiassed mind can study any living creature, however humble, without being struck with enthusiasm at its marvellous structure and properties.

CHAPTER VII

On the Races of Man

IT is not my intention here to describe the several so-called races of men; but to inquire what is the value of the differences between them under a classificatory point of view, and how they have originated. In determining whether two or more allied forms ought to be ranked as species or varieties, naturalists are practically guided by the following considerations; namely, the amount of difference between them, and whether such differences relate to few or many points of structure, and whether they are of physiological importance; but more especially whether they are constant. Constancy of character is what is chiefly valued and sought for by naturalists. Whenever it can be shewn, or rendered probable, that the forms in question have remained distinct for a long period, this becomes an argument of much weight in favour of treating them as species. Even a slight degree of sterility between any two forms when first crossed, or in their offspring, is generally considered as a decisive test of their specific distinctness; and their continued persistence without blending within the same area, is usually accepted as sufficient evidence, either of some degree of mutual sterility, or in the case of animals of some repugnance to mutual pairing.

Independently of blending from intercrossing, the complete absence, in a well-investigated region, of varieties linking together any two closely-allied forms, is probably the most important of all the criterions of their specific distinctness; and this is a somewhat different consideration from mere constancy of character, for two forms may be highly variable and yet not yield intermediate varieties. Geographical distribution is often unconsciously and sometimes consciously brought into play; so that forms living in two widely separated areas, in which most of the other inhabitants are specifically distinct, are themselves usually looked at as distinct; but in truth this affords no aid in distinguishing geographical races from so-called good or true species.

Now let us apply these generally-admitted principles to the races of man, viewing him in the same spirit as a naturalist would any other animal. In regard to the amount of difference between the races, we must make some allowance for our nice powers of discrimination gained by the long habit of observing ourselves. In India, as Elphinstone remarks, although a newly-arrived European cannot at first distinguish the various native races, yet they soon appear to him extremely dissimilar; and the Hindoo cannot at first perceive any difference between the several European nations. Even the most distinct races of man, with the exception of certain negro tribes, are much more like each other in form than would at first be supposed. This is well shewn by the French photographs in the Collection Anthropologique du Muséum of the men belonging to various races, the greater number of which, as many persons to whom I have shown them have remarked, might pass for Europeans. Nevertheless, these men if seen alive would undoubtedly appear very distinct, so that we are clearly much influenced in our judgment by the mere colour of the skin and hair, by slight differences in the features, and by expression.

There is, however, no doubt that the various races, when carefully compared and measured, differ much from each other,—as in the texture of the hair, the relative proportions of all parts of the body, the capacity of the lungs, the form and capacity of the skull, and even in the convolutions of the brain. But it would be an endless task to specify the numerous points of structural difference. The races differ also in constitution, in acclimatisation, and in liability to certain diseases. Their mental characteristics are likewise very distinct; chiefly as it would appear in their emotional, but partly in their intellectual, faculties. Every one who has had the opportunity of comparison, must have been struck with the contrast between the taciturn, even morose, aborigines of S. America and the light-hearted, talkative negroes. There is a nearly similar contrast between the Malays and the Papuans, who live under the same physical conditions, and are separated from each other only by a narrow space of sea.

[*Darwin summarizes first the case for, then against, classing man as a single species.*]

The question whether mankind consists of one or several species has of late years been much agitated by anthropologists, who are divided into two schools of monogenists and polygenists. Those who do not admit the principle of evolution, must look at species either as separate creations or as in some manner distinct entities; and they must decide what forms to rank as species by the analogy of other organic beings which are commonly thus received. But it is a hopeless endeavour to decide this point on sound grounds, until some definition of the term "species" is generally accepted; and the definition must not include an element which cannot possibly be ascertained, such as an act of creation. We might as well attempt without any definition to decide whether a certain number of houses should be called a village, or town, or city. We have a practical illustration of the difficulty in the never-ending doubts whether many closely-allied mammals, birds, insects, and plants, which represent each other in North America and Europe, should be ranked species or geographical races; and so it is with the productions of many islands situated at some little distance from the nearest continent.

Those naturalists, on the other hand, who admit the principle of evolution, and this is now admitted by the greater number of rising men, will feel no doubt that all the races of man are descended from a single primitive stock; whether or not they think fit to designate them as distinct species, for the sake of expressing their amount of difference. With our domestic animals the question whether the various races have arisen from one or more species is different. Although all such races, as well as all the natural species within the same genus, have undoubtedly sprung from the same primitive stock, yet it is a fit subject for discussion, whether, for instance, all the domestic races of the dog have acquired their present differences since some one species was first domesticated and bred by man; or whether they owe some of their characters to inheritance from distinct species, which had already been modified in a state of nature. With mankind no such question can arise, for he cannot be said to have been domesticated at any particular period.

When the races of man diverged at an extremely remote epoch from their common progenitor, they will have differed but little from each other, and been few in number; consequently they will then, as far as their distinguishing characters are concerned, have had less claim to rank as distinct species, than the existing so-called races. Nevertheless such early races would perhaps have been ranked by some naturalists as distinct species, so arbitrary is the term, if their differences, although extremely slight, had been more constant than at present, and had not graduated into each other.

It is, however, possible, though far from probable, that the early progenitors of man might at first have diverged much in character, until they became more unlike each other than are any existing races; but that subsequently, as suggested by Vogt, they converged in character. When man selects for the same object the offspring of two distinct species, he sometimes induces, as far as general appearance is concerned, a considerable amount of convergence. This is the case, as shewn by Von Nathusius, with the improved breeds of pigs, which are descended from two distinct species; and in a less well-marked manner with the improved breeds of cattle. A great anatomist, Gratiolet, maintains that the anthropomorphous apes do not form a natural sub-group; but that the orang is a highly developed gibbon or semnopithecus; the chimpanzee a highly developed macacus; and the gorilla a highly developed mandrill.* If this conclusion, which rests almost exclusively on brain-characters, be admitted, we should have a case of convergence at least in external characters, for the anthropomorphous apes are certainly more like each other in many points than they are to other apes. All analogical resemblances,* as of a whale to a fish, may indeed be said to be cases of convergence; but this term has never been applied to superficial and adaptive resemblances. It would be extremely rash in most cases to attribute to convergence close similarity in many points of structure in beings which had once been widely different. The form of a crystal is determined solely by the molecular forces, and it is not surprising that dissimilar substances should sometimes assume the same form; but with organic beings we should bear in mind that the form of each depends on an infinitude of complex relations, namely on the variations which have arisen, these being due to causes far too intricate to be followed out,—on the nature of the variations which have been preserved, and this depends on the surrounding physical conditions, and in a still higher degree on the surrounding organisms with which each has come into competition,—and lastly, on inheritance (in itself a fluctuating element) from innumerable progenitors, all of which have had their forms determined through equally complex relations. It appears utterly incredible that two organisms, if differing in a marked manner, should ever afterwards converge so closely as to lead to a near approach to identity throughout their whole organisation. In the case of the convergent pigs above referred to, evidence of their descent from two primitive stocks is still plainly retained, according to Von Nathusius, in certain bones of their skulls. If the races of man were descended, as supposed by some naturalists, from two or more distinct species, which had differed as much, or nearly as much, from each other, as the orang differs from the gorilla, it can hardly be doubted that marked differences in the structure of certain bones would still have been discoverable in man as he now exists.

Although the existing races of man differ in many respects, as in colour, hair, shape of skull, proportions of the body, &c., yet if their whole organisation be taken into consideration they are found to resemble each other closely in a multitude of points. Many of these points are of so unimportant or of so singular a nature, that it is extremely improbable that they should have been independently acquired by aboriginally distinct species or races. The same remark holds good with equal or greater force with respect to the numerous points of mental similarity between the most distinct races of man. The American aborigines, Negroes and Europeans differ as much from each other in mind as any three races that can be named; yet I was incessantly struck, whilst living with the Fuegians on board the "Beagle," with the many little traits of character, shewing how similar their minds were to ours; and so it was with a full-blooded negro* with whom I happened once to be intimate.

He who will carefully read Mr. Tylor's and Sir J. Lubbock's interesting works can hardly fail to be deeply impressed with the close similarity between the men of all races in tastes, dispositions and habits. This is shewn by the pleasure which they all take in dancing, rude music, acting, painting, tattooing, and otherwise decorating themselves,—in their mutual comprehension of gesture-language—and, as I shall be able to shew in a future essay, by the same expression in their features, and by the same inarticulate cries, when they are excited by various emotions. This similarity, or rather identity, is striking, when contrasted with the different expressions which may be observed in distinct species of monkeys. There is good evidence that the art of shooting with bows and arrows has not been handed down from any common progenitor of mankind, yet the stone arrow-heads, brought from the most distant parts of the world and manufactured at the most remote periods, are, as Nilsson has shewn, almost identical; and this fact can only be accounted for by the various races having similar inventive or mental powers. The same observation has been made by archæologists with respect to certain widely-prevalent ornaments, such as zigzags, &c.; and with respect to various simple beliefs and customs, such as the burying of the dead under megalithic structures. I remember observing in South America, that there, as in so many other parts of the world, man has generally chosen the summits of lofty hills, on which to throw up piles of stones, either for the sake of recording some remarkable event, or for burying his dead.

Now when naturalists observe a close agreement in numerous small details of habits, tastes and dispositions between two or more domestic races, or between nearly-allied natural forms, they use this fact as an argument that all are descended from a common progenitor who was thus endowed; and consequently that all should be classed under the same species. The same argument may be applied with much force to the races of man.

As it is improbable that the numerous and unimportant points of resemblance between the several races of man in bodily structure and mental faculties (I do not here refer to similar customs) should all have been independently acquired, they must have been inherited from progenitors who were thus characterised. We thus gain some insight into the early state of man, before he had spread step by step over the face of the earth. The spreading of man to regions widely separated by the sea, no doubt, preceded any considerable amount of divergence of character in the several races; for otherwise we should sometimes meet with the same race in distinct continents; and this is never the case. Sir J. Lubbock, after comparing the arts now prac- tised by savages in all parts of the world, specifies those which man could not have known, when he first wandered from his original birth-place; for if once learnt they would never have been forgotten. He thus shews that "the spear, which is but a development of the knife-point, and the club, which is but a long hammer, are the only things left." He admits, however, that the art of making fire probably had already been discovered, for it is common to all the races now existing, and was known to the ancient cave-inhabitants of Europe. Perhaps the art of making rude canoes or rafts was likewise known; but as man existed at a remote epoch, when the land in many places stood at a very different level, he would have been able, without the aid of canoes, to have spread widely. Sir J. Lubbock further remarks how improbable it is that our earliest ancestors could have "counted as high as ten, considering that so many races now in existence cannot get beyond four." Nevertheless, at this early period, the intellectual and social faculties of man could hardly have been inferior in any extreme degree to those now possessed by the lowest savages; otherwise primeval man could not have been so eminently successful in the struggle for life, as proved by his early and wide diffusion.

From the fundamental differences between certain languages, some philologists have inferred that when man first became widely diffused he was not a speaking animal; but it may be suspected that languages, far less perfect than any now spoken, aided by gestures, might have been used, and yet have left no traces on subsequent and more highly-developed tongues. Without the use of some language, however imperfect, it appears doubtful whether man's intellect could have risen to the standard implied by his dominant position at an early period.

Whether primeval man, when he possessed very few arts of the rudest kind, and when his power of language was extremely imperfect, would have deserved to be called man, must depend on the definition which we employ. In a series of forms graduating insensibly from some ape-like creature to man as he now exists, it would be impossible to fix on any definite point when the term "man" ought to be used. But this is a matter of very little importance. So again it is almost a matter of indifference whether the so-called races of man are thus designated, or are ranked as species or sub-species; but the latter term appears the most appropriate. Finally, we may conclude that when the principles of evolution are generally accepted, as they surely will be before long, the dispute between the monogenists and the polygenists will die a silent and unobserved death.

One other question ought not to be passed over without notice, namely, whether, as is sometimes assumed, each sub-species or race of man has sprung from a single pair of progenitors. With our domestic animals a new race can readily be formed from a single pair possessing some new character, or even from a single individual thus characterised, by carefully matching the varying offspring; but most of our races have been formed, not intentionally from a selected pair, but unconsciously by the preservation of many individuals which have varied, however slightly, in some useful or desired manner. If in one country stronger and heavier horses, and in another country lighter and fleeter horses, were habitually preferred, we may feel sure that two distinct sub-breeds would, in the course of time, be produced, without any particular pairs or individuals having been separated and bred from in either country. Many races have been thus formed, and their manner of formation is closely analogous with that of natural species. We know, also, that the horses which have been brought to the Falkland Islands have become, during successive generations, smaller and weaker, whilst those which have run wild on the Pampas have acquired larger and coarser heads; and such changes are manifestly due, not to any one pair, but to all the individuals having been subjected to the same conditions, aided, perhaps, by the principle of reversion.* The new sub-breeds in none of these cases are descended from any single pair, but from many individuals which have varied in different degrees, but in the same general manner; and we may conclude that the races of man have been similarly produced, the modifications being either the direct result of exposure to different conditions, or the indirect result of some form of selection. But to this latter subject we shall presently return.

On the Extinction of the Races of Man.—The partial and complete extinction of many races and sub-races of man are historically known events. Humboldt saw in South America a parrot which was the sole living creature that could speak the language of a lost tribe. Ancient monuments and stone implements found in all parts of the world, of which no tradition is preserved by the present inhabitants, indicate much extinction. Some small and broken tribes, remnants of former races, still survive in isolated and generally mountainous districts. In Europe the ancient races were all, according to Schaaffhausen, "lower in the scale than the rudest living savages;" they must therefore have differed, to a certain extent, from any existing race. The remains described by Professor Broca from Les Eyzies,* though they unfortunately appear to have belonged to a single family, indicate a race with a most singular combination of low or simious and high characteristics, and is "entirely different from any other race, ancient or modern, that we have ever heard of." It differed, therefore, from the quaternary race of the caverns of Belgium.

Unfavourable physical conditions appear to have had but little effect in the extinction of races. Man has long lived in the extreme regions of the North, with no wood wherewith to make his canoes or other implements, and with blubber alone for burning and giving him warmth, but more especially for melting the snow. In the Southern extremity of America the Fuegians survive without the protection of clothes, or of any building worthy to be called a hovel. In South Africa the aborigines wander over the most arid plains, where dangerous beasts abound. Man can withstand the deadly influence of the Terai* at the foot of the Himalaya, and the pestilential shores of tropical Africa.

Extinction follows chiefly from the competition of tribe with tribe, and race with race. Various checks are always in action, as specified in a former chapter, which serve to keep down the numbers of each savage tribe,—such as periodical famines, the wandering of the parents and the consequent deaths of infants, prolonged suckling, the stealing of women, wars, accidents, sickness, licentiousness, especially infanticide, and, perhaps, lessened fertility from less nutritious food, and many hardships. If from any cause any one of these checks is lessened, even in a slight degree, the tribe thus favoured will tend to increase; and when one of two adjoining tribes becomes more numerous and powerful than the other, the contest is soon settled by war, slaughter, cannibalism, slavery, and absorption. Even when a weaker tribe is not thus abruptly swept away, if it once begins to decrease, it generally goes on decreasing until it is extinct.

When civilised nations come into contact with barbarians the struggle is short, except where a deadly climate gives its aid to the native race. Of the causes which lead to the victory of civilised nations, some are plain and some very obscure. We can see that the cultivation of the land will be fatal in many ways to savages, for they cannot, or will not, change their habits. New diseases and vices are highly destructive; and it appears that in every nation a new disease causes much death, until those who are most susceptible to its destructive influence are gradually weeded out; and so it may be with the evil effects from spirituous liquors, as well as with the unconquerably strong taste for them shewn by so many savages. It further appears, mysterious as is the fact, that the first meeting of distinct and separated people generates disease. Mr. Sproat, who in Vancouver Island closely attended to the subject of extinction, believes that changed habits of life, which always follow from the advent of Europeans, induces much ill-health. He lays, also, great stress on so trifling a cause as that the natives become "bewildered and dull by the new life around them; they lose the motives for exertion, and get no new ones in their place."

The grade of civilisation seems a most important element in the success of nations which come in competition. A few centuries ago Europe feared the inroads of Eastern barbarians; now, any such fear would be ridiculous. It is a more curious fact, that savages did not formerly waste away, as Mr. Bagehot has remarked, before the classical nations, as they now do before modern civilised nations; had they done so, the old moralists would have mused over the event; but there is no lament in any writer of that period over the perishing barbarians.

Although the gradual decrease and final extinction of the races of man is an obscure problem, we can see that it depends on many causes, differing in different places and at different times. It is the same difficult problem as that presented by the extinction of one of the higher animals—of the fossil horse, for instance, which disappeared from South America, soon afterwards to be replaced, within the same districts, by countless troops of the Spanish horse. The New Zealander seems conscious of this parallelism, for he compares his future fate with that of the native rat almost exterminated by the European rat. The difficulty, though great to our imagination, and really great if we wish to ascertain the precise causes, ought not to be so to our reason, as long as we keep steadily in mind that the increase of each species and each race is constantly hindered by various checks; so that if any new check, or cause of destruction, even a slight one, be superadded, the race will surely decrease in number; and as it has everywhere been observed that savages are much opposed to any change of habits, by which means injurious checks could be counterbalanced, decreasing numbers will sooner or later lead to extinction; the end, in most cases, being promptly determined by the inroads of increasing and conquering tribes.

On the Formation of the Races of Man.—It may be premised that when we find the same race, though broken up into distinct tribes, ranging over a great area, as over America, we may attribute their general resemblance to descent from a common stock. In some cases the crossing of races already distinct has led to the formation of new races. The singular fact that Europeans and Hindoos, who belong to the same Aryan stock* and speak a language fundamentally the same, differ widely in appearance, whilst Europeans differ but little from Jews, who belong to the Semitic stock* and speak quite another language, has been accounted for by Broca through the Aryan branches having been largely crossed during their wide diffusion by various indigenous tribes. When two races in close contact cross, the first result is a heterogeneous mixture: thus Mr. Hunter, in describing the Santali or hill-tribes of India, says that hundreds of imperceptible gradations may be traced "from the black, squat tribes of the mountains to the tall olive-coloured Brahman, with his intellectual brow, calm eyes, and high but narrow head;" so that it is necessary in courts of justice to ask the witnesses whether they are Santalis or Hindoos. Whether a heterogeneous people, such as the inhabitants of some of the Polynesian islands, formed by the crossing of two distinct races, with few or no pure members left, would ever become homogeneous, is not known from direct evidence. But as with our domesticated animals, a crossed breed can certainly, in the course of a few generations, be fixed and made uniform by careful selection, we may infer that the free and prolonged intercrossing during many generations of a heterogeneous mixture would supply the place of selection, and overcome any tendency to reversion, so that a crossed race would ultimately become homogeneous, though it might not partake in an equal degree of the characters of the two parent-races.

Of all the differences between the races of man, the colour of the skin is the most conspicuous and one of the best marked. Differences of this kind, it was formerly thought, could be accounted for by long exposure under different climates; but Pallas first shewed that this view is not tenable, and he has been followed by almost all anthropologists. The view has been rejected chiefly because the distribution of the variously coloured races, most of whom must have long inhabited their present homes, does not coincide with corresponding differences of climate. Weight must also be given to such cases as that of the Dutch families, who, as we hear on excellent authority, have not undergone the least change of colour, after residing for three centuries in South Africa. The uniform appearance in various parts of the world of gypsies and Jews, though the uniformity of the latter has been somewhat exaggerated, is likewise an argument on the same side. A very damp or a very dry atmosphere has been supposed to be more influential in modifying the colour of the skin than mere heat; but as D'Orbigny in South America, and Livingstone in Africa, arrived at diametrically opposite conclusions with respect to dampness and dryness, any conclusion on this head must be considered as very doubtful.

Various facts, which I have elsewhere given, prove that the colour of the skin and hair is sometimes correlated in a surprising manner with a complete immunity from the action of certain vegetable poisons and from the attacks of certain parasites. Hence it occurred to me, that negroes and other dark races might have acquired their dark tints by the darker individuals escaping during a long series of generations from the deadly influence of the miasmas of their native countries.

I afterwards found that the same idea had long ago occurred to Dr. Wells. That negroes, and even mulattoes, are almost completely exempt from the yellow-fever, which is so destructive in tropical America, has long been known. They likewise escape to a large extent the fatal intermittent fevers that prevail along, at least, 2600 miles of the shores of Africa, and which annually cause one-fifth of the white settlers to die, and another fifth to return home invalided. This immunity in the negro seems to be partly inherent, depending on some unknown peculiarity of constitution, and partly the result of acclimatisation. Pouchet states that the negro regiments, borrowed from the Viceroy of Egypt for the Mexican war, which had been recruited near the Soudan, escaped the yellow-fever almost equally well with the negroes originally brought from various parts of Africa, and accustomed to the climate of the West Indies. That acclimatisation plays a part is shewn by the many cases in which negroes, after having resided for some time in a colder climate, have become to a certain extent liable to tropical fevers. The nature of the climate under which the white races have long resided, likewise has some influence on them; for during the fearful epidemic of yellow-fever in Demerara during 1837, Dr. Blair found that the death-rate of the immigrants was proportional to the latitude of the country whence they had come. With the negro the immunity, as far as it is the result of acclimatisation, implies exposure during a prodigious length of time; for the aborigines of tropical America, who have resided there from time immemorial, are not exempt from yellow-fever; and the Rev. B. Tristram states, that there are districts in Northern Africa which the native inhabitants are compelled annually to leave, though the negroes can remain with safety.

That the immunity of the negro is in any degree correlated with the colour of his skin is a mere conjecture: it may be correlated with some difference in his blood, nervous system, or other tissues. Nevertheless, from the facts above alluded to, and from some connection apparently existing between complexion and a tendency to consumption, the conjecture seemed to me not improbable. Consequently I endeavoured, with but little success, to ascertain how far it held good. The late Dr. Daniell, who had long lived on the West Coast of Africa, told me that he did not believe in any such relation. He was himself unusually fair, and had withstood the climate in a wonderful manner. When he first arrived as a boy on the coast, an old and experienced negro chief predicted from his appearance that this would prove the case. Dr. Nicholson, of Antigua, after having attended to this subject, wrote to me that he did not think that dark-coloured Europeans escaped the yellow-fever better than those that were light-coloured. Mr. J. M. Harris altogether denies that Europeans with dark hair withstand a hot climate better than other men; on the contrary, experience has taught him in making a selection of men for service on the coast of Africa, to choose those with red hair. As far, therefore, as these slight indications serve, there seems no foundation for the hypothesis, which has been accepted by several writers, that the colour of the black races may have resulted from darker and darker individuals having survived in greater numbers, during their exposure to the fever-generating miasmas of their native countries.

Although with our present knowledge we cannot account for the strongly-marked differences in colour between the races of man, either through correlation with constitutional peculiarities, or through the direct action of climate; yet we must not quite ignore the latter agency, for there is good reason to believe that some inherited effect is thus produced.

We have seen in our third chapter that the conditions of life, such as abundant food and general comfort, affect in a direct manner the development of the bodily frame, the effects being transmitted. Through the combined influences of climate and changed habits of life, European settlers in the United States undergo, as is generally admitted, a slight but extraordinarily rapid change of appearance. There is, also, a considerable body of evidence shewing that in the Southern States the house-slaves of the third generation present a markedly different appearance from the field-slaves.

If, however, we look to the races of man, as distributed over the world, we must infer that their characteristic differences cannot be accounted for by the direct action of different conditions of life, even after exposure to them for an enormous period of time. The Esquimaux live exclusively on animal food; they are clothed in thick fur, and are exposed to intense cold and to prolonged darkness; yet they do not differ in any extreme degree from the inhabitants of Southern China, who live entirely on vegetable food and are exposed almost naked to a hot, glaring climate. The unclothed Fuegians live on the marine productions of their inhospitable shores; the Botocudos of Brazil wander about the hot forests of the interior and live chiefly on vegetable productions; yet these tribes resemble each other so closely that the Fuegians on board the "Beagle" were mistaken by some Brazilians for Botocudos. The Botocudos again, as well as the other inhabitants of tropical America, are wholly different from the Negroes who inhabit the opposite shores of the Atlantic, are exposed to a nearly similar climate, and follow nearly the same habits of life.

Nor can the differences between the races of man be accounted for, except to a quite insignificant degree, by the inherited effects of the increased or decreased use of parts. Men who habitually live in canoes, may have their legs somewhat stunted; those who inhabit lofty regions have their chests enlarged; and those who constantly use certain sense-organs have the cavities in which they are lodged somewhat increased in size, and their features consequently a little modified. With civilised nations, the reduced size of the jaws from lessened use, the habitual play of different muscles serving to express different emotions, and the increased size of the brain from greater intellectual activity, have together produced a considerable effect on their general appearance in comparison with savages. It is also possible that increased bodily stature, with no corresponding increase in the size of the brain, may have given to some races (judging from the previously adduced cases of the rabbits) an elongated skull of the dolichocephalic type.*

Lastly, the little-understood principle of correlation will almost certainly have come into action, as in the case of great muscular development and strongly projecting supra-orbital ridges. It is not improbable that the texture of the hair, which differs much in the different races, may stand in some kind of correlation with the structure of the skin; for the colour of the hair and skin are certainly correlated, as is its colour and texture with the Mandans. The colour of the skin and the odour emitted by it are likewise in some manner connected. With the breeds of sheep the number of hairs within a given space and the number of the excretory pores stand in some relation to each other. If we may judge from the analogy of our domesticated animals, many modifications of structure in man probably come under this principle of correlated growth.

We have now seen that the characteristic differences between the races of man cannot be accounted for in a satisfactory manner by the direct action of the conditions of life, nor by the effects of the continued use of parts, nor through the principle of correlation. We are therefore led to inquire whether slight individual differences, to which man is eminently liable, may not have been preserved and augmented during a long series of generations through natural selection. But here we are at once met by the objection that beneficial variations alone can be thus preserved; and as far as we are enabled to judge (although always liable to error on this head) not one of the external differences between the races of man are of any direct or special service to him. The intellectual and moral or social faculties must of course be excepted from this remark; but differences in these faculties can have had little or no influence on external characters. The variability of all the characteristic differences between the races, before referred to, likewise indicates that these differences cannot be of much importance; for, had they been important, they would long ago have been either fixed and preserved, or eliminated. In this respect man resembles those forms, called by naturalists protean or polymorphic, which have remained extremely variable, owing, as it seems, to their variations being of an indifferent nature, and consequently to their having escaped the action of natural selection.

We have thus far been baffled in all our attempts to account for the differences between the races of man; but there remains one important agency, namely Sexual Selection, which appears to have acted as powerfully on man, as on many other animals. I do not intend to assert that sexual selection will account for all the differences between the races. An unexplained residuum is left, about which we can in our ignorance only say, that as individuals are continually born with, for instance, heads a little rounder or narrower, and with noses a little longer or shorter, such slight differences might become fixed and uniform, if the unknown agencies which induced them were to act in a more constant manner, aided by long-continued intercrossing. Such modifications come under the provisional class, alluded to in our fourth chapter, which for the want of a better term have been called spontaneous variations. Nor do I pretend that the effects of sexual selection can be indicated with scientific precision; but it can be shewn that it would be an inexplicable fact if man had not been modified by this agency, which has acted so powerfully on innumerable animals, both high and low in the scale. It can further be shewn that the differences between the races of man, as in colour, hairyness, form of features, &c., are of the nature which it might have been expected would have been acted on by sexual selection. But in order to treat this subject in a fitting manner, I have found it necessary to pass the whole animal kingdom in review; I have therefore devoted to it the Second Part of this work. At the close I shall return to man, and, after attempting to shew how far he has been modified through sexual selection, will give a brief summary of the chapters in this First Part.

PART II.—SEXUAL SELECTION

CHAPTER VIII

Principles of Sexual Selection

WITH animals which have their sexes separated, the males necessarily differ from the females in their organs of reproduction; and these afford the primary sexual characters. But the sexes often differ in what Hunter has called secondary sexual characters, which are not directly connected with the act of reproduction; for instance, in the male possessing certain organs of sense or locomotion, of which the female is quite destitute, or in having them more highly-developed, in order that he may readily find or reach her; or again, in the male having special organs of prehension so as to hold her securely. These latter organs of infinitely diversified kinds graduate into, and in some cases can hardly be distinguished from, those which are commonly ranked as primary, such as the complex appendages at the apex of the abdomen in male insects. Unless indeed we confine the term "primary" to the reproductive glands, it is scarcely possible to decide, as far as the organs of prehension are concerned, which ought to be called primary and which secondary.

The female often differs from the male in having organs for the nourishment or protection of her young, as the mammary glands of mammals, and the abdominal sacks of the marsupials. The male, also, in some few cases differs from the female in possessing analogous organs, as the receptacles for the ova possessed by the males of certain fishes, and those temporarily developed in certain male frogs. Female bees have a special apparatus for collecting and carrying pollen, and their ovipositor* is modified into a sting for the defence of their larvæ and the community. In the females of many insects the ovipositor is modified in the most complex manner for the safe placing of the eggs. Numerous similar cases could be given, but they do not here concern us. There are, however, other sexual differences quite disconnected with the primary organs with which we are more especially concerned—such as the greater size, strength, and pugnacity of the male, his weapons of offence or means of defence against rivals, his gaudy colouring and various ornaments, his power of song, and other such characters.

Besides the foregoing primary and secondary sexual differences, the male and female sometimes differ in structures connected with different habits of life, and not at all, or only indirectly, related to the reproductive functions. Thus the females of certain flies (Culicidæ and Tabanidæ) are blood-suckers, whilst the males live on flowers and have their mouths destitute of mandibles.* The males alone of certain moths and of some crustaceans (*e.g.* Tanais) have imperfect, closed mouths, and cannot feed. The Complemental males of certain cirripedes* live like epiphytic* plants either on the female or hermaphrodite form, and are destitute of a mouth and prehensile* limbs. In these cases it is the male which has been modified and has lost certain important organs, which the other members of the same group possess. In other cases it is the female which has lost such parts; for instance, the female glowworm is destitute of wings, as are many female moths, some of which never leave their cocoons. Many female parasitic crustaceans have lost their natatory* legs. In some weevil-beetles (Curculionidæ) there is a great difference between the male and female in the length of the rostrum or snout; but the meaning of this and of many analogous differences, is not at all understood. Differences of structure between the two sexes in relation to different habits of life are generally confined to the lower animals; but with some few birds the beak of the male differs from that of the female. No doubt in most, but apparently not in all these cases, the differences are indirectly connected with the propagation of the species: thus a female which has to nourish a multitude of ova will require more food than the male, and consequently will require special means for procuring it. A male animal which lived for a very short time might without detriment lose through disuse its organs for procuring food; but he would retain his locomotive organs in a perfect state, so that he might reach the female. The female, on the other hand, might safely lose her organs for flying, swimming, or walking, if she gradually acquired habits which rendered such powers useless.

We are, however, here concerned only with that kind of selection, which I have called sexual selection. This depends on the advantage which certain individuals have over other individuals of the same sex and species, in exclusive relation to reproduction. When the two sexes differ in structure in relation to different habits of life, as in the cases above mentioned, they have no doubt been modified through natural selection, accompanied by inheritance limited to one and the same sex. So again the primary sexual organs, and those for nourishing or protecting the young, come under this same head; for those individuals which generated or nourished their offspring best, would leave, *cæteris paribus*,* the greatest number to inherit their superiority; whilst those which generated or nourished their offspring badly, would leave but few to inherit their weaker powers. As the male has to search for the female, he requires for this purpose organs of sense and locomotion, but if these organs are necessary for the other purposes of life, as is generally the case, they will have been developed through natural selection. When the male has found the female he sometimes absolutely requires prehensile organs to hold her; thus Dr. Wallace informs me that the males of certain moths cannot unite with the females if their tarsi* or feet are broken. The males of many oceanic crustaceans have their legs and antennæ modified in an extraordinary manner for the prehension* of the female; hence we may suspect that owing to these animals being washed about by the waves of the open sea, they absolutely require these organs in order to propagate their kind, and if so their development will have been the result of ordinary or natural selection.

When the two sexes follow exactly the same habits of life, and the male has more highly developed sense or locomotive organs than the female, it may be that these in their perfected state are indispensable to the male for finding the female; but in the vast majority of cases, they serve only to give one male an advantage over another, for the less well-endowed males, if time were allowed them, would succeed in pairing with the females; and they would in all other respects, judging from the structure of the female, be equally well adapted for their ordinary habits of life. In such cases sexual selection must have come into action, for the males have acquired their present structure, not from being better fitted to survive in the struggle for existence, but from having gained an advantage over other males, and from having transmitted this advantage to their male offspring alone. It was the importance of this distinction which led me to designate this form of selection as sexual selection. So again, if the chief service rendered to the male by his prehensile organs is to prevent the escape of the female before the arrival of other males, or when assaulted by them, these organs will have been perfected through sexual selection, that is by the advantage acquired by certain males over their rivals. But in most cases it is scarcely possible to distinguish between the effects of natural and sexual selection. Whole chapters could easily be filled with details on the differences between the sexes in their sensory, locomotive, and prehensile organs. As, however, these structures are not more interesting than others adapted for the ordinary purposes of life, I shall almost pass them over, giving only a few instances under each class.

There are many other structures and instincts which must have been developed through sexual selection—such as the weapons of offence and the means of defence possessed by the males for fighting with and driving away their rivals—their courage and pugnacity—their ornaments of many kinds—their organs for producing vocal or instrumental music—and their glands for emitting odours; most of these latter structures serving only to allure or excite the female.

That these characters are the result of sexual and not of ordinary selection is clear, as unarmed, unornamented, or unattractive males would succeed equally well in the battle for life and in leaving a numerous progeny, if better endowed males were not present. We may infer that this would be the case, for the females, which are unarmed and unornamented, are able to survive and procreate their kind. Secondary sexual characters of the kind just referred to, will be fully discussed in the following chapters, as they are in many respects interesting, but more especially as they depend on the will, choice, and rivalry of the individuals of either sex. When we behold two males fighting for the possession of the female, or several male birds displaying their gorgeous plumage, and performing the strangest antics before an assembled body of females, we cannot doubt that, though led by instinct, they know what they are about, and consciously exert their mental and bodily powers.

In the same manner as man can improve the breed of his game-cocks by the selection of those birds which are victorious in the cockpit, so it appears that the strongest and most vigorous males, or those provided with the best weapons, have prevailed under nature, and have led to the improvement of the natural breed or species. Through repeated deadly contests, a slight degree of variability, if it led to some advantage, however slight, would suffice for the work of sexual selection; and it is certain that secondary sexual characters are eminently variable. In the same manner as man can give beauty, according to his standard of taste, to his male poultry—can give to the Sebright bantam a new and elegant plumage, an erect and peculiar carriage—so it appears that in a state of nature female birds, by having long selected the more attractive males, have added to their beauty. No doubt this implies powers of discrimination and taste on the part of the female which will at first appear extremely improbable; but I hope hereafter to shew that this is not the case.

From our ignorance on several points, the precise manner in which sexual selection acts is to a certain extent uncertain. Nevertheless if those naturalists who already believe in the mutability of species, will read the following chapters, they will, I think, agree with me that sexual selection has played an important part in the history of the organic world. It is certain that with almost all animals there is a struggle between the males for the possession of the female. This fact is so notorious that it would be superfluous to give instances.

Hence the females, supposing that their mental capacity sufficed for the exertion of a choice, could select one out of several males. But in numerous cases it appears as if it had been specially arranged that there should be a struggle between many males. Thus with migratory birds, the males generally arrive before the females at their place of breeding, so that many males are ready to contend for each female. The bird-catchers assert that this is invariably the case with the nightingale and blackcap,* as I am informed by Mr. Jenner Weir, who confirms the statement with respect to the latter species.

Mr. Swaysland of Brighton, who has been in the habit, during the last forty years, of catching our migratory birds on their first arrival, writes to me that he has never known the females of any species to arrive before their males. During one spring he shot thirty-nine males of Ray's wagtail (*Budytes Raii*) before he saw a single female. Mr. Gould has ascertained by dissection, as he informs me, that male snipes arrive in this country before the females. In the case of fish, at the period when the salmon ascend our rivers, the males in large numbers are ready to breed before the females. So it apparently is with frogs and toads. Throughout the great class of insects the males almost always emerge from the pupal state before the other sex, so that they generally swarm for a time before any females can be seen. The cause of this difference between the males and females in their periods of arrival and maturity is sufficiently obvious. Those males which annually first migrated into any country, or which in the spring were first ready to breed, or were the most eager, would leave the largest number of offspring; and these would tend to inherit similar instincts and constitutions. On the whole there can be no doubt that with almost all animals, in which the sexes are separate, there is a constantly recurrent struggle between the males for the possession of the females.

Our difficulty in regard to sexual selection lies in understanding how it is that the males which conquer other males, or those which prove the most attractive to the females, leave a greater number of offspring to inherit their superiority than the beaten and less attractive males. Unless this result followed, the characters which gave to certain males an advantage over others, could not be perfected and augmented through sexual selection. When the sexes exist in exactly equal numbers, the worst-endowed males will ultimately find females (excepting where polygamy prevails), and leave as many offspring, equally well fitted for their general habits of life, as the best-endowed males. From various facts and considerations, I formerly inferred that with most animals, in which secondary sexual characters were well developed, the males considerably exceeded the females in number; and this does hold good in some few cases. If the males were to the females as two to one, or as three to two, or even in a somewhat lower ratio, the whole affair would be simple; for the better-armed or more attractive males would leave the largest number of offspring. But after investigating, as far as possible, the numerical proportions of the sexes, I do not believe that any great inequality in number commonly exists. In most cases sexual selection appears to have been effective in the following manner.

Let us take any species, a bird for instance, and divide the females inhabiting a district into two equal bodies: the one consisting of the more vigorous and better-nourished individuals, and the other of the less vigorous and healthy. The former, there can be little doubt, would be ready to breed in the spring before the others; and this is the opinion of Mr. Jenner Weir, who has during many years carefully attended to the habits of birds. There can also be no doubt that the most vigorous, healthy, and best-nourished females would on an average succeed in rearing the largest number of offspring. The males, as we have seen, are generally ready to breed before the females; of the males the strongest, and with some species the best armed, drive away the weaker males; and the former would then unite with the more vigorous and best-nourished females, as these are the first to breed. Such vigorous pairs would surely rear a larger number of offspring than the retarded females, which would be compelled, supposing the sexes to be numerically equal, to unite with the con-quered and less powerful males; and this is all that is wanted to add, in the course of successive generations, to the size, strength and courage of the males, or to improve their weapons.

But in a multitude of cases the males which conquer other males, do not obtain possession of the females, independently of choice on the part of the latter. The courtship of animals is by no means so simple and short an affair as might be thought. The females are most excited by, or prefer pairing with, the more ornamented males, or those which are the best songsters, or play the best antics; but it is obvi-ously probable, as has been actually observed in some cases, that they would at the same time prefer the more vigorous and lively males.

Thus the more vigorous females, which are the first to breed, will have the choice of many males; and though they may not always select the strongest or best armed, they will select those which are vigorous and well armed, and in other respects the most attractive. Such early pairs would have the same advantage in rearing offspring on the female side as above explained, and nearly the same advantage on the male side. And this apparently has sufficed during a long course of generations to add not only to the strength and fighting-powers of the males, but likewise to their various ornaments or other attractions.

In the converse and much rarer case of the males selecting particular females, it is plain that those which were the most vigorous and had conquered others, would have the freest choice; and it is almost certain that they would select vigorous as well as attractive females. Such pairs would have an advantage in rearing offspring, more especially if the male had the power to defend the female during the pairing-season, as occurs with some of the higher animals, or aided in providing for the young. The same principles would apply if both sexes mutually preferred and selected certain individuals of the opposite sex; supposing that they selected not only the more attractive, but likewise the more vigorous individuals.

[Darwin had expected that in those species in which the action of sexual selection was most evident, males would greatly outnumber females. His researches, however, suggested that the proportions were nearly equal: thus not only do males compete, but females choose, which is why in most cases the males possess much more marked characters.]

Sexual selection acts in a less rigorous manner than natural selection. The latter produces its effects by the life or death at all ages of the more or less successful individuals. Death, indeed, not rarely ensues from the conflicts of rival males. But generally the less successful male merely fails to obtain a female, or obtains later in the season a retarded and less vigorous female, or, if polygamous, obtains fewer females; so that they leave fewer, or less vigorous, or no offspring. In regard to structures acquired through ordinary or natural selection, there is in most cases, as long as the conditions of life remain the same, a limit to the amount of advantageous modification in relation to certain special ends; but in regard to structures adapted to make one male victorious over another, either in fighting or in charming the female, there is no definite limit to the amount of advantageous modification; so that as long as the proper variations arise the work of sexual selection will go on. This circumstance may partly account for the frequent and extraordinary amount of variability presented by secondary sexual characters. Nevertheless, natural selection will determine that characters of this kind shall not be acquired by the victorious males, which would be injurious to them in any high degree, either by expending too much of their vital powers, or by exposing them to any great danger. The development, however, of certain structures—of the horns, for instance, in certain stags—has been carried to a wonderful extreme; and in some instances to an extreme which, as far as the general conditions of life are concerned, must be slightly injurious to the male. From this fact we learn that the advantages which favoured males have derived from conquering other males in battle or courtship, and thus leaving a numerous progeny, have been in the long run greater than those derived from rather more perfect adaptation to the external conditions of life. We shall further see, and this could never have been anticipated, that the power to charm the female has been in some few instances more important than the power to conquer other males in battle.

Laws of Inheritance

In order to understand how sexual selection has acted, and in the course of ages has produced conspicuous results with many animals of many classes, it is necessary to bear in mind the laws of inheritance, as far as they are known. Two distinct elements are included under the term "inheritance," namely the transmission and the development of characters; but as these generally go together, the distinction is often over-looked. We see this distinction in those characters which are transmitted through the early years of life, but are developed only at maturity or during old age. We see the same distinction more clearly with secondary sexual characters, for these are transmitted through both sexes, though developed in one alone. That they are present in both sexes, is manifest when two species, having strongly-marked sexual characters, are crossed, for each transmits the characters proper to its own male and female sex to the hybrid offspring of both sexes. The same fact is likewise manifest, when characters proper to the male are occasionally developed in the female when she grows old or becomes diseased; and so conversely with the male. Again, characters occasionally appear, as if transferred from the male to the female, as when, in certain breeds of the fowl, spurs* regularly appear in the young and healthy females; but in truth they are simply developed in the female; for in every breed each detail in the structure of the spur is transmitted through the female to her male offspring. In all cases of reversion, characters are transmitted through two, three, or many generations, and are then under certain unknown favourable conditions developed. This important distinction between transmission and development will be easiest kept in mind by the aid of the hypothesis of pangenesis,* whether or not it be accepted as true. According to this hypothesis, every unit or cell of the body throws off gemmules or undeveloped atoms, which are transmitted to the offspring of both sexes, and are multiplied by self-division. They may remain undeveloped during the early years of life or during successive generations; their development into units or cells, like those from which they were derived, depending on their affinity for, and union with, other units or cells previously developed in the due order of growth.

[*The rest of the chapter shows how these laws apply to the case of sexual selection.*]

Summary and concluding remarks.—From the foregoing discussion on the various laws of inheritance, we learn that characters often or even generally tend to become developed in the same sex, at the same age, and periodically at the same season of the year, in which they first appeared in the parents. But these laws, from unknown causes, are very liable to change. Hence the successive steps in the modification of a species might readily be transmitted in different ways; some of the steps being transmitted to one sex, and some to both; some to the offspring at one age, and some at all ages. Not only are the laws of inheritance extremely complex, but so are the causes which induce and govern variability. The variations thus caused are preserved and accumulated by sexual selection, which is in itself an extremely complex affair, depending, as it does, on ardour in love, courage, and the rivalry of the males, and on the powers of perception, taste, and will of the female. Sexual selection will also be dominated by natural selection for the general welfare of the species. Hence the manner in which the individuals of either sex or of both sexes are affected through sexual selection cannot fail to be complex in the highest degree.

When variations occur late in life in one sex, and are transmitted to the same sex at the same age, the other sex and the young are necessarily left unmodified. When they occur late in life, but are transmitted to both sexes at the same age, the young alone are left unmodified. Variations, however, may occur at any period of life in one sex or in both, and be transmitted to both sexes at all ages, and then all the individuals of the species will be similarly modified. In the following chapters it will be seen that all these cases frequently occur under nature.

Sexual selection can never act on any animal whilst young, before the age for reproduction has arrived. From the great eagerness of the male it has generally acted on this sex and not on the females. The males have thus become provided with weapons for fighting with their rivals, or with organs for discovering and securely holding the female, or for exciting and charming her. When the sexes differ in these respects, it is also, as we have seen, an extremely general law that the adult male differs more or less from the young male; and we may conclude from this fact that the successive variations, by which the adult male became modified, cannot have occurred much before the age for reproduction. How then are we to account for this general and remarkable coincidence between the period of variability and that of sexual selection,—principles which are quite independent of each other? I think we can see the cause: it is not that the males have never varied at an early age, but that such variations have commonly been lost, whilst those occurring at a later age have been preserved.

All animals produce more offspring than can survive to maturity; and we have every reason to believe that death falls heavily on the weak and inexperienced young. If then a certain proportion of the offspring were to vary at birth or soon afterwards, in some manner which at this age was of no service to them, the chance of the preservation of such variations would be small. We have good evidence under domestication how soon variations of all kinds are lost, if not selected. But variations which occurred at or near maturity, and which were of immediate service to either sex, would probably be preserved; as would similar variations occurring at an earlier period in any individuals which happened to survive. As this principle has an important bearing on sexual selection, it may be advisable to give an imaginary illustration. We will take a pair of animals, neither very fertile nor the reverse, and assume that after arriving at maturity they live on an average for five years, producing each year five young. They would thus produce 25 offspring; and it would not, I think, be an unfair estimate to assume that 18 or 20 out of the 25 would perish before maturity, whilst still young and inexperienced; the remaining seven or five sufficing to keep up the stock of mature individuals. If so, we can see that variations which occurred during youth, for instance in brightness, and which were not of the least service to the young, would run a good chance of being utterly lost. Whilst similar variations, which occurring at or near maturity in the comparatively few individuals surviving to this age, and which immediately gave an advantage to certain males, by rendering them more attractive to the females, would be likely to be preserved. No doubt some of the variations in brightness which occurred at an earlier age would by chance be preserved, and eventually give to the male the same advantage as those which appeared later; and this will account for the young males commonly partaking to a certain extent (as may be observed with many birds) of the bright colours of their adult male parents. If only a few of the successive variations in brightness were to occur at a late age, the adult male would be only a little brighter than the young male; and such cases are common.

In this illustration I have assumed that the young varied in a manner which was of no service to them; but many characters proper to the adult male would be actually injurious to the young,—as bright colours from making them conspicuous, or horns of large size from expending much vital force. Such variations in the young would promptly be eliminated through natural selection. With the adult and experienced males, on the other hand, the advantage thus derived in their rivalry with other males would often more than counter-balance exposure to some degree of danger. Thus we can understand how it is that variations which must originally have appeared rather late in life have alone or in chief part been preserved for the development of secondary sexual characters; and the remarkable coincidence between the periods of variability and of sexual selection is intelligible.

As variations which give to the male an advantage in fighting with other males, or in finding, securing, or charming the female, would be of no use to the female, they will not have been preserved in this sex either during youth or maturity. Consequently such variations would be extremely liable to be lost; and the female, as far as these characters are concerned, would be left unmodified, excepting in so far as she may have received them by transference from the male. No doubt if the female varied and transferred serviceable characters to her male offspring, these would be favoured through sexual selection; and then both sexes would thus far be modified in the same manner. But I shall hereafter have to recur to these more intricate contingencies.

In the following chapters, I shall treat of the secondary sexual characters in animals of all classes, and shall endeavour in each case to apply the principles explained in the present chapter. The lowest classes will detain us for a very short time, but the higher animals, especially birds, must be treated at considerable length. It should be borne in mind that for reasons already assigned, I intend to give only a few illustrative instances of the innumerable structures by the aid of which the male finds the female, or, when found, holds her. On the other hand, all structures and instincts by which the male conquers other males, and by which he allures or excites the female, will be fully discussed, as these are in many ways the most interesting.

[*Chapters IX–XVIII catalogue the differences between the sexes in the lower animals, from insects, molluscs, birds, and mammals to the non-human primates, or quadrumana.*]

CHAPTER XIX

Secondary Sexual Characters of Man

WITH mankind the differences between the sexes are greater than in most species of Quadrumana, but not so great as in some, for instance, the mandrill. Man on an average is considerably taller, heavier, and stronger than woman, with squarer shoulders and more plainly-pronounced muscles. Owing to the relation which exists between muscular development and the projection of the brows, the superciliary ridge* is generally more strongly marked in man than in woman. His body, and especially his face, is more hairy, and his voice has a different and more powerful tone. In certain tribes the women are said, whether truly I know not, to differ slightly in tint from the men; and with Europeans, the women are perhaps the more brightly coloured of the two, as may be seen when both sexes have been equally exposed to the weather.

Man is more courageous, pugnacious, and energetic than woman, and has a more inventive genius. His brain is absolutely larger, but whether relatively to the larger size of his body, in comparison with that of woman, has not, I believe been fully ascertained. In woman the face is rounder; the jaws and the base of the skull smaller; the outlines of her body rounder, in parts more prominent; and her pelvis is broader than in man; but this latter character may perhaps be considered rather as a primary than a secondary sexual character. She comes to maturity at an earlier age than man.

As with animals of all classes, so with man, the distinctive characters of the male sex are not fully developed until he is nearly mature; and if emasculated they never appear. The beard, for instance, is a secondary sexual character, and male children are beardless, though at an early age they have abundant hair on their heads. It is probably due to the rather late appearance in life of the successive variations, by which man acquired his masculine characters, that they are transmitted to the male sex alone. Male and female children resemble each other closely, like the young of so many other animals in which the adult sexes differ; they likewise resemble the mature female much more closely, than the mature male. The female, however, ultimately assumes certain distinctive characters, and in the formation of her skull, is said to be intermediate between the child and the man. Again, as the young of closely allied though distinct species do not differ nearly so much from each other as do the adults, so it is with the children of the different races of man. Some have even maintained that race-differences cannot be detected in the infantile skull. In regard to colour, the new-born negro child is reddish nut-brown, which soon becomes slaty-grey; the black colour being fully developed within a year in the Sudan, but not until three years in Egypt. The eyes of the negro are at first blue, and the hair chesnut-brown rather than black, being curled only at the ends. The children of the Australians immediately after birth are yellowish-brown, and become dark at a later age. Those of the Guaranys of Paraguay* are whitish-yellow, but they acquire in the course of a few weeks the yellowish-brown tint of their parents. Similar observations have been made in other parts of America.

I have specified the foregoing familiar differences between the male and female sex in mankind, because they are curiously the same as in the Quadrumana. With these animals the female is mature at an earlier age than the male; at least this is certainly the case with the *Cebus azaræ*.* With most of the species the males are larger and much stronger than the females, of which fact the gorilla offers a well-known instance. Even in so trifling a character as the greater prominence of the superciliary ridge, the males of certain monkeys differ from the females, and agree in this respect with man-kind. In the gorilla and certain other monkeys, the cranium of the adult male presents a strongly-marked sagittal crest,* which is absent in the female; and Ecker found a trace of a similar difference between the two sexes in the Australians. With monkeys when there is any difference in the voice, that of the male is the more powerful. We have seen that certain male monkeys, have a well-developed beard, which is quite deficient, or much less developed in the female. No instance is known of the beard, whiskers, or moustache being larger in a female than in the male monkey. Even in the colour of the beard there is a curious parallelism between man and the Quadrumana, for when in man the beard differs in colour from the hair of the head, as is often the case, it is, I believe, invariably of a lighter tint, being often reddish. I have observed this fact in England, and Dr. Hooker, who attended to this little point for me in Russia, found no exception to the rule. In Calcutta, Mr. J. Scott, of the Botanic Gardens, was so kind as to observe with care the many races of men to be seen there, as well as in some other parts of India, namely, two races in Sikhim, the Bhoteas, Hindoos, Burmese, and Chinese. Although most of these races have very little hair on the face, yet he always found that when there was any difference in colour between the hair of the head and the beard, the latter was invariably of a lighter tint. Now with monkeys, as has already been stated, the beard frequently differs in a striking manner in colour from the hair of the head, and in such cases it is invariably of a lighter hue, being often pure white, sometimes yellow or reddish.

In regard to the general hairyness of the body, the women in all races are less hairy than the men, and in some few Quadrumana the under side of the body of the female is less hairy than that of the male. Lastly, male monkeys, like men, are bolder and fiercer than the females. They lead the troop, and when there is danger, come to the front. We thus see how close is the parallelism between the sexual differences of man and the Quadrumana. With some few species, however, as with certain baboons, the gorilla and orang, there is a considerably greater difference between the sexes, in the size of the canine teeth, in the development and colour of the hair, and especially in the colour of the naked parts of the skin, than in the case of mankind.

[*The development of the beard and other bodily hair is typical of the variability of secondary sexual characters between different races. Some characters, however, are virtually universal: the most significant is the 'law of battle' in which men compete for women.*]

There can be little doubt that the greater size and strength of man, in comparison with woman, together with his broader shoulders, more developed muscles, rugged outline of body, his greater courage and pugnacity, are all due in chief part to inheritance from some early male progenitor, who, like the existing anthropoid apes, was thus characterised. These characters will, however, have been preserved or even augmented during the long ages whilst man was still in a barbarous condition, by the strongest and boldest men having succeeded best in the general struggle for life, as well as in securing wives, and thus having left a large number of offspring. It is not probable that the greater strength of man was primarily acquired through the inherited effects of his having worked harder than woman for his own subsistence and that of his family; for the women in all barbarous nations are compelled to work at least as hard as the men. With civilised people the arbitrament of battle for the possession of the women has long ceased; on the other hand, the men, as a general rule, have to work harder than the women for their mutual subsistence; and thus their greater strength will have been kept up.

Difference in the Mental Powers of the two Sexes.—With respect to differences of this nature between man and woman, it is probable that sexual selection has played a very important part. I am aware that some writers doubt whether there is any inherent difference; but this is at least probable from the analogy of the lower animals which present other secondary sexual characters. No one will dispute that the bull differs in disposition from the cow, the wild-boar from the sow, the stallion from the mare, and, as is well known to the keepers of menageries, the males of the larger apes from the females. Woman seems to differ from man in mental disposition, chiefly in her greater tenderness and less selfishness; and this holds good even with savages, as shewn by a well-known passage in Mungo Park's

Travels,* and by statements made by many other travellers. Woman, owing to her maternal instincts, displays these qualities towards her infants in an eminent degree; therefore it is likely that she should often extend them towards her fellow-creatures. Man is the rival of other men; he delights in competition, and this leads to ambition which passes too easily into selfishness. These latter qualities seem to be his natural and unfortunate birthright. It is generally admitted that with woman the powers of intuition, of rapid perception, and perhaps of imitation, are more strongly marked than in man; but some, at least, of these faculties are characteristic of the lower races, and therefore of a past and lower state of civilisation.

The chief distinction in the intellectual powers of the two sexes is shewn by man attaining to a higher eminence, in whatever he takes up, than woman can attain—whether requiring deep thought, reason, or imagination, or merely the use of the senses and hands. If two lists were made of the most eminent men and women in poetry, painting, sculpture, music,—comprising composition and performance, history, science, and philosophy, with half-a-dozen names under each subject, the two lists would not bear comparison. We may also infer, from the law of the deviation of averages, so well illustrated by Mr. Galton, in his work on 'Hereditary Genius,' that if men are capable of decided eminence over women in many subjects, the average standard of mental power in man must be above that of woman.

The half-human male progenitors of man, and men in a savage state, have struggled together during many generations for the possession of the females. But mere bodily strength and size would do little for victory, unless associated with courage, perseverance, and determined energy. With social animals, the young males have to pass through many a contest before they win a female, and the older males have to retain their females by renewed battles. They have, also, in the case of man, to defend their females, as well as their young, from enemies of all kinds, and to hunt for their joint subsistence. But to avoid enemies, or to attack them with success, to capture wild animals, and to invent and fashion weapons, requires the aid of the higher mental faculties, namely, observation, reason, invention, or imagination. These various faculties will thus have been continually put to the test, and selected during manhood; they will, moreover, have been strengthened by use during this same period of life. Consequently, in accordance with the principle often alluded to, we might expect that they would at least tend to be transmitted chiefly to the male offspring at the corresponding period of manhood.

Now, when two men are put into competition, or a man with a woman, who possess every mental quality in the same perfection, with the exception that the one has higher energy, perseverance, and courage, this one will generally become more eminent, whatever the object may be, and will gain the victory.* He may be said to possess genius—for genius has been declared by a great authority to be patience; and patience, in this sense, means unflinching, undaunted perseverance. But this view of genius is perhaps deficient; for without the higher powers of the imagination and reason, no eminent success in many subjects can be gained. But these latter as well as the former faculties will have been developed in man, partly through sexual selection,—that is, through the contest of rival males, and partly through natural selection,—that is, from success in the general struggle for life; and as in both cases the struggle will have been during maturity, the characters thus gained will have been transmitted more fully to the male than to the female offspring. Thus man has ultimately become superior to woman. It is, indeed, fortunate that the law of the equal transmission of characters to both sexes has commonly prevailed throughout the whole class of mammals; otherwise it is probable that man would have become as superior in mental endowment to woman, as the peacock is in ornamental plumage to the peahen.

It must be borne in mind that the tendency in characters acquired at a late period of life by either sex, to be transmitted to the same sex at the same age, and of characters acquired at an early age to be transmitted to both sexes, are rules which, though general, do not always hold good. If they always held good, we might conclude (but I am here wandering beyond my proper bounds) that the inherited effects of the early education of boys and girls would be transmitted equally to both sexes; so that the present inequality between the sexes in mental power could not be effaced by a similar course of early training; nor can it have been caused by their dissimilar early training. In order that woman should reach the same standard as man, she ought, when nearly adult, to be trained to energy and perseverance, and to have her reason and imagination exercised to the highest point; and then she would probably transmit these qualities chiefly to her adult daughters. The whole body of women, however, could not be thus raised, unless during many generations the women who excelled in the above robust virtues were married, and produced offspring in larger numbers than other women. As before remarked with respect to bodily strength, although men do not now fight for the sake of obtaining wives, and this form of selection has passed away, yet they generally have to undergo, during manhood, a severe struggle in order to maintain themselves and their families; and this will tend to keep up or even increase their mental powers, and, as a consequence, the present inequality between the sexes.

[*After showing that music has its origins among the lower animals, and continues to play a role in human courtship, Darwin examines the remarkable range of ideas of physical attractiveness among different human races.*]

On the influence of beauty in determining the marriages of mankind.— In civilised life man is largely, but by no means exclusively, influenced in the choice of his wife by external appearance; but we are chiefly concerned with primeval times, and our only means of forming a judgment on this subject is to study the habits of existing semi-civilised and savage nations. If it can be shewn that the men of different races prefer women having certain characteristics, or conversely that the women prefer certain men, we have then to enquire whether such choice, continued during many generations, would produce any sensible effect on the race, either on one sex or both sexes; this latter circumstance depending on the form of inheritance which prevails.

It will be well first to shew in some detail that savages pay the greatest attention to their personal appearance. That they have a passion for ornament is notorious; and an English philosopher goes so far as to maintain that clothes were first made for ornament and not for warmth. As Professor Waitz remarks, "however poor and miserable man is, he finds a pleasure in adorning himself." The extravagance of the naked Indians of South America in decorating themselves is shewn "by a man of large stature gaining with difficulty enough by the labour of a fortnight to procure in exchange the *chica* necessary to paint himself red." The ancient barbarians of Europe during the Reindeer period* brought to their caves any brilliant or singular objects which they happened to find. Savages at the present day everywhere deck themselves with plumes, necklaces, armlets, earrings, &c. They paint themselves in the most diversified manner. "If painted nations," as Humboldt observes, "had been examined with the same attention as clothed nations, it would have been perceived that the most fertile imagination and the most mutable caprice have created the fashions of painting, as well as those of garments."

In one part of Africa the eyelids are coloured black; in another the nails are coloured yellow or purple. In many places the hair is dyed of various tints. In different countries the teeth are stained black, red, blue, &c., and in the Malay Archipelago it is thought shameful to have white teeth like those of a dog. Not one great country can be named, from the Polar regions in the north to New Zealand in the south, in which the aborigines do not tattoo themselves. This practice was followed by the Jews of old and by the ancient Britons. In Africa some of the natives tattoo themselves, but it is much more common to raise protuberances by rubbing salt into incisions made in various parts of the body; and these are considered by the inhabitants of Kordofan and Durfur "to be great personal attractions."* In the Arab countries no beauty can be perfect until the cheeks "or temples have been gashed."* In South America, as Humboldt remarks, "a mother would be accused of culpable indifference towards her children, if she did not employ artificial means to shape the calf of the leg after the fashion of the country." In the Old and New World the shape of the skull was formerly modified during infancy in the most extraordinary manner, as is still the case in many places, and such deformities are considered ornamental. For instance, the savages of Colombia deem a much flattened head "an essential point of beauty."

[*Further instances of bodily decoration and mutilation are considered, with examples drawn from around the world.*]

In the fashions of our own dress we see exactly the same principle and the same desire to carry every point to an extreme; we exhibit, also, the same spirit of emulation. But the fashions of savages are far more permanent than ours; and whenever their bodies are artificially modified this is necessarily the case. The Arab women of the Upper Nile occupy about three days in dressing their hair; they never imitate other tribes, "but simply vie with each other in the superlativeness of their own style."* Dr. Wilson, in speaking of the compressed skulls of various American races, adds, "such usages are among the least eradicable, and long survive the shock of revolutions that change dynasties and efface more important national peculiarities." The same principle comes largely into play in the art of selection; and we can thus understand, as I have elsewhere explained, the wonderful development of all the races of animals and plants which are kept merely for ornament. Fanciers always wish each character to be somewhat increased; they do not admire a medium standard; they certainly do not desire any great and abrupt change in the character of their breeds; they admire solely what they are accustomed to behold, but they ardently desire to see each characteristic feature a little more developed.

No doubt the perceptive powers of man and the lower animals are so constituted that brilliant colours and certain forms, as well as harmonious and rhythmical sounds, give pleasure and are called beautiful; but why this should be so, we know no more than why certain bodily sensations are agreeable and others disagreeable. It is certainly not true that there is in the mind of man any universal standard of beauty with respect to the human body. It is, however, possible that certain tastes may in the course of time become inherited, though I know of no evidence in favour of this belief; and if so, each race would possess its own innate ideal standard of beauty. It has been argued that ugliness consists in an approach to the structure of the lower animals, and this no doubt is true with the more civilised nations, in which intellect is highly appreciated; but a nose twice as prominent, or eyes twice as large as usual, would not be an approach in structure to any of the lower animals, and yet would be utterly hideous. The men of each race prefer what they are accustomed to behold; they cannot endure any great change; but they like variety, and admire each characteristic point carried to a moderate extreme. Men accustomed to a nearly oval face, to straight and regular features, and to bright colours, admire, as we Europeans know, these points when strongly developed. On the other hand, men accustomed to a broad face, with high cheek-bones, a depressed nose, and a black skin, admire these points strongly developed. No doubt characters of all kinds may easily be too much developed for beauty. Hence a perfect beauty, which implies many characters modified in a particular manner, will in every race be a prodigy. As the great anatomist Bichat long ago said, if every one were cast in the same mould, there would be no such thing as beauty. If all our women were to become as beautiful as the Venus de Medici, we should for a time be charmed; but we should soon wish for variety; and as soon as we had obtained variety, we should wish to see certain characters in our women a little exaggerated beyond the then existing common standard.

CHAPTER XX

Secondary Sexual Characters of Man—continued

WE have seen in the last chapter that with all barbarous races ornaments, dress, and external appearance are highly valued; and that the men judge of the beauty of their women by widely different standards. We must next inquire whether this preference and the consequent selection during many generations of those women, which appear to the men of each race the most attractive, has altered the character either of the females alone or of both sexes. With mammals the general rule appears to be that characters of all kinds are inherited equally by the males and females; we might therefore expect that with mankind any characters gained through sexual selection by the females would commonly be transferred to the offspring of both sexes. If any change has thus been effected it is almost certain that the different races will have been differently modified, as each has its own standard of beauty.

With mankind, especially with savages, many causes interfere with the action of sexual selection as far as the bodily frame is concerned. Civilised men are largely attracted by the mental charms of women, by their wealth, and especially by their social position; for men rarely marry into a much lower rank of life. The men who succeed in obtaining the more beautiful women, will not have a better chance of leaving a long line of descendants than other men with plainer wives, with the exception of the few who bequeath their fortunes according to primogeniture. With respect to the opposite form of selection, namely of the more attractive men by the women, although in civilised nations women have free or almost free choice, which is not the case with barbarous races, yet their choice is largely influenced by the social position and wealth of the men; and the success of the latter in life largely depends on their intellectual powers and energy, or on the fruits of these same powers in their forefathers.

There is, however, reason to believe that sexual selection has effected something in certain civilised and semi-civilised nations. Many persons are convinced, as it appears to me with justice, that the members of our aristocracy, including under this term all wealthy families in which primogeniture has long prevailed, from having chosen during many generations from all classes the more beautiful women as their wives, have become handsomer, according to the European standard of beauty, than the middle classes; yet the middle classes are placed under equally favourable conditions of life for the perfect development of the body. Cook remarks that the superiority in personal appearance "which is observable in the erees or nobles in all the other islands (of the Pacific) is found in the Sandwich islands;" but this may be chiefly due to their better food and manner of life.

[*Those whom Darwin sees as primitive peoples appear to be relatively immune from the action of sexual selection. Reasons for this in such societies include the predominance of polygamy and promiscuity; infanticide, especially of girls; the early commencement of sexual activity; and the low esteem—akin to the status of slaves—in which women are held.*]

We thus see that several customs prevail with savages which would greatly interfere with, or completely stop, the action of sexual selection. On the other hand, the conditions of life to which savages are exposed, and some of their habits, are favourable to natural selection; and this always comes into play together with sexual selection. Savages are known to suffer severely from recurrent famines; they do not increase their food by artificial means; they rarely refrain from marriage, and generally marry young. Consequently they must be subjected to occasional hard struggles for existence, and the favoured individuals will alone survive.

Turning to primeval times when men had only doubtfully attained the rank of manhood, they would probably have lived, as already stated, either as polygamists or temporarily as monogamists. Their intercourse, judging from analogy, would not then have been promiscuous. They would, no doubt, have defended their females to the best of their power from enemies of all kinds, and would probably have hunted for their subsistence, as well as for that of their offspring. The most powerful and able males would have succeeded best in the struggle for life and in obtaining attractive females. At this early period the progenitors of man, from having only feeble powers of reason, would not have looked forward to distant contingencies. They would have been governed more by their instincts and even less by their reason than are savages at the present day. They would not at that period have partially lost one of the strongest of all instincts, common to all the lower animals, namely the love of their young offspring; and consequently they would not have practised infanticide. There would have been no artificial scarcity of women, and polyandry would not have been followed; there would have been no early betrothals; women would not have been valued as mere slaves; both sexes, if the females as well as the males were permitted to exert any choice, would have chosen their partners, not for mental charms, or property, or social position, but almost solely from external appearance. All the adults would have married or paired, and all the offspring, as far as that was possible, would have been reared; so that the struggle for existence would have been periodically severe to an extreme degree. Thus during these primordial times all the conditions for sexual selection would have been much more favourable than at a later period, when man had advanced in his intellectual powers, but had retrograded in his instincts. Therefore, whatever influence sexual selection may have had in producing the differences between the races of man, and between man and the higher Quadrumana, this influence would have been much more powerful at a very remote period than at the present day.

On the Manner of Action of Sexual Selection with mankind.—With primeval men under the favourable conditions just stated, and with those savages who at the present time enter into any marriage tie (but subject to greater or less interference according as the habits of female infanticide, early betrothals, &c., are more or less practised), sexual selection will probably have acted in the following manner. The strongest and most vigorous men,—those who could best defend and hunt for their families, and during later times the chiefs or head-men,—those who were provided with the best weapons and who possessed the most property, such as a larger number of dogs or other animals, would have succeeded in rearing a greater average number of offspring, than would the weaker, poorer and lower members of the same tribes. There can, also, be no doubt that such men would generally have been able to select the more attractive women. At present the chiefs of nearly every tribe throughout the world succeed in obtaining more than one wife. Until recently, as I hear from Mr. Mantell, almost every girl in New Zealand, who was pretty, or promised to be pretty, was *tapu* to some chief. With the Kafirs, as Mr. C. Hamilton states, "the chiefs generally have the pick of the women for many miles round, and are most persevering in establishing or confirming their privilege." We have seen that each race has its own style of beauty, and we know that it is natural to man to admire each characteristic point in his domestic animals, dress, ornaments, and personal appearance, when carried a little beyond the common standard. If then the several foregoing propositions be admitted, and I cannot see that they are doubtful, it would be an inexplicable circumstance, if the selection of the more attractive women by the more powerful men of each tribe, who would rear on an average a greater number of children, did not after the lapse of many generations modify to a certain extent the character of the tribe.

With our domestic animals, when a foreign breed is introduced into a new country, or when a native breed is long and carefully attended to, either for use or ornament, it is found after several generations to have undergone, whenever the means of comparison exist, a greater or less amount of change. This follows from unconscious selection during a long series of generations—that is, the preservation of the most approved individuals—without any wish or expectation of such a result on the part of the breeder. So again, if two careful breeders rear during many years animals of the same family, and do not compare them together or with a common standard, the animals are found after a time to have become to the surprise of their owners slightly different. Each breeder has impressed, as Von Nathusius well expresses it, the character of his own mind—his own taste and judgment—on his animals. What reason, then, can be assigned why similar results should not follow from the long-continued selection of the most admired women by those men of each tribe, who were able to rear to maturity the greater number of children? This would be unconscious selection, for an effect would be produced, independently of any wish or expectation on the part of the men who preferred certain women to others.

Let us suppose the members of a tribe, in which some form of marriage was practised, to spread over an unoccupied continent; they would soon split up into distinct hordes, which would be separated from each other by various barriers, and still more effectually by the incessant wars between all barbarous nations. The hordes would thus be exposed to slightly different conditions and habits of life, and would sooner or later come to differ in some small degree. As soon as this occurred, each isolated tribe would form for itself a slightly different standard of beauty; and then unconscious selection would come into action through the more powerful and leading savages preferring certain women to others. Thus the differences between the tribes, at first very slight, would gradually and inevitably be increased to a greater and greater degree.

With animals in a state of nature, many characters proper to the males, such as size, strength, special weapons, courage and pugnacity, have been acquired through the law of battle. The semi-human progenitors of man, like their allies the Quadrumana, will almost certainly have been thus modified; and, as savages still fight for the possession of their women, a similar process of selection has probably gone on in a greater or less degree to the present day. Other characters proper to the males of the lower animals, such as bright colours and various ornaments, have been acquired by the more attractive males having been preferred by the females. There are, however, exceptional cases in which the males, instead of having been the selected, have been the selectors. We recognise such cases by the females having been rendered more highly ornamented than the males,—their ornamental characters having been transmitted exclusively or chiefly to their female offspring. One such case has been described in the order to which man belongs, namely, with the Rhesus monkey.

Man is more powerful in body and mind than woman, and in the savage state he keeps her in a far more abject state of bondage than does the male of any other animal; therefore it is not surprising that he should have gained the power of selection. Women are everywhere conscious of the value of their beauty; and when they have the means, they take more delight in decorating themselves with all sorts of ornaments than do men. They borrow the plumes of male birds, with which nature decked this sex in order to charm the females. As women have long been selected for beauty, it is not surprising that some of the successive variations should have been transmitted in a limited manner; and consequently that women should have transmitted their beauty in a somewhat higher degree to their female than to their male offspring. Hence women have become more beautiful, as most persons will admit, than men. Women, however, certainly transmit most of their characters, including beauty, to their offspring of both sexes; so that the continued preference by the men of each race of the more attractive women, according to their standard of taste, would tend to modify in the same manner all the individuals of both sexes belonging to the race.

With respect to the other form of sexual selection (which with the lower animals is much the most common), namely, when the females are the selectors, and accept only those males which excite or charm them most, we have reason to believe that it formerly acted on the progenitors of man. Man in all probability owes his beard, and perhaps some other characters, to inheritance from an ancient progenitor who gained in this manner his ornaments. But this form of selection may have occasionally acted during later times; for in utterly barbarous tribes the women have more power in choosing, rejecting, and tempting their lovers, or of afterwards changing their husbands, than might have been expected. As this is a point of some importance, I will give in detail such evidence as I have been able to collect.

Hearne describes how a woman in one of the tribes of Arctic America repeatedly ran away from her husband and joined a beloved man; and with the Charruas* of S. America, as Azara states, the power of divorce is perfectly free. With the Abipones, when a man chooses a wife he bargains with the parents about the price. But "it frequently happens that the girl rescinds what has been agreed upon between the parents and the bridegroom, obstinately rejecting the very mention of marriage."* She often runs away, hides herself, and thus eludes the bridegroom. In the Fiji Islands the man seizes on the woman whom he wishes for his wife by actual or pretended force; but "on reaching the home of her abductor, should she not approve of the match, she runs to some one who can protect her; if, however, she is satisfied, the matter is settled forthwith."* In Tierra del Fuego a young man first obtains the consent of the parents by doing them some service, and then he attempts to carry off the girl; "but if she is unwilling, she hides herself in the woods until her admirer is heartily tired of looking for her, and gives up the pursuit; but this seldom happens."* With the Kalmucks there is a regular race between the bride and bridegroom, the former having a fair start; and Clarke "was assured that no instance occurs of a girl being caught, unless she has a partiality to the pursuer." So with the wild tribes of the Malay archipelago there is a similar racing match; and it appears from M. Bourien's account,* as Sir J. Lubbock remarks, that "the race 'is not to the swift, nor the battle to the strong,' but to the young man who has the good fortune to please his intended bride."

Turning to Africa: the Kafirs buy their wives, and girls are severely beaten by their fathers if they will not accept a chosen husband; yet it is manifest from many facts given by the Rev. Mr. Shooter, that they have considerable power of choice. Thus very ugly, though rich men, have been known to fail in getting wives. The girls, before consenting to be betrothed, compel the men to shew themselves off, first in front and then behind, and "exhibit their paces." They have been known to propose to a man, and they not rarely run away with a favoured lover. With the degraded bush-women of S. Africa, "when a girl has grown up to womanhood without having been betrothed, which, however, does not often happen, her lover must gain her approbation, as well as that of the parents."* Mr. Winwood Reade made inquiries for me with respect to the negroes of Western Africa, and he informs me that "the women, at least among the more intelligent Pagan tribes, have no difficulty in getting the husbands whom they may desire, although it is considered unwomanly to ask a man to marry them. They are quite capable of falling in love, and of forming tender, passionate, and faithful attachments."

We thus see that with savages the women are not in quite so abject a state in relation to marriage as has often been supposed. They can tempt the men whom they prefer, and can sometimes reject those whom they dislike, either before or after marriage. Preference on the part of the women, steadily acting in any one direction, would ultimately affect the character of the tribe; for the women would generally choose not merely the handsomer men, according to their standard of taste, but those who were at the same time best able to defend and support them. Such well-endowed pairs would commonly rear a larger number of offspring than the less well endowed. The same result would obviously follow in a still more marked manner if there was selection on both sides; that is if the more attractive, and at the same time more powerful men were to prefer, and were preferred by, the more attractive women. And these two forms of selection seem actually to have occurred, whether or not simultaneously, with mankind, especially during the earlier periods of our long history.

We will now consider in a little more detail, relatively to sexual selection, some of the characters which distinguish the several races of man from each other and from the lower animals, namely, the more or less complete absence of hair from the body and the colour of the skin. We need say nothing about the great diversity in the shape of the features and of the skull between the different races, as we have seen in the last chapter how different is the standard of beauty in these respects. These characters will therefore probably have been acted on through sexual selection; but we have no means of judging, as far as I can see, whether they have been acted on chiefly through the male or female side.

[*The first case to be considered is the absence of hair on the human body, and its development on the face and head. That hairiness varies not only between the sexes, but between races, indicates that it is a secondary sexual character.*]

Colour of the Skin.—The best kind of evidence that the colour of the skin has been modified through sexual selection is wanting in the case of mankind; for the sexes do not differ in this respect, or only slightly and doubtfully. On the other hand we know from many facts already given that the colour of the skin is regarded by the men of all races as a highly important element in their beauty; so that it is a character which would be likely to be modified through selection, as has occurred in innumerable instances with the lower animals. It seems at first sight a monstrous supposition that the jet blackness of the negro has been gained through sexual selection; but this view is supported by various analogies, and we know that negroes admire their own blackness. With mammals, when the sexes differ in colour, the male is often black or much darker than the female; and it depends merely on the form of inheritance whether this or any other tint shall be transmitted to both sexes or to one alone. The resemblance of *Pithecia satanas** with his jet black skin, white rolling eyeballs, and hair parted on the top of the head, to a negro in miniature, is almost ludicrous.

The colour of the face differs much more widely in the various kinds of monkeys than it does in the races of man; and we have good reason to believe that the red, blue, orange, almost white and black tints of their skin, even when common to both sexes, and the bright colours of their fur, as well as the ornamental tufts of hair about the head, have all been acquired through sexual selection. As the newly-born infants of the most distinct races do not differ nearly as much in colour as do the adults, although their bodies are completely destitute of hair, we have some slight indication that the tints of the different races were acquired subsequently to the removal of the hair, which, as before stated, must have occurred at a very early period.

Summary.—We may conclude that the greater size, strength, courage, pugnacity, and even energy of man, in comparison with the same qualities in woman, were acquired during primeval times, and have subsequently been augmented, chiefly through the contests of rival males for the possession of the females. The greater intellectual vigour and power of invention in man is probably due to natural selection combined with the inherited effects of habit, for the most able men will have succeeded best in defending and providing for themselves, their wives and offspring. As far as the extreme intricacy of the subject permits us to judge, it appears that our male ape-like progenitors acquired their beards as an ornament to charm or excite the opposite sex, and transmitted them to man as he now exists. The females apparently were first denuded of hair in like manner as a sexual ornament; but they transmitted this character almost equally to both sexes. It is not improbable that the females were modified in other respects for the same purpose and through the same means; so that women have acquired sweeter voices and become more beautiful than men.

It deserves particular attention that with mankind all the conditions for sexual selection were much more favourable, during a very early period, when man had only just attained to the rank of manhood, than during later times. For he would then, as we may safely conclude, have been guided more by his instinctive passions, and less by foresight or reason. He would not then have been so utterly licentious as many savages now are; and each male would have jealously guarded his wife or wives. He would not then have practised infanticide; nor valued his wives merely as useful slaves; nor have been betrothed to them during infancy. Hence we may infer that the races of men were differentiated, as far as sexual selection is concerned, in chief part during a very remote epoch; and this conclusion throws light on the remarkable fact that at the most ancient period, of which we have as yet obtained any record, the races of man had already come to differ nearly or quite as much as they do at the present day.

The views here advanced, on the part which sexual selection has played in the history of man, want scientific precision. He who does not admit this agency in the case of the lower animals, will properly disregard all that I have written in the later chapters on man. We cannot positively say that this character, but not that, has been thus modified; it has, however, been shewn that the races of man differ from each other and from their nearest allies amongst the lower animals, in certain characters which are of no service to them in their ordinary habits of life, and which it is extremely probable would have been modified through sexual selection. We have seen that with the lowest savages the people of each tribe admire their own characteristic qualities,—the shape of the head and face, the squareness of the cheek-bones, the prominence or depression of the nose, the colour of the skin, the length of the hair on the head, the absence of hair on the face and body, or the presence of a great beard, and so forth. Hence these and other such points could hardly fail to have been slowly and gradually exaggerated from the more powerful and able men in each tribe, who would succeed in rearing the largest number of offspring, having selected during many generations as their wives the most strongly characterised and therefore most attractive women. For my own part I conclude that of all the causes which have led to the differences in external appearance between the races of man, and to a certain extent between man and the lower animals, sexual selection has been by far the most efficient.

CHAPTER XXI

General Summary and Conclusion

A BRIEF summary will here be sufficient to recall to the reader's mind the more salient points in this work. Many of the views which have been advanced are highly speculative, and some no doubt will prove erroneous; but I have in every case given the reasons which have led me to one view rather than to another. It seemed worth while to try how far the principle of evolution would throw light on some of the more complex problems in the natural history of man.

False facts are highly injurious to the progress of science, for they often long endure; but false views, if supported by some evidence, do little harm, as every one takes a salutary pleasure in proving their falseness; and when this is done, one path towards error is closed and the road to truth is often at the same time opened.

The main conclusion arrived at in this work, and now held by many naturalists who are well competent to form a sound judgment, is that man is descended from some less highly organised form. The grounds upon which this conclusion rests will never be shaken, for the close similarity between man and the lower animals in embryonic development, as well as in innumerable points of structure and constitution, both of high and of the most trifling importance,—the rudiments which he retains, and the abnormal reversions to which he is occasionally liable,—are facts which cannot be disputed. They have long been known, but until recently they told us nothing with respect to the origin of man. Now when viewed by the light of our knowledge of the whole organic world, their meaning is unmistakeable. The great principle of evolution stands up clear and firm, when these groups of facts are considered in connection with others, such as the mutual affinities of the members of the same group, their geographical distribution in past and present times, and their geological succession. It is incredible that all these facts should speak falsely. He who is not content to look, like a savage, at the phenomena of nature as disconnected, cannot any longer believe that man is the work of a separate act of creation. He will be forced to admit that the close resemblance of the embryo of man to that, for instance, of a dog—the construction of his skull, limbs, and whole frame, independently of the uses to which the parts may be put, on the same plan with that of other mammals—the occasional reappearance of various structures, for instance of several distinct muscles, which man does not normally possess, but which are common to the Quadrumana—and a crowd of analogous facts—all point in the plainest manner to the conclusion that man is the co-descendant with other mammals of a common progenitor.

We have seen that man incessantly presents individual differences in all parts of his body and in his mental faculties. These differences or variations seem to be induced by the same general causes, and to obey the same laws as with the lower animals. In both cases similar laws of inheritance prevail. Man tends to increase at a greater rate than his means of subsistence; consequently he is occasionally sub-jected to a severe struggle for existence, and natural selection will have effected whatever lies within its scope. A succession of strongly-marked variations of a similar nature are by no means requisite; slight fluctuating differences in the individual suffice for the work of natural selection. We may feel assured that the inherited effects of the long-continued use or disuse of parts will have done much in the same direction with natural selection. Modifications formerly of importance, though no longer of any special use, will be long inher-ited. When one part is modified, other parts will change through the principle of correlation, of which we have instances in many curious cases of correlated monstrosities. Something may be attributed to the direct and definite action of the surrounding conditions of life, such as abundant food, heat, or moisture; and lastly, many characters of slight physiological importance, some indeed of considerable import-ance, have been gained through sexual selection.

No doubt man, as well as every other animal, presents structures, which as far as we can judge with our little knowledge, are not now of any service to him, nor have been so during any former period of his existence, either in relation to his general conditions of life, or of one sex to the other. Such structures cannot be accounted for by any form of selection, or by the inherited effects of the use and disuse of parts. We know, however, that many strange and strongly-marked peculiarities of structure occasionally appear in our domesticated productions, and if the unknown causes which produce them were to act more uniformly, they would probably become common to all the individuals of the species. We may hope hereafter to understand something about the causes of such occasional modifications, especially through the study of monstrosities: hence the labours of experimentalists, such as those of M. Camille Dareste, are full of promise for the future. In the greater number of cases we can only say that the cause of each slight variation and of each monstrosity lies much more in the nature or constitution of the organism, than in the nature of the surrounding conditions; though new and changed conditions certainly play an important part in exciting organic changes of all kinds.

Through the means just specified, aided perhaps by others as yet undiscovered, man has been raised to his present state. But since he attained to the rank of manhood, he has diverged into distinct races, or as they may be more appropriately called sub-species. Some of these, for instance the Negro and European, are so distinct that, if specimens had been brought to a naturalist without any further information, they would undoubtedly have been considered by him as good and true species. Nevertheless all the races agree in so many unimportant details of structure and in so many mental peculiarities, that these can be accounted for only through inheritance from a common progenitor; and a progenitor thus characterised would probably have deserved to rank as man.

It must not be supposed that the divergence of each race from the other races, and of all the races from a common stock, can be traced back to any one pair of progenitors. On the contrary, at every stage in the process of modification, all the individuals which were in any way best fitted for their conditions of life, though in different degrees, would have survived in greater numbers than the less well fitted. The process would have been like that followed by man, when he does not intentionally select particular individuals, but breeds from all the superior and neglects all the inferior individuals. He thus slowly but surely modifies his stock, and unconsciously forms a new strain. So with respect to modifications, acquired independently of selection, and due to variations arising from the nature of the organism and the action of the surrounding conditions, or from changed habits of life, no single pair will have been modified in a much greater degree than the other pairs which inhabit the same country, for all will have been continually blended through free intercrossing.

By considering the embryological structure of man,—the homologies which he presents with the lower animals,—the rudiments which he retains,—and the reversions to which he is liable, we can partly recall in imagination the former condition of our early progenitors; and can approximately place them in their proper position in the zoological series. We thus learn that man is descended from a hairy quadruped, furnished with a tail and pointed ears, probably arboreal* in its habits, and an inhabitant of the Old World. This creature, if its whole structure had been examined by a naturalist, would have been classed amongst the Quadrumana, as surely as would the common and still more ancient progenitor of the Old and New World monkeys. The Quadrumana and all the higher mammals are probably derived from an ancient marsupial animal, and this through a long line of diversified forms, either from some reptile-like or some amphibian-like creature, and this again from some fish-like animal. In the dim obscurity of the past we can see that the early progenitor of all the Vertebrata must have been an aquatic animal, provided with branchiæ* with the two sexes united in the same individual, and with the most important organs of the body (such as the brain and heart) imperfectly developed. This animal seems to have been more like the larvæ of our existing marine Ascidians* than any other known form.

The greatest difficulty which presents itself, when we are driven to the above conclusion on the origin of man, is the high standard of intellectual power and of moral disposition which he has attained. But every one who admits the general principle of evolution, must see that the mental powers of the higher animals, which are the same in kind with those of mankind, though so different in degree, are capable of advancement. Thus the interval between the mental powers of one of the higher apes and of a fish, or between those of an ant and scale-insect, is immense. The development of these powers in animals does not offer any special difficulty; for with our domesticated animals, the mental faculties are certainly variable, and the variations are inherited. No one doubts that these faculties are of the utmost importance to animals in a state of nature. Therefore the conditions are favourable for their development through natural selection. The same conclusion may be extended to man; the intellect must have been all-important to him, even at a very remote period, enabling him to use language, to invent and make weapons, tools, traps, &c.; by which means, in combination with his social habits, he long ago became the most dominant of all living creatures.

A great stride in the development of the intellect will have followed, as soon as, through a previous considerable advance, the half-art and half-instinct of language came into use; for the continued use of language will have reacted on the brain, and produced an inherited effect; and this again will have reacted on the improvement of language. The large size of the brain in man, in comparison with that of the lower animals, relatively to the size of their bodies, may be attributed in chief part, as Mr. Chauncey Wright has well remarked, to the early use of some simple form of language,—that wonderful engine which affixes signs to all sorts of objects and qualities, and excites trains of thought which would never arise from the mere impression of the senses, and if they did arise could not be followed out. The higher intellectual powers of man, such as those of ratiocination, abstraction, self-consciousness, &c., will have followed from the continued improvement of other mental faculties; but without considerable culture of the mind, both in the race and in the individual, it is doubtful whether these high powers would be exercised, and thus fully attained.

The development of the moral qualities is a more interesting and difficult problem. Their foundation lies in the social instincts, including in this term the family ties. These instincts are of a highly complex nature, and in the case of the lower animals give special tendencies towards certain definite actions; but the more important elements for us are love, and the distinct emotion of sympathy. Animals endowed with the social instincts take pleasure in each other's company, warn each other of danger, defend and aid each other in many ways. These instincts are not extended to all the individuals of the species, but only to those of the same community. As they are highly beneficial to the species, they have in all probability been acquired through natural selection.

A moral being is one who is capable of comparing his past and future actions and motives,—of approving of some and disapproving of others; and the fact that man is the one being who with certainty can be thus designated makes the greatest of all distinctions between him and the lower animals. But in our third chapter I have endeavoured to shew that the moral sense follows, firstly, from the enduring and always present nature of the social instincts, in which respect man agrees with the lower animals; and secondly, from his mental faculties being highly active and his impressions of past events extremely vivid, in which respects he differs from the lower animals. Owing to this condition of mind, man cannot avoid looking backwards and comparing the impressions of past events and actions. He also continually looks forward. Hence after some temporary desire or passion has mastered his social instincts, he will reflect and compare the now weakened impression of such past impulses, with the ever present social instinct; and he will then feel that sense of dissatisfaction which all unsatisfied instincts leave behind them. Consequently he resolves to act differently for the future—and this is conscience. Any instinct which is permanently stronger or more enduring than another, gives rise to a feeling which we express by saying that it ought to be obeyed. A pointer dog, if able to reflect on his past conduct, would say to himself, I ought (as indeed we say of him) to have pointed at that hare and not have yielded to the passing temptation of hunting it.

Social animals are partly impelled by a wish to aid the members of the same community in a general manner, but more commonly to perform certain definite actions. Man is impelled by the same general wish to aid his fellows, but has few or no special instincts. He differs also from the lower animals in being able to express his desires by words, which thus become the guide to the aid required and bestowed. The motive to give aid is likewise somewhat modified in man: it no longer consists solely of a blind instinctive impulse, but is largely influenced by the praise or blame of his fellow men. Both the appreciation and the bestowal of praise and blame rest on sympathy; and this emotion, as we have seen, is one of the most important elements of the social instincts. Sympathy, though gained as an instinct, is also much strengthened by exercise or habit. As all men desire their own happiness, praise or blame is bestowed on actions and motives, according as they lead to this end; and as happiness is an essential part of the general good, the greatest-happiness principle indirectly serves as a nearly safe standard of right and wrong. As the reasoning powers advance and experience is gained, the more remote effects of certain lines of conduct on the character of the individual, and on the general good, are perceived; and then the self-regarding virtues, from coming within the scope of public opinion, receive praise, and their opposites receive blame. But with the less civilised nations reason often errs, and many bad customs and base superstitions come within the same scope, and consequently are esteemed as high virtues, and their breach as heavy crimes.

The moral faculties are generally esteemed, and with justice, as of higher value than the intellectual powers. But we should always bear in mind that the activity of the mind in vividly recalling past impressions is one of the fundamental though secondary bases of conscience. This fact affords the strongest argument for educating and stimulating in all possible ways the intellectual faculties of every human being. No doubt a man with a torpid mind, if his social affections and sympathies are well developed, will be led to good actions, and may have a fairly sensitive conscience. But whatever renders the imagination of men more vivid and strengthens the habit of recalling and comparing past impressions, will make the conscience more sensitive, and may even compensate to a certain extent for weak social affections and sympathies.

The moral nature of man has reached the highest standard as yet attained, partly through the advancement of the reasoning powers and consequently of a just public opinion, but especially through the sympathies being rendered more tender and widely diffused through the effects of habit, example, instruction, and reflection. It is not improbable that virtuous tendencies may through long practice be inherited. With the more civilised races, the conviction of the existence of an all-seeing Deity has had a potent influence on the advancement of morality. Ultimately man no longer accepts the praise or blame of his fellows as his chief guide, though few escape this influence, but his habitual convictions controlled by reason afford him the safest rule. His conscience then becomes his supreme judge and monitor. Nevertheless the first foundation or origin of the moral sense lies in the social instincts, including sympathy; and these instincts no doubt were primarily gained, as in the case of the lower animals, through natural selection.

The belief in God has often been advanced as not only the greatest, but the most complete of all the distinctions between man and the lower animals. It is however impossible, as we have seen, to maintain that this belief is innate or instinctive in man. On the other hand a belief in all-pervading spiritual agencies seems to be universal; and apparently follows from a considerable advance in the reasoning powers of man, and from a still greater advance in his faculties of imagination, curiosity and wonder. I am aware that the assumed instinctive belief in God has been used by many persons as an argument for His existence. But this is a rash argument, as we should thus be compelled to believe in the existence of many cruel and malignant spirits, possessing only a little more power than man; for the belief in them is far more general than of a beneficent Deity. The idea of a universal and beneficent Creator of the universe does not seem to arise in the mind of man, until he has been elevated by long-continued culture.

He who believes in the advancement of man from some lowly-organised form, will naturally ask how does this bear on the belief in the immortality of the soul. The barbarous races of man, as

Sir J. Lubbock has shewn, possess no clear belief of this kind; but arguments derived from the primeval beliefs of savages are, as we have just seen, of little or no avail. Few persons feel any anxiety from the impossibility of determining at what precise period in the development of the individual, from the first trace of the minute germinal vesicle to the child either before or after birth, man becomes an immortal being; and there is no greater cause for anxiety because the period in the gradually ascending organic scale cannot possibly be determined.

I am aware that the conclusions arrived at in this work will be denounced by some as highly irreligious; but he who thus denounces them is bound to shew why it is more irreligious to explain the origin of man as a distinct species by descent from some lower form, through the laws of variation and natural selection, than to explain the birth of the individual through the laws of ordinary reproduction. The birth both of the species and of the individual are equally parts of that grand sequence of events, which our minds refuse to accept as the result of blind chance. The understanding revolts at such a conclusion, whether or not we are able to believe that every slight variation of structure,—the union of each pair in marriage,—the dissemination of each seed,—and other such events, have all been ordained for some special purpose.

Sexual selection has been treated at great length in these volumes; for, as I have attempted to shew, it has played an important part in the history of the organic world. As summaries have been given to each chapter, it would be superfluous here to add a detailed summary. I am aware that much remains doubtful, but I have endeavoured to give a fair view of the whole case. In the lower divisions of the animal kingdom, sexual selection seems to have done nothing: such animals are often affixed for life to the same spot, or have the two sexes combined in the same individual, or what is still more important, their perceptive and intellectual faculties are not sufficiently advanced to allow of the feelings of love and jealousy, or of the exertion of choice. When, however, we come to the Arthropoda* and Vertebrata, even to the lowest classes in these two great Sub-Kingdoms, sexual selection has effected much; and it deserves notice that we here find the intellectual faculties developed, but in two very distinct lines, to the highest standard, namely in the Hymenoptera (ants, bees, &c.)

amongst the Arthropoda, and in the Mammalia, including man, amongst the Vertebrata.

In the most distinct classes of the animal kingdom, with mammals, birds, reptiles, fishes, insects, and even crustaceans, the differences between the sexes follow almost exactly the same rules. The males are almost always the wooers; and they alone are armed with special weapons for fighting with their rivals. They are generally stronger and larger than the females, and are endowed with the requisite qualities of courage and pugnacity. They are provided, either exclusively or in a much higher degree than the females, with organs for producing vocal or instrumental music, and with odoriferous glands. They are ornamented with infinitely diversified appendages, and with the most brilliant or conspicuous colours, often arranged in elegant patterns, whilst the females are left unadorned. When the sexes differ in more important structures, it is the male which is provided with special sense-organs for discovering the female, with locomotive organs for reaching her, and often with prehensile organs for holding her. These various structures for securing or charming the female are often developed in the male during only part of the year, namely the breeding season. They have in many cases been transferred in a greater or less degree to the females; and in the latter case they appear in her as mere rudiments. They are lost by the males after emasculation. Generally they are not developed in the male during early youth, but appear a short time before the age for reproduction. Hence in most cases the young of both sexes resemble each other; and the female resembles her young offspring throughout life. In almost every great class a few anomalous cases occur in which there has been an almost complete transposition of the characters proper to the two sexes; the females assuming characters which properly belong to the males. This surprising uniformity in the laws regulating the differences between the sexes in so many and such widely separated classes, is intelligible if we admit the action throughout all the higher divisions of the animal kingdom of one common cause, namely sexual selection.

Sexual selection depends on the success of certain individuals over others of the same sex in relation to the propagation of the species; whilst natural selection depends on the success of both sexes, at all ages, in relation to the general conditions of life. The sexual struggle is of two kinds; in the one it is between the individuals of the same sex, generally the male sex, in order to drive away or kill their rivals, the females remaining passive; whilst in the other, the struggle is likewise between the individuals of the same sex, in order to excite or charm those of the opposite sex, generally the females, which no longer remain passive, but select the more agreeable partners. This latter kind of selection is closely analogous to that which man unintentionally, yet effectually, brings to bear on his domesticated productions, when he continues for a long time choosing the most pleasing or useful individuals, without any wish to modify the breed.

The laws of inheritance determine whether characters gained through sexual selection by either sex shall be transmitted to the same sex, or to both sexes; as well as the age at which they shall be developed. It appears that variations which arise late in life are commonly transmitted to one and the same sex. Variability is the necessary basis for the action of selection, and is wholly independent of it. It follows from this, that variations of the same general nature have often been taken advantage of and accumulated through sexual selection in relation to the propagation of the species, and through natural selection in relation to the general purposes of life. Hence secondary sexual characters, when equally transmitted to both sexes can be distinguished from ordinary specific characters only by the light of analogy. The modifications acquired through sexual selection are often so strongly pronounced that the two sexes have frequently been ranked as distinct species, or even as distinct genera. Such strongly-marked differences must be in some manner highly important; and we know that they have been acquired in some instances at the cost not only of inconvenience, but of exposure to actual danger.

The belief in the power of sexual selection rests chiefly on the following considerations. The characters which we have the best reason for supposing to have been thus acquired are confined to one sex; and this alone renders it probable that they are in some way connected with the act of reproduction. These characters in innumerable instances are fully developed only at maturity; and often during only a part of the year, which is always the breeding-season. The males (passing over a few exceptional cases) are the most active in courtship; they are the best armed, and are rendered the most attractive in various ways. It is to be especially observed that the males display their attractions with elaborate care in the presence of the females; and that they rarely or never display them excepting during the season of love. It is incredible that all this display should be purposeless. Lastly we have distinct evidence with some quadrupeds and birds that the individuals of the one sex are capable of feeling a strong antipathy or preference for certain individuals of the opposite sex.

Bearing these facts in mind, and not forgetting the marked results of man's unconscious selection, it seems to me almost certain that if the individuals of one sex were during a long series of generations to prefer pairing with certain individuals of the other sex, characterised in some peculiar manner, the offspring would slowly but surely become modified in this same manner. I have not attempted to conceal that, excepting when the males are more numerous than the females, or when polygamy prevails, it is doubtful how the more attractive males succeed in leaving a larger number of offspring to inherit their superiority in ornaments or other charms than the less attractive males; but I have shewn that this would probably follow from the females,—especially the more vigorous females which would be the first to breed, preferring not only the more attractive but at the same time the more vigorous and victorious males.

Although we have some positive evidence that birds appreciate bright and beautiful objects, as with the Bower-birds of Australia, and although they certainly appreciate the power of song, yet I fully admit that it is an astonishing fact that the females of many birds and some mammals should be endowed with sufficient taste for what has apparently been effected through sexual selection; and this is even more astonishing in the case of reptiles, fish, and insects. But we really know very little about the minds of the lower animals. It cannot be supposed that male Birds of Paradise or Peacocks, for instance, should take so much pains in erecting, spreading, and vibrating their beautiful plumes before the females for no purpose. We should remember the fact given on excellent authority in a former chapter, namely that several peahens, when debarred from an admired male, remained widows during a whole season rather than pair with another bird.

Nevertheless I know of no fact in natural history more wonderful than that the female Argus pheasant should be able to appreciate the exquisite shading of the ball-and-socket ornaments and the elegant patterns on the wing-feathers of the male. He who thinks that the male was created as he now exists must admit that the great plumes, which prevent the wings from being used for flight, and which, as well as the primary feathers, are displayed in a manner quite peculiar to this one species during the act of courtship, and at no other time, were given to him as an ornament. If so, he must likewise admit that the female was created and endowed with the capacity of appreciating such ornaments. I differ only in the conviction that the male Argus pheasant acquired his beauty gradually, through the females having preferred during many generations the more highly ornamented males; the æsthetic capacity of the females having been advanced through exercise or habit in the same manner as our own taste is gradually improved. In the male, through the fortunate chance of a few feathers not having been modified, we can distinctly see how simple spots with a little fulvous shading on one side might have been developed by small and graduated steps into the wonderful ball-and-socket ornaments; and it is probable that they were actually thus developed.

Everyone who admits the principle of evolution, and yet feels great difficulty in admitting that female mammals, birds, reptiles, and fish, could have acquired the high standard of taste which is implied by the beauty of the males, and which generally coincides with our own standard, should reflect that in each member of the vertebrate series the nerve-cells of the brain are the direct offshoots of those possessed by the common progenitor of the whole group. It thus becomes intelligible that the brain and mental faculties should be capable under similar conditions of nearly the same course of development, and consequently of performing nearly the same functions.

The reader who has taken the trouble to go through the several chapters devoted to sexual selection, will be able to judge how far the conclusions at which I have arrived are supported by sufficient evidence. If he accepts these conclusions, he may, I think, safely extend them to mankind; but it would be superfluous here to repeat what I have so lately said on the manner in which sexual selection has apparently acted on both the male and female side, causing the two sexes of man to differ in body and mind, and the several races to differ from each other in various characters, as well as from their ancient and lowly-organised progenitors.

He who admits the principle of sexual selection will be led to the remarkable conclusion that the cerebral system not only regulates most of the existing functions of the body, but has indirectly influenced the progressive development of various bodily structures and of certain mental qualities. Courage, pugnacity, perseverance, strength and size of body, weapons of all kinds, musical organs, both vocal and instrumental, bright colours, stripes and marks, and ornamental appendages, have all been indirectly gained by the one sex or the other, through the influence of love and jealousy, through the appreciation of the beautiful in sound, colour or form, and through the exertion of a choice; and these powers of the mind manifestly depend on the development of the cerebral system.

Man scans with scrupulous care the character and pedigree of his horses, cattle, and dogs before he matches them; but when he comes to his own marriage he rarely, or never, takes any such care. He is impelled by nearly the same motives as are the lower animals when left to their own free choice, though he is in so far superior to them that he highly values mental charms and virtues. On the other hand he is strongly attracted by mere wealth or rank. Yet he might by selection do something not only for the bodily constitution and frame of his offspring, but for their intellectual and moral qualities. Both sexes ought to refrain from marriage if in any marked degree inferior in body or mind; but such hopes are Utopian and will never be even partially realised until the laws of inheritance are thoroughly known. All do good service who aid towards this end. When the principles of breeding and of inheritance are better understood, we shall not hear ignorant members of our legislature rejecting with scorn a plan for ascertaining by an easy method whether or not consanguineous marriages are injurious to man.

The advancement of the welfare of mankind is a most intricate problem: all ought to refrain from marriage who cannot avoid abject poverty for their children; for poverty is not only a great evil, but tends to its own increase by leading to recklessness in marriage. On the other hand, as Mr. Galton has remarked, if the prudent avoid marriage, whilst the reckless marry, the inferior members will tend to supplant the better members of society. Man, like every other animal, has no doubt advanced to his present high condition through a struggle for existence consequent on his rapid multiplication; and if he is to advance still higher he must remain subject to a severe struggle. Otherwise he would soon sink into indolence, and the more highly-gifted men would not be more successful in the battle of life than the less gifted. Hence our natural rate of increase, though leading to many and obvious evils, must not be greatly diminished by any means. There should be open competition for all men; and the most able should not be prevented by laws or customs from succeeding best and rearing the largest number of offspring. Important as the struggle for existence has been and even still is, yet as far as the highest part of man's nature is concerned there are other agencies more important. For the moral qualities are advanced, either directly or indirectly, much more through the effects of habit, the reasoning powers, instruction, religion, &c., than through natural selection; though to this latter agency the social instincts, which afforded the basis for the development of the moral sense, may be safely attributed.

The main conclusion arrived at in this work, namely that man is descended from some lowly-organised form, will, I regret to think, be highly distasteful to many persons. But there can hardly be a doubt that we are descended from barbarians. The astonishment which I felt on first seeing a party of Fuegians on a wild and broken shore will never be forgotten by me, for the reflection at once rushed into my mind—such were our ancestors. These men were absolutely naked and bedaubed with paint, their long hair was tangled, their mouths frothed with excitement, and their expression was wild, startled, and distrustful. They possessed hardly any arts, and like wild animals lived on what they could catch; they had no government, and were merciless to every one not of their own small tribe. He who has seen a savage in his native land will not feel much shame, if forced to acknowledge that the blood of some more humble creature flows in his veins. For my own part I would as soon be descended from that heroic little monkey, who braved his dreaded enemy in order to save the life of his keeper; or from that old baboon, who, descending from the mountains, carried away in triumph his young comrade from a crowd of astonished dogs—as from a savage who delights to torture his enemies, offers up bloody sacrifices, practises infanticide without remorse, treats his wives like slaves, knows no decency, and is haunted by the grossest superstitions.

Man may be excused for feeling some pride at having risen, though not through his own exertions, to the very summit of the organic scale; and the fact of his having thus risen, instead of having been aboriginally placed there, may give him hopes for a still higher destiny in the distant future. But we are not here concerned with hopes or fears, only with the truth as far as our reason allows us to discover it. I have given the evidence to the best of my ability; and we must acknowledge, as it seems to me, that man with all his noble qualities, with sympathy which feels for the most debased, with benevolence which extends not only to other men but to the humblest living creature, with his god-like intellect which has penetrated into the movements and constitution of the solar system—with all these exalted powers—Man still bears in his bodily frame the indelible stamp of his lowly origin.

REVIEWS AND RESPONSES

Thanks for your noble work on 'The Descent of Man' which reached me this morning. It is strong as iron and clear as crystal so fascinating that it has kept me from church and even from dinner. It is beyond all praise and I can only ask you to accept my very sincere thanks for it and my assurance that it will henceforth be one of the most honoured and familiar inhabitants of my library.

Have you met with rudimentary nipples? I have a woman under my care now, who has a second small, *milk giving* nipple, with a distinct areola on one breast. I could send you a photograph if you desired it. I shall also endeavour to get you a photograph of the scars of a man whom I saw some time ago which were covered, with hair, not down but long strong *bristles* particularly long towards the top of the helix, & coming somewhat to a point.

British physician James Crichton-Browne in a letter to Darwin, 19 Feb. 1871, CUL: DAR 161: 313

I am delighted to realize that the artisans and mill-hands of Manchester and Oldham *read* and *understand* and believe in your books. They club together to buy them.

Geologist and curator of the Manchester Museum, W. Boyd Dawkins, in a letter to Darwin, 23 Feb. 1871, CUL: DAR 162: 126

Your book has been already of use to me in showing me how to write on subjects where you have old fashioned prejudices—instincts as you well define them—against you. You have succeeded in a very wonderful way in combining the gentle & resolute treatment. The frog is transfixed but in such a delicate and sympathising manner that no well-conditioned frog can feel angry. Huxley seems to enjoy pelting them with stones; and that is amusing to look on at when it is remembered how they have croaked down science so many centuries.

Do not trouble to answer this note: I shall of course send you my volume on African His[ty.] which will be out I trust in the autumn. It would be mere platitude to wish yours success. That is assured.

English traveller in Africa W. Winwood Reade in a letter to Darwin, 4 Mar. 1871, CUL: DAR 176: 46

The translation and the press of the Russian edition is progressing but provisionally I cannot print more than the first volume, as our civilised Government has prohibited Your new work. After the first volume of the translation will be printed it will be seized by the Ministry of the Interior and I shall to appeal to a Court of Law against the seizure, I have a strong hope in the success as the Courts of Law give very often decisions against the shamefull arbitrary measure of the Minister. But at all events the result is doubtful and I dare not print the whole work as my losses, in case the Court confirms the prohibition of the Minister, will be to heavy. I case of final prohibition I will have another fight to obtain a respite in such sense thant the whole edition should not be burned (as they always make such auto da fe's) but stored under my responsability awaiting better times.— I hope at the meeting of our Association of Naturalist in August of this year to start up a protest against this realy shamefull act of the Ministry, still there is hope left that the Court will overrule it.

Russian palaeontologist Vladimir Kovalevsky in a letter to Darwin, 14 Mar. 1871, CUL: DAR 169: 88

We wish we could think that these speculations were as innocuous as they are unpractical and unscientific, but it is too probable that if unchecked they might exert a very mischievous influence. We abstain from noticing their bearings on religious thought, although it is hard to see how, on Mr. Darwin's hypothesis, it is possible to ascribe to Man any other immortality or any other spiritual existence, than that possessed by the Brutes. But, apart from these considerations, if such views as he advances on the nature of the Moral Sense were generally accepted, it seems evident that morality would lose all elements of stable authority, and the "ever fixed marks" around which the tempests of human passion now break themselves would cease to exert their guiding and controlling influence. . . . Men, unfortunately, have the power of acting not according to what is their ultimate social interest, but according to their ideas of it; and if the doctrine could be impressed on them that right and wrong have no other meaning than the pursuit or the neglect of that ultimate interest, Conscience would cease to be a check upon the wildest, or, as Mr. Darwin's own illustration allows us to add, the most murderous revolutions. At a moment when every artificial principle of authority seems undermined, we have no other guarantee for the order and peace of life except in the eternal authority of those elementary principles of duty which are independent of all times and all circumstances. There is much reason to fear that loose philosophy, stimulated by an irrational religion, has done not a little to weaken the force of these principles in France, and that this is, at all events, one potent element in the disorganization of French society. A man incurs a grave responsibility who, with the authority of a well-earned reputation, advances at such a time the disintegrating speculations of this book. He ought to be capable of supporting them by the most conclusive evidence of facts. To put them forward on such incomplete evidence, such cursory investigation, such hypothetical arguments as we have exposed, is more than unscientific—it is reckless.

The Times of London (8 Apr. 1871) condemning the publication of *Descent* two weeks after the revolutionary Commune seized control of Paris

The few passages devoted to sexual selection in the *Origin of Species*, led many persons to suppose that it was but a vague hypothesis almost unsupported by direct evidence; and most of its opponents have shown an utter ignorance of, or disbelief in, the whole matter. . . . From the reticence with which the sexual relations of animals have been treated in popular works, most of the readers of this book will be astonished to find that a new and inner world of animal life exists, of which they had hitherto had no conception; and that a considerable portion of the form and structure, the weapons, the ornaments, and the colouring of animals, owes its very existence to the separation of the sexes. This new branch of natural history is one of the most striking creations of Mr. Darwin's genius, and it is all his own; and although we believe he imputes far too much to its operation, it must be admitted to have exerted a most powerful influence over the higher forms of life. . . . Considerable space is devoted to prove that savages think much of personal appearance, admire certain types of form and complexion, and that probably selection of wives and husbands has been an important agent in determining both the racial and the sexual differences of mankind. The evidence adduced, however, seems only to show that the men as a rule ornament themselves more than the women, and that they do so to be admired by their fellow-men quite as much as by the women; and also that men of each race admire all the characteristic features of their own race, and abhor any wide departure from it; the natural effect of which would be to keep the race true, not to favour the production of new races. . . .

We can hardly therefore impute much influence to sexual selection in the case of man, even as regards less important characters than the loss of hair, because it requires the very same tastes to persist in the majority of the race during a period of long and unknown duration. All analogy teaches us that there would be no such identity of taste in successive generations; and this seems a fatal objection to the belief that any fixed and definite characters could have been produced in man by sexual selection alone.

> English naturalist Alfred Russel Wallace, reviewing *Descent* in the
> *Academy*, 2 (15 Mar. 1871), 177–83, at 179–80

Man is not merely an intellectual animal, but he is also a free moral agent, and, as such—and with the infinite future such freedom opens out before him—differs from all the rest of the visible universe by a distinction so profound that none of those which separate other visible beings is comparable with it. The gulf which lies between his being as a whole, and that of the highest brute, marks off vastly more than a mere kingdom of material beings; and man, so considered, differs far more from an elephant or a gorilla than do these from the dust of the earth on which they tread.

> English comparative anatomist St George Jackson Mivart,
> writing anonymously in the *Quarterly Review*, 131 (July 1871),
> 47–90, at 89

We are going through that change in regard to Mr. Darwin's speculations which has occurred so often in regard to scientific theories. When first propounded, divines regarded them with horror, and declared them to be radically opposed not only to the Book of Genesis, but to all the religious beliefs which elevate us above the brutes. The opinions have gained wider acceptance; and whatever may be the ultimate verdict as to their soundness, it certainly cannot be doubted that they are destined profoundly to modify the future current of thought. As Darwinism has won its way to respectability, as it has ceased to be the rash conjecture of some hasty speculator, and is received with all the honours of grave scientific discussion, divines have naturally come to look upon it with different eyes. They have gradually sidled up towards the object which at first struck them as so dark and portentous a phenomenon, and discovered that after all it is not of so diabolic a nature as they had imagined. . . . Darwinists are not necessarily hoofed and horned monsters, but are occasionally of pacific habits, and may even be detected in the act of going to church.

> English critic Leslie Stephen, 'Darwinism and Divinity', *Fraser's Magazine*, NS 5 (Apr. 1872), 409–21, at 409

Darwin & Co.'s hypothesis that I was once an oyster or an ape, and that Almighty God is mere Caloric does not seem a useful result!

> Scottish essayist Thomas Carlyle to John Aitken Carlyle, 9 Dec. 1873, National Library of Scotland, MS 527, No. 103

Nay, we might even sufficiently represent the general manner of conclusion in the Darwinian system by the statement that if you fasten a hair-brush to a mill-wheel, with the handle forward, so as to develop itself into a neck by moving always in the same direction, and within continual hearing of a steam-whistle, after a certain number of revolutions the hair-brush will fall in love with the whistle; they will marry, lay an egg, and the produce will be a nightingale.

> Art critic John Ruskin, *Love's Meinie: Lectures on Greek and English Birds. Given before the University of Oxford* (Keston, Kent, 1873–81), i, 30

All which Darwinism declares is, that creation is not to be explained through a miracle but through the natural law of progressive development of life under favorable circumstances. . . . Does not this view of life . . . harmonize perfectly with our comprehension of religion, which we do not recognize in form, but in reform, which has its living power in the internal remodeling of Judaism and its Messianic mission as progress towards completed humanity?

> 'Science and Religion', *Jewish Times* (20 Feb. 1874), 821, reporting Rabbi Kaufmann Kohler's first Sunday sermon at Sinai Congregation in Chicago

We have thus arrived at the answer to our question, What is Darwinism? It is Atheism. This does not mean, as before said, that Mr. Darwin himself and all who adopt his views are atheists; but it means that his theory is atheistic; that the exclusion of design from nature is, as Dr. Gray says, tantamount to atheism.

American Calvinist theologian Charles Hodge, *What is Darwinism?* (New York, 1874), 176–7

Without a doubt, when looked at from the point of view of science, this question of [the descent of] man, and of his soul in particular, is one of the most difficult and uncertain of questions. A purely rational mind cannot judge this problem on the basis of one of the theories of the scientific and philosophical schools. Anyone who wants to look at Darwin's argumentation on the power of man vis-à-vis his intellect and ethics would find it affected and far-fetched, unlike any of Darwin's other research, let alone the fact that he himself admits that there is a lack of knowledge, difficulty in research and feeble evidence [on this subject]. And anyone who looks at the [various] philosophical schools will find inconsistencies, and aberrations and many long and troublesome conflicts, and they will become convinced, after researching them, that between all these different schools, the truth is lost and unknown. So the wise one will hold on to what God has revealed and will take from science [only] the clearest of truths. Whatever mistakes and missing links there are in Darwin's theory or whatever errors were added to it, there is no doubt that despite these limitations, it now includes established truths and that it has given scientists many benefits and opened for them paths to [uncover] unsolved problems in a number of ways. And so it should be said that the just will be pleased with the truth wherever they see it and accept it as a gift from the Lord however it comes.

'Darwinism' (*'al-Madhab al-Dārwinī'*), *al-Muqtaṭaf*, 7.3 (1882), 63; trans. Marwa Elshakry

In the Animal Kingdom, we have the Indian.
Is his cause just?
To put the matter bluntly, the Indian is defending his land, which we have usurped, and so he hurts us, robs us, and kills us.
But does he do right? It's not always clear. This is the struggle for life. The broad brushstrokes of nature's laws are most apparent when applied to large groups of people, rather than individuals. It is then that the belief in Providence, justice, equality, fraternity, diverse

opinions so entrenched in every one of us, collides with the most universal manifestation of natural law—and then we whites, we civilized Christians, armed with our Remingtons, shall do away with the Indians, because the Law of Malthus stands above all individual opinion. No matter how excellent these may be, they remain intangible, perhaps because humanity is still far short of being civilized, perhaps for some other reason. Armed with good ideas, weapons, and resources, we too struggle for life, playing our advantages to the hilt.

"Do we do right?" That's a question.

"We struggle for life." That's an answer.

Argentine naturalist and author Eduardo L. Holmberg, *Carlos Roberto Darwin* (Buenos Aires, 1882), 52–68, at 65–6, trans. in A. Novoa and A. Levine, *¡Darwinistas!* (forthcoming)

I do not know that I am an evolutionist, but to this extent I am one. I certainly have more patience with those who trace mankind upward from a low condition, even from the lower animals, than with those that start him at a high point of perfection and conduct him to the level with the brutes. I have no sympathy with a theory that starts man in heaven and stops him in hell.

American abolitionist Frederick Douglass, '"It Moves," or the Philosophy of Reform: An Address Delivered in Washington, D.C., on 20 November 1883', in J. W. Blassingame and J. R. McKivigan (eds.), *The Frederick Douglass Papers*, ser. 1, vol. v (New Haven, 1992), 124–45, at 129

The conception of struggle for existence as a factor of evolution, introduced into science by Darwin and Wallace, has permitted us to embrace an immensely wide range of phenomena in one single generalisation, which soon became the very basis of our philosophical, biological, and sociological speculations. . . . While he himself was chiefly using the term in its narrow sense for his own special purpose, he warned his followers against committing the error (which he seems once to have committed himself) of overrating its narrow meaning. In the *Descent of Man* he gave some powerful pages to illustrate its proper, wide sense. He pointed out how, in numberless animal societies, the struggle between separate individuals for the means of existence disappears, how *struggle* is replaced by *co-operation*, and how that substitution results in the development of intellectual and moral faculties which secure to the species the best conditions for survival. He intimated that in such cases the fittest are not the physically strongest, nor the cunningest, but those who learn to combine so as mutually to support each other, strong and weak alike, for the welfare of the community. 'Those communities,' he wrote, 'which included the greatest number of the most sympathetic members would flourish best, and rear the greatest number of off-spring'. . . . The term, which originated from the narrow Malthusian conception of competition between each and all, thus lost its narrow-ness in the mind of one who knew nature.

Unhappily, these remarks, which might have become the basis of most fruitful researches, were overshadowed by the masses of facts gathered for the purpose of illustrating the consequences of a real competition for life. . . . Nay, on the very pages just mentioned, amidst data disproving the narrow Malthusian conception of struggle, the old Malthusian leaven reappeared—namely, in Darwin's remarks as to the alleged inconveniences of maintaining the 'weak in mind and body' in our civilised societies (ch. v.). As if thousands of weak-bodied and infirm poets, scientists, inventors, and reformers, together with other thousands of so-called 'fools' and 'weak-minded enthusiasts,' were not the most precious weapons used by humanity in its struggle for existence by intellectual and moral arms, which Darwin himself emphasised in those same chapters of the *Descent of Man*.

It happened with Darwin's theory as it always happens with theories having any bearing upon human relations. Instead of widen-ing it according to his own hints, his followers narrowed it still more. . . . They came to conceive the animal world as a world of perpetual struggle among half-starved individuals, thirsting for one another's blood.

Russian anarchist Prince Peter Kropotkin, 'Mutual Aid Among Animals', *Nineteenth Century*, 28 (Sept. 1890), 337–54, at 337–8

By the way, remember the current theories of Darwin and others concerning the descent of man from monkeys. Without going into theory of any kind, Christ explicitly states that in man, besides the animal world, there is also a spiritual world. And what of it—so let man be descended from anywhere at all (in the Bible it's not explained at all how God moulded him out of clay, summoned him from stone), but still God breathed the breath of life into him. (But on the negative side, through sin man could revert to a bestial state.)

Russian novelist Fyodor Dostoevsky to V. A. Alekseev in a letter dated 7 June 1876, in A. S. Dolinia (ed.), *Pis'ma*, iii (Moscow, 1934), 212–13

'Speaking seriously, we know that a really good book will more likely than not receive fair treatment from two or three reviewers; yes, but also more likely than not it will be swamped in the flood of literature that pours forth week after week, and won't have attention fixed long enough upon it to establish its repute. The struggle for existence among books is nowadays as severe as among men. If a writer has friends connected with the press, it is the plain duty of those friends to do their utmost to help him. What matter if they exaggerate, or even lie?'

The novelist Jaspar Milvain in George Gissing's *New Grub Street: A Novel* (London, 1891), iii, 225

Yet, just as in the establishment of the white man's supremacy in the Cape Colony, the aboriginal black races have either been displaced or reduced to a state of submission to the white man's rule at the cost of much blood and injustice to the black man, so also will it be in Matabeleland, and so must it ever be in any country where the European comes into contact with native races, and where at the same time the climate is such that the more highly organised and intelligent race can live and thrive, as it can do in Matabeleland; whilst the presence of valuable minerals or anything else that excites the greed of the stronger race will naturally hasten the process. Therefore Matabeleland is doomed by what seems a law of nature to be ruled by the white man, and the black man must go, or conform to the white man's laws, or die in resisting them. It seems a hard and cruel fate for the black man, but it is a destiny which the broadest philanthropy cannot avert, whilst the British colonist is but the irresponsible atom employed in carrying out a preordained law—the law which has ruled upon this planet ever since, in the far-off misty depths of time, organic life was first evolved upon the earth—the inexorable law which Darwin has aptly termed the 'Survival of the Fittest'.

Traveller and hunter Frederick Courteney Selous, *Sunshine and Storm in Rhodesia* (London, 1896), 66–7

Competition among males, with selection by the female of the superior male, is the process of sexual selection, and works to racial
improvement. So far as the human male competes freely with his
peers in higher and higher activities, and the female chooses the win-
ner, so far we are directly benefited. But there is a radical distinction
between sex-competition and marriage by purchase. . . . To make the
sexual gain of the male rest on his purchasing power puts the
immense force of sex-competition into the field of social economics,
not only as an incentive to labor and achievement, which is good, but
as an incentive to individual gain, however obtained, which is bad;
thus accounting for our multiplied and intensified desire to get,—the
inordinate greed of our industrial world.

Novelist and social reformer Charlotte Perkins Gilman, *Women
and Economics* (Boston, 1898), 110–11

The real difficulty in woman's case is that the whole foundation of
the Christian religion rests on her temptation and man's fall, hence
the necessity of a Redeemer and a plan of salvation. As the chief
cause of this dire calamity, woman's degradation and subordination
were made a necessity. If, however, we accept the Darwinian theory,
that the race has been a gradual growth from the lower to a higher
form of life, and that the story of the fall is a myth, we can exonerate
the snake, emancipate the woman, and reconstruct a more rational
religion for the nineteenth century, and thus escape all the perplex-
ities of the Jewish mythology as of no more importance than those of
the Greek, Persian and Egyptian.

American social activist Elizabeth Cady Stanton, *The Woman's
Bible. Part II. Comments on the Old and New Testaments from Joshua
to Revelation* (New York, 1898), 214

What matters it then that some millionaires are idle, or silly, or vulgar,
that their ideas are sometimes futile, and their plans grotesque,
when they turn aside from money-making? How do they differ in
this from any other class? The millionaires are a product of natural
selection, acting on the whole body of men, to pick out those who can
meet the requirement of certain work to be done. In this respect they
are just like the great statesmen, or scientific men, or military men.
It is because they are thus selected that wealth aggregates under their
hands—both their own and that intrusted to them. Let one of them

make a mistake and see how quickly the concentration gives way to dispersion. They may fairly be regarded as the naturally selected agents of society for certain work. They get high wages and live in luxury, but the bargain is a good one for society. There is the intensest competition for their place and occupation. This assures us that all who are competent for this function will be employed in it, so that the cost of it will be reduced to the lowest terms, and furthermore that the competitors will study the proper conduct to be observed in their occupation. This will bring discipline and the correction of arrogance and masterfulness.

Yale economist William Graham Sumner, a leading American Social Darwinist, in 'Economic Aspects: Its Justification', *The Independent*, 54 (1 May 1902), 1036–40, at 1040

If a country can strengthen itself and make itself one of the fittest, then, even if it annihilates the unfit and the weak, it can still not be said to be immoral. Why? Because it is a law of evolution. Even if we do not extinguish a country that is weak and unfit, it will be unable to survive in the end anyway. That is why violent aggression, which used to be viewed as an act of barbarism, is now viewed as a normal rule of civilization.

Chinese Social Darwinist Liang Qichao, writing in the journal *Xinmin congbao* (Journal of A New People), 1/2 (1902), 34, trans. in J. R. Pusey, *China and Charles Darwin* (Cambridge, Mass., 1983), 311

We both, friend Bölsche, are idealists in that we ascribe a psychical, spiritual character to all of nature. At the same time we profess Darwinism because in spite of its gaps it is a purely reasonable, clearly intelligible, and in a certain sense irrefutable theory. On our walks in the woods we have often sketched Darwin's theory into our panpsychic picture of nature.

The German philosopher Bruno Wille addressing the novelist and popular science writer Wilhelm Bölsche in *Darwins Weltanschauung* (Heilbronn, 1906), p. xii, trans. in A. Kelly, *The Descent of Darwin* (Chapel Hill, NC, 1981), 40

Few people seem to perceive fully as yet that the most far-reaching consequence of the establishment of the common origin of all species is ethical; that it logically involved a re-adjustment of altruistic morals by enlarging as a *necessity of rightness* the application of what has been called 'The Golden Rule' beyond the area of mere mankind to that of the whole animal kingdom. Possibly Darwin himself did not wholly perceive it, though he alluded to it. While man was deemed to be a creation apart from all other creations, a secondary or tertiary morality was considered good enough toward the 'inferior' races; but no person who reasons nowadays can escape the trying conclusion that this is not maintainable.

English poet and novelist Thomas Hardy, writing to the Secretary of the Humanitarian League, 10 Apr. 1910, in M. Millgate (ed.), *The Life and Work of Thomas Hardy* (London, 1985), 376–7

When in 1871—while I was studying medicine at the Syrian Protestant College—I heard—I don't know how—that someone was claiming that man is descended from the ape. I could not determine the truth of this claim, as there was nothing in the school curriculum that allowed me to make it out. And when I heard it, I was greatly disgusted by it and by whoever said it, who I thought was simply trying to be shocking. No wonder, given the way it was mentioned to me and the way it is always described by his opponents—that the origin of man is an ape—such that anyone who hears this for the first time and is full of different beliefs will disregard it, even if we know that there are certain types of men who are much lower than apes. And this was the weapon that the detractors of this theory used to disparage it. . . .

Then months passed and I don't recall learning anything new about this theory, such that I forgot about it. Strangely enough, after a while, at my graduation, my thesis was on the following: 'The Difference between Man and Animal from the Perspective of Climate, Nutrition and Breeding.' And I came up with many ideas which supported this theory, without realizing it, and I was like one who spoke of something without knowing it.

But what I hadn't intended, soon became my way of thinking and the focus of my thought and all my speeches and writings after my graduation from the college and after my trip to Europe and after acquainting myself with the theory through the works of its authors. . . . And when the idea of the materiality of the universe settled in my mind, the advantages of the theory of evolution and progress became clear to me.

Shiblī Shumayyil, *The Philosophy of Evolution and Progress* (*Kitāb falsafat al-nushū' wairtiqā'*) (Cairo, 2nd edn., 1910), 26–7; trans. Marwa Elshakry

The views you have acquired about Darwinism, evolution and the struggle for existence won't explain to you the meaning of your life and won't give you guidance in your actions, and a life without an explanation of its meaning and importance, and without the unfailing guidance that stems from it is a pitiful existence. Think about it. I say it, probably on the eve of my death, because I love you.

> Russian novelist Leo Tolstoy in a letter dictated on his deathbed to his children, 1 Nov. 1910, in *Tolstoy's Letters*, ed. R. F. Christian (London, 1978), ii, 717

It seems to me that the question of evolution is very largely a question for experts and I am willing to leave it to them to settle.

As to the general ideas of progress of mankind, which have been associated with the name of Darwin in most of our minds, I confess that I share them, though I have never taken much stock in the theories that people have advanced from time to time as to the way in which this progress has come about. As far as my experience goes, I have seen but very little progress either moral or material that was not pretty closely associated with work. It is the evolution brought about by human beings who work that I believe in.

> Booker T. Washington, writing from Tuskegee, Alabama, in answer to an enquiry about his beliefs, 29 Nov. 1911, in L. Harlan, R. W. Smock, and G. McTigue (eds.), *The Booker T. Washington Papers*, xi (1981), 378–9

Women must consider themselves the main agents for the continuity and evolution of the race towards a higher physical, intellectual and spiritual level. . . . Education of girls and young women must prepare them for this great mission. Upon reaching marital age they must have such an elevated and clear notion about it that they should refuse to wed men with inferior physical, intellectual and moral conditions.

. . . Natural selection would not be wholly incompatible with love if these tremendous issues were studied and better understood throughout the whole world.

The improvement of race dreamed by philosophers and preached by biologists would not be a monstrous violation of affection, if we became accustomed to edifying our love on solid moral and religious foundations.

> Cuban writer Blanche Z. de Baralt, 'El Feminismo eugénico', *El Diario* (Mexico City), 24 Dec. 1911, trans. in L. Suárez y

Lopez-Guazo, 'The Mexican Eugenics Society', in T. Glick, M. A. Puig-Samper, and R. Ruiz (eds.), *The Reception of Darwinism in the Iberian World* (Dordrecht, 2001), 143–51, at 146

As I reflected upon the intensive application of man to war in cold, rain, and mud; in rivers, canals, and lakes; underground, in the air, and under the sea; infected with vermin, covered with scabs, adding the stench of his own filthy body to that of his decomposing comrades; hairy, begrimed, bedraggled, yet with unflagging zeal striving eagerly to kill his fellows; and as I felt within myself the mystical urge of the sound of great cannon I realized that war is a normal state of man.

In taking into account the training and the education of the men now at war it is obvious that although this war was precipitated by certain nations, its fundamental cause is to be found in no one nation alone; for every nation, race, or tribe has waged war. The impulse to war is stronger than the desire to live; it is stronger than the fear of death. Those who believe that man is a mechanism evolved through an endless struggle for existence, and that the struggle among men differs only in kind and not in principle from the struggle among other animals or from the equally fierce struggle among plants, will turn for the explanation of war among men to the principles of evolution. . . .

My aim is to make an analysis of war; to point out the probability that these phenomena are explainable on a mechanistic basis; to seek its origin and inherent force in man; and to suggest means by which the very forces which have made cycles of war inevitable may be utilized for the evolution of longer and more secure cycles of peace.

American army surgeon and physiologist George W. Crile writing from the front during the Great War in *A Mechanistic View of War and Peace* (New York, 1915), 3–5

AUTOBIOGRAPHIES

Life. Written August — 1838

My earliest recollection, the date of which I can approximately tell, and which must have been before I [was] four years old, was when sitting on Carolines* knee in the dining room, whilst she was cutting an orange for me, a cow run by the window, which made me jump; so that I received a bad cut of which I bear the scar to this day. Of this scene I recollect the place where I sat & the cause of the fright, but not the cut itself.— & I think my memory is real, & not as often happens in similar cases, from hearing the thing so often repeated, one obtains so vivid an image, that it cannot be separated from memory, because I clearly remember which way the cow ran, which would not probably have been told me. My memory here is an obscure picture, in which from not recollecting any pain I am scarcely conscious of its reference to myself.—

1813 summer.— When I was four year & a half old went the sea & staid there some weeks— I remember many things, but with the exception of the maid servants (& these are not individualised) I recollect none of my family, who were there.— I remember either myself or Catherine* being naughty, & being shut up in a room & trying to break the windows.— I have obscure picture of house before my eyes, & of a neighbouring small shop, where the owner gave me one fig, but which to my great joy turned out to be two:— this fig was given me that this man might kiss the maidservant:— I remember a common walk to a kind of well, on the road to which was a cottage shaded with damascene trees, inhabited by old man, called a hermit, with white hair, used to give us damascenes*— I know not whether the damascenes, or the reverence & indistinct fear for this old man produced the greatest effect, on my memory.— I remember, when going there crossing in the carriage a broad ford, & fear & astonishment of white foaming water has made vivid impression.— I think memory of events commences abruptly, that is I remember these earliest things quite as clearly as others very much later in life, which were equally impressed on me.— Some very early recollections are connected with fear, at Parkfields with poor Betty Harvey* I remember with horror her story of people being pushed into the canal by the towing rope, by going wrong side of the horse..—I had greatest horror of this story.—keen instinct against death.—Some other recollections are those of vanity, & what is odder a consciousness, as if instinctive, & contempt of myself that I was vain—namely thinking that people were admiring me in one instance for perseverance & another for boldness in climbing a low tree.—My supposed admirer was old Peter Hailes the bricklayer,* & the tree the Mountain Ash on the lawn.

All my recollections seem to be connected most closely with self.—now Catherine seems to recollect scenes, where others were chief actors.—When my mother died, I was 8 & ½ old.—& she one year less, yet she remember all particular & events of each day, whilst I scarcely recollect anything, except being sent for—memory of going into her room, my Father meeting us crying afterwards.—She remembers my mother crying, when she heard of my grandmother's death.—Also when at Parkfields, how Aunt Sarah & Kitty* used to receive her—& so with very many other cases

Susan* like me, only remember affairs personal—It is sufficiently odd, this difference in subjects remembered. Catherine says she does not remember the impression made upon her by external things as scenery., but things which she reads she has excellent memory—ie for *ideas*. now her sympathy being ideal, it is part of her character, & shows how early her kind of memory was stamped. A vivid thought is repeated, a vivid impression forgotten.—

I recollect my mother's gown & scarcely anything of her appearance, except one or two walks with her I have no distinct remembrance of any conversations, & those only of very trivial nature.—I remember her saying "if she did ask me to do something, which I said she had, it was solely for my good.".—

I remember oscurely the illumination after the Battle of Waterloo, & the militia exercising, about that period, in the field opposite our House.—

1817. 8½ old went to Mr Cases school.—I remember how very much I was afraid of meeting the dogs in Barker St & how at school I could not get up my courage to fight.—I was very timid by nature. I remember I took great delight at school in fishing for newts in the quary pool.—I had thus young formed a strong taste for collecting, chiefly seals, franks & but also pebbles & minerals,—one which was given me by some boy, decided this taste.—I believe shortly after this or before I had smattered in botany, & certainly when at Mr Case's school I was very fond of gardening, & invented some great falsehoods about being able to colour crocuses as I liked.—At this time I felt strong friendships for some boys.—It was soon after I began collecting stones, ie when 9 or 10 I distinctly recollect the desire I had of being able to know something about every pebble in front of the Hall door—it was my earliest—only geological aspiration at that time.—I was in these days a very great story teller,—for the pure pleasure of exciting attention & surprise. I stole fruit & hid it for these same motives, & injured trees by barking them for similar ends.—I scarcely ever went out walking without saying I had seen a pheasant or some strange bird, (natural History taste). these lies, when not detected, I presume excited my attention, as I recollect them vividly,.—not connected with shame, though some I do,—but as something which by having produced great effect on my mind, gave pleasure, like a tragedy.—

I recollect when at Mr Cases, inventing a whole fabric to show how fond I was of speaking the **truth!**—my invention is still so vivid in my mind, that I could almost fancy it was true did not memory of former shame tell me it was false.—I have no particularly happy or unhappy recollections of this time or earlier periods of my life.—

I remember well a walk I took with boy named Ford across some fields to a farmhouse on Church Stretton Road.—

I do not remember any mental pursuits excepting those of collecting stones &c.—gardening, & about this time often going with my father in his carriage, telling him of my lessons, & seeing game & other wild birds, which was a great delight to me.—I was born a naturalist.—

When I was 9 & ½ years old (July 1818) I went with Erasmus* to see Liverpool.—it has left no impression in my mind, except most trifling ones.—fear of the coach upsetting, a good dinner, & an extremely vague memory of ships.

In midsummer of this year I went to D^r. Butlers school.—I well recollect the first going there, which oddly enough I cannot of going to Mr Cases, the first school of all.—I remember the year 1818 well, not from having first gone to a Public school, but from writing those figures in my school book, accompanied with obscure thoughts, now fullfilled, whether I should recollect in future life that year.—

In September (1818) I was ill with the Scarlet Fever I well remember the wretched feeling of being delirious.—

1819. July. (10 & ½ years old) Went to sea at Plas Edwards* & staid there three weeks, which now appears to me like three months.—I remember a certain shady green road (where I saw a snake) & a waterfall with a degree of pleasure, which must be connected with the pleasure from scenery, though not directly recognized as such.—The sandy plain before the house has left a strong impression, which is obscurely connected with indistinct remembrance of curious insects—probably a Cimex mottled with red—the Zygena.*—I was at that time very passionate, (when I swore like a trooper) & quarrelsome,—the former passion has I think nearly wholly, but slowly died away.—When journeying there by stage Coach I remember a recruiting officer (I think I should know his face to this day) at tea time, asking the maid servant for *toasted* bread butter.—I was convulsed with laughter, & thought it the quaintest & wittiest speech, that ever passed from the mouth of man.—Such is wit at 10 & ½ years old.—

The memory now flashes across me, of the pleasure I had in the evening or on blowy day walking along the beach by myseelf, & seeing the gulls & cormorants wending their way home in a wild & irregular course.—Such poetic pleasures, felt so keenly in after years,, I should not have expected so early, in life.—

1820 July. Went riding tour (on old Dobbin) with Erasmus to Pistol Rhyadwr.*—of this I recollect little.—an indistinct picture of the fall.—but I well remember my astonishment on hearing that fishes could jump up it.—

RECOLLECTIONS

OF THE

DEVELOPMENT OF MY MIND AND CHARACTER

———

*Table of Contents**

1876. May 31ˢᵗ Recollections of the Development of my mind & character. C. Darwin

A German editor* having written to me to ask for an account of the development of my mind & character with some sketch of my auto-biography, I have thought that the attempt would amuse me, & might possibly interest my children or their children. I know that it would have interested me greatly to have read even so short & dull a sketch of the mind of my grandfather written by himself,* & what he thought & did & how he worked. I have attempted to write the following account of myself, as if I were a dead man in another world looking back at my own life. Nor have I found this difficult, for life is nearly over with me. I have taken no pains about my style of writing.

I was born at Shrewsbury on Feb. 12ᵗʰ 1809. I have heard my Father say that he believed that persons with powerful minds generally had memories extending back to a very early period of life. This is not my case, for my earliest recollection goes back only to when I was a few months over four years old, when we went to near Abergele for sea-bathing, & I recollect some events & places there with some little distinctness.

My mother* died in July 1817, when I was a little over eight years old, & it is odd that I can remember hardly anything about her, except her death-bed, her black velvet gown & her curiously constructed work-table. I believe that my forgetfulness is partly due to my sisters owing to their great grief never being able to speak about her or mention her name; & partly to her previous invalid state. In the spring of this same year I was sent to a day-school in Shrewsbury, where I staid a year. Before going to school I was educated by my sister Caroline, but I doubt whether this plan answered. I have been told that I was much slower in learning than my younger sister Catherine, & I believe that I was in many ways a naughty boy. Caroline was extremely kind, clever & zealous; but she was too zealous in trying to improve me; for I clearly remember after this long interval of years, saying to myself when about to enter a room where she was—"what will she blame me for now."; & I made myself dogged so as not to care what she might say.

By the time I went to this day-school my taste for natural-history, & more especially for collecting, was well developed. I tried to make out the names of plants, & collected all sorts of things, shells, seals, franks, coins & minerals. The passion for collecting, which leads a man to be a systematic naturalist, a virtuoso or a miser was very strong in me, & was clearly innate as none of my sisters or brother ever had this taste. One little event during this year has fixed itself very firmly in my mind, & I hope that it has done so from my conscience having been afterwards sorely troubled by it: it is curious as showing that apparently I was interested at this early age in the variability of plants! I told another little boy (I believe it was Leighton who afterwards became a well-known Lichenologist & botanist) that I could produce variously coloured Polyanthuses & Primroses by watering them with certain coloured fluids, which was of course a monstrous fable, & had never been tried by me. I may here also confess that as a little boy I was much given to inventing deliberate falsehoods & this was always done for the sake of causing excitement. For instance I once gathered much valuable fruit from my Father's trees & hid them in the shrubbery, & then ran in breathless haste to spread the news that I had discovered a hoard of stolen fruit.

About this time, or as I hope at a somewhat earlier age, I sometimes stole fruit for the sake of eating it; & one of my schemes was ingenious. The kitchen garden was kept locked in the evening & was surrounded by a high wall, but by the aid of neighbouring trees I could easily get on the coping. I then fixed a long stick into the hole at the bottom of a rather large flower-pot, & by dragging this upwards pulled off peaches & plums, which fell into the pot & the prizes were thus secured. When a very little boy I remember stealing apples from the orchard, for the sake of giving away to some boys & young men, who lived in a cottage not far off, but before I gave them the fruit I showed off how quickly I could run & it is wonderful that I did not perceive that the surprise & admiration which they expressed at my powers of running, was given for the sake of the apples. But I well remember that I was delighted at them declaring that they had never seen a boy run so fast!

I remember clearly only one other incident during the year whilst at Mr Case's daily school,—namely the burial of a dragoon-soldier;* & it is surprising how clearly I can still see the horse with the man's empty boots & carbine suspended to the saddle & the firing over the grave. This scene deeply stirred whatever poetic fancy there was in me.

In the summer of 1818 I went to Dr Butlers great school in Shrewsbury,* & remained there for 7 years till midsummer of 1825, when I was 16 years old. I boarded at this school, so that I had the great advantage of living the life of a true school-boy, but as the distance was hardly more than a mile to my home, I very often ran there in the longer intervals between the callings over & before locking up at night. This I think was in many ways advantageous to me by keeping up home affections & interests. I remember in the early part of my school life that I often had to run very quickly to be in time, & from being a fleet runner was generally successful; but when in doubt I prayed earnestly to God to help me, & I well remember that I attributed my success to the prayers & not to my quick running, & marvelled how generally I was aided. I have heard my Father & elder sisters say that I had as a very young boy a strong taste for long solitary walks; but what I thought about I know not. I often became quite absorbed, & once whilst returning to school on the summit of the old fortifications round Shrewsbury, which have been converted into a public foot-path with no parapet on one side, I walked off & fell to the ground, but the height was only 7 or 8 feet. Nevertheless the number of thoughts which passed through my mind during this very short, but sudden & wholly unexpected fall was astonishing, & seems hardly compatible with what physiologists have, I believe, proved about each thought requiring quite an appreciable amount of time.

I must have been a very simple little fellow when I first went to the school. A boy of the name of Garnett took me into a cake-shop one day, & bought some cakes for which he did not pay, as the shop-man trusted him. When we came out, I asked him why he did not pay for them, & he instantly answered, "why do you not know that my uncle left a great sum of money to the Town on condition that every tradesman should give whatever was wanted without payment to anyone who wore his old hat & moved it in a particular manner"; & he then showed me how it was moved. He then went into another shop where he was trusted, & asked for some small article, moving his hat in the proper manner, & of course obtained it without payment. When we came out he said "now if you like to go by yourself into that cake-shop (how well I remember its exact position) I will lend you my hat & you can get whatever you like if you move the hat on your head properly". I gladly accepted the generous offer, & went in & asked for some cakes, moved the old hat & was walking out of the shop, when the shop-man made a rush at me, so I dropped the cakes & ran away for dear life, & was astonished by being greeted by shouts of laughter by my false friend Garnett.

I can say in my own favour that I was as a boy humane, but I owed this entirely to the instruction & example of my sisters. I doubt indeed whether humanity is a natural or innate quality. I was very fond of collecting eggs, but I never took more than a single egg out of a bird's nest, except on one single occasion, when I took all, not for their value, but from a sort of bravado. I had a strong taste for angling & would sit for any number of hours on the bank of a river or pond watching the float; & when at Maer* I was told that I could kill the worms with salt & water, & from that day I never spitted a living worm, though at the expense, probably, of some loss of success. Once as a very little boy, whilst at the day-school or before that time I acted cruelly, for I beat a puppy I believe, simply from enjoying the sense of power; but the beating could not have been severe, for the puppy did not howl, of which I feel sure as the spot was near to the house. This act lay heavy on my conscience, as is shown by my remembering the exact spot where the crime was committed. It probably lay all the heavier from my love of dogs being then & for a long time afterwards a passion. Dogs seemed to know this, for I was an adept in robbing their love from their masters.

Nothing could have been worse for the development of my mind than D^r. Butler's school, as it was strictly classical, nothing else being taught except a little ancient geography & history. The school as a means of education to me was simply a blank. During my whole life I have been singularly incapable of mastering any language. Especial attention was paid to verse-making, & this I could never do well. I had many friends & got together a grand collection of old verses, which by patching together, sometimes aided by other boys, I could work into any subject. Much attention was paid to learning by heart the lessons of the previous day; this I could effect with great facility learning 40 or 50 lines of Virgil or Homer, whilst I was in morning chapel; but this exercise was utterly useless for every verse was forgotten in 48 hours. I was not idle, & with the exception of versification generally worked conscientiously at my classics, not using cribs. The sole pleasure I ever received from such studies, was from some of the odes of Horace, which I admired greatly. When I left the school I was for my age neither high nor low in it; & I believe that I was considered by all my masters & by my Father as a very ordinary boy, rather below the common standard in intellect. To my deep mortification my Father once said to me "you care for nothing but shooting, dogs & rat-catching, & you will be a disgrace to yourself & all your family". But my Father who was the kindest man I ever knew & whose memory I love with all my heart, must have been angry & somewhat unjust when he used such words.

I may here add a few pages about my Father, who was in many ways a remarkable man.* He was about 6 ft 2 inches in height, with broad shoulders & very corpulent, so that he was the largest man, whom I ever saw. When he last weighed himself, he was 24 stone,* but afterwards increased much in weight. His chief mental characteristics were his powers of observation & his sympathy, neither of which have I ever seen exceeded or even equalled. His sympathy was not only with the distresses of others, but in a greater degree with the pleasures of all around him. This led him to be always scheming to give pleasure to others, & though hating extravagance to perform many generous actions. For instance, M^r B, a small manufacturer in

Shrewsbury* came to him one day, & said he sh^d. be bankrupt unless he could at once borrow £10,000, but that he was unable to give any legal security. My Father heard his reasons for believing that he could ultimately repay the money, & from my father's intuitive perception of character felt sure that he was to be trusted. So he advanced this sum, which was a very large one for him while young, & was after a time repaid.

I suppose that it was his sympathy which gave him unbounded power of winning confidence, & as a consequence made him highly successful as a physician. He began to practice before he was 21 years old, & his fees during the first year paid for the keep of two horses & a servant. On the following year his practice was larger & so continued for above 60 years, when he ceased to attend on anyone. His great success as a doctor was the more remarkable as he told me that he at first hated his profession so much that if he had been sure of the smallest pittance, or if his Father had given him any choice, nothing sh^d. have induced him to follow it. To the end of his life, the thought of an operation almost sickened him, & he could scarcely endure to see a person bled,—a horror which he has transmitted to me, & I remember the horror which I felt as a schoolboy in reading about Pliny (I think) bleeding to death in a warm bath. My father told me two odd stories about bleeding: one was that as a very young man he became a Free-mason. A friend of his who was a free-mason & who pretended not to know about his strong feeling with respect to blood, remarked casually to him, as they walked to the meeting, "I suppose that you do not care about losing a few drops of blood". It seems that when he was received as a member, his eyes were bandaged & his coat-sleeves turned up. Whether any such ceremony is now performed I know not, but my Father mentioned the case as an excellent instance of the power of imagination, for he distinctly felt the blood trickling down his arm, & could hardly believe his own eyes, when he afterwards could not find the smallest prick on his arm. A great slaughtering butcher from London once consulted my grandfather, when another man very ill was brought in, & my grandfather wished to have him instantly bled by the accompanying apothecary. The butcher was asked to hold the patient's arm, but he made some excuse & left the room. Afterwards he explained to my grandfather that although he believed that he had killed with his own hands more animals than any other man in London, yet absurd as it might seem he assuredly should have fainted if he had seen the patient bled.

Owing to my father's power of winning confidence, many patients, especially ladies, consulted him when suffering from any misery, as a sort of Father-Confessor. He told me that they always began by complaining in a vague manner about their health, & by practice he soon guessed what was really the matter. He then suggested that they had been suffering in their minds, & now they would pour out their troubles & he heard nothing more about the body. Family quarrels were a common subject. When gentlemen complained to him about their wives, & the quarrel seemed serious, my father advised them to act in the following manner; & his advice always succeeded, if the gentleman followed it to the letter, which was not always the case. The husband was to say to his wife that he was very sorry that they could not live happily together,—that he felt sure that she would be happier if separated from him—that he did not blame her in the least (this was the point on which the man oftenest failed),—that he would not blame her to any of her relations or friends,—& lastly that he would settle on her as large a provision as he could afford. She was then asked to deliberate on this proposal. As no fault had been found, her temper was unruffled, & she soon felt what an awkward position she would be in, with no accusation to rebut, & with her husband & not herself proposing a separation. Invariably the lady begged her husband not to think of separation, & usually behaved much better ever afterwards. Owing to my fathers skill in winning confidence he received many strange confessions of misery & guilt. He often remarked how many miserable wives he had known. In several instances husbands & wives had gone on pretty well together for between 20 & 30 years, & then hated each other bitterly: this he attributed to their having lost a common bond in their young children having grown up.

But the most remarkable power which my father possessed was that of reading the characters, & even the thoughts of those whom he saw even for a short time. We had many instances of this power some of which seemed almost supernatural. It saved my father from ever making (with one exception & the character of this man was soon discovered) an unworthy friend. A strange clergyman came to Shrewsbury & seemed to be a rich man: everybody called on him & he was invited to many houses. My father called, & on his return home told my sisters on no account to invite him or his family to our house; for he felt sure that the man was not to be trusted. After a few months he suddenly bolted being heavily in debt, & was found out to be little better than an habitual swindler.

Here is a case of trust-fulness, which not many men would have ventured on. An Irish gentleman, a complete stranger, called on my father one day, & said that he had lost his purse, & that it would be a serious inconvenience to him to wait in Shrewsbury until he could receive a remittance from Ireland. He then asked my father to lend him 20 £, which was immediately done, as my father felt certain that the story was a true one. As soon as a letter could arrive from Ireland, one came with the most profuse thanks, & enclosing, as he said, a 20 £ Bank of England note; but no note was enclosed. I asked my father whether this did not stagger him, but he answered "not in the least". On the next day another letter came with many apologies for having forgotten (like a true Irishman) to put the note into his letter of the day before.

A connection of my Father's* consulted him about his son who was strangely idle & would settle to no work. My father said "I believe that the foolish young man thinks that I shall bequeath him a large sum of money. Tell him that I have declared to you that I shall not leave him a penny." The father of the youth owned with shame that this preposterous idea had taken possession of his son's mind; & he asked my father how he could possibly have discovered it, but my father said that he not in the least knew. The Earl of ——* brought his nephew, who was insane but quite gentle, to my father; & the young man's insanity led him to accuse himself of all the crimes under heaven. When my father afterwards talked about the case with the uncle, he said "I am sure that your nephew is really guilty of - - - - -, a heinous crime". Whereupon the Earl exclaimed, "Good God, Dr Darwin who told you, we thought that no human being knew the fact except ourselves." My father told me this story many years after the event, & I asked him how he distinguished the true from the false self-accusations; & it was very characteristic of my Father, that he said he could not explain how it was.

The following story shows what good guesses my father could make. Lord Sherburn,* afterwards the first Marquis of Landsdown, was famous (as Macaulay somewhere remarks) for his knowledge of the affairs of Europe, on which he greatly prided himself. He consulted my father medically, & afterwards harangued him on the state of Holland. My father had studied medicine at Leyden, & one day went a long walk into the country with a friend, who took him to the house of a clergyman (we will say the Rev^d M^r. A., for I have forgotten his name) who had married an English woman. My father was very hungry, & there was little for luncheon except cheese, which he could never eat. The old lady was surprised & grieved at this, & assured my father that it was an excellent cheese & had been sent her from Bowood, the seat of L^d. Sherburn. My father wondered why a cheese should have been sent her from Bowood, but thought nothing more about it, until it flashed across his mind many years afterwards whilst L^d. Sherburn was talking about Holland. So he answered, "I should think from what I saw of the Rev^d M^r. A. that he was a very able man & well acquainted with the state of Holland.["] My father saw that the Earl, who immediately changed the conversation, was much startled. On the next morning my father received a note from the Earl, saying that he had delayed starting on his journey, & wished particularly to see my father. When he called the Earl said, "D^r. Darwin it is of the utmost importance to me & to the Rev. M^r. A. to learn how you have discovered that he is the source of my information about Holland." So my father had to explain the state of the case, & he supposed that Ld Sherburn was much struck with his diplomatic skill in guessing, for during many years afterwards he received many kind messages from him through various friends. I think that he must have told the story to his children; for Sir C. Lyell asked me many years ago why the Marquis of Lansdown (the son or grandson of the first Marquis) felt so much interest about me, whom he had never seen, & my family. When 40 new members (the 40 thieves as they were then called) were added to the Athenæum Club,* there was much canvassing to be one of them; & without my having asked anyone, L^d. Lansdown proposed me & got me elected. If I am right in my supposition, it was a queer concatanation of events that my father not eating cheese half-a-century before in Holland led to my election as a member of the Athenæum.

Early in life my father occasionally wrote down a short account of some curious events & conversations, which are enclosed in a separate envelope.

The sharpness of his observation led him to predict with remarkable skill the course of any illness, & he suggested endless small details of relief. I was told that a young Doctor in Shrewsbury* who disliked my father, used to say that he was wholly unscientific, but owned that his power of predicting the end of an illness was unparalleled. Formerly when he thought that I sh^d be a doctor, he talked much to me about his patients. In the old days the practice of bleeding largely was universal, but my father maintained that far more evil was thus caused than good done; & he advised me, if ever I was myself ill not to allow any doctor to take from me more than extremely small quantity of blood. Long before typhoid fever was recognised as distinct, my father told me that two utterly distinct kinds of illness were confounded under the name of typhus fever. He was vehement against drinking, & was convinced of both the direct & inherited evil effects of alcohol, when habitually taken even in moderate quantity, in a very large majority of cases. But he admitted & advanced instances of certain persons, who could drink largely during their whole lives, without apparently suffering any evil effects; & he believed that he could often beforehand tell who would not thus suffer. He himself never drank a drop of any alcoholic fluid.

This remark reminds me of a case showing how a witness under the most favourable circumstances may be wholly mistaken. A gentleman-farmer was strongly urged by my father not to drink, & was encouraged by being told that he himself never touched any spirituous liquor. Whereupon the gentleman said, "come, come Doctor, that won't do,—though it is very kind of you to say so for my sake,—for I know that you take a very large glass of hot gin & water every evening after your dinner." So my father asked him how he knew this. The man answered my cook was your kitchen-maid for 2 or 3 years, & she saw the butler every day prepare & take to you the gin & water. The explanation was that my father had the odd habit of drinking hot water in a very tall & large glass after his dinner; & the butler used first to put some cold water in the glass, which the girl mistook for gin, & then filled it up with boiling water from the kitchen-boiler.

My father used to tell me many little things which he had found useful in his medical practice. Thus ladies often cried much while telling him their troubles, & thus caused much loss of his precious time. He soon found that begging them to command & restrain themselves, always made them weep the more, so that afterwards he always encouraged them to go on crying, saying that this would relieve them more than anything else, with the invariable result that they soon ceased to cry, & he could hear what they had to say & give his advice. When patients, who were very ill, craved for some strange & unnatural food, my father asked them what had put such an idea into their heads: if they answered that they did not know, he would allow them to try the food & often with success, as he trusted to their having a kind of instinctive desire; but if they answered that they had heard that the food in question had done good to some one else, he firmly refused his assent.

He gave one day an odd little specimen of human nature. When a very young man he was called in to consult with the family physician in the case of a gentleman of much distinction in Shropshire. The old doctor told the wife that the illness was of such a nature that it must end fatally. My father took a different view & maintained that the gentleman would recover; he was proved quite wrong in all respects (I think by autopsy) & he owned his error. He was then convinced that he sh^d never again be consulted by this family; but after a few months the widow sent for him, having dismissed the old family doctor. My father was so much surprised at this, that he asked a friend of the widow to find out why he was again consulted. The widow answered her friend, that she would never again see that odious old doctor who said from the first that her husband would die, while Dr Darwin always maintained that he would recover! In another case my father told a lady that her husband would certainly die. Some months afterward he saw the widow who was a very sensible woman & she said "you are a very young man & allow me to advise you always to give, as long as you possibly can, hope to any near relation nursing a patient. You made me despair, & from that moment I lost strength." My father said that he had often since seen the paramount importance for the sake of the patient of keeping up the hope & with it the strength of the nurse in charge. This he sometimes found it difficult to do compatibly with truth. One old gentleman, however, Mr Pemberton, caused him no such perplexity. He was sent for by M$^{r.}$ P. who said, "from all that I have seen & heard of you, I believe you are the sort of man who will speak the truth, & if I ask you will tell me when I am dying. Now I much desire that you should attend me, if you will promise, whatever I may say, always to declare that I am not going to die." My father acquiesced on this understanding that his words should in fact have no meaning.

My father possessed an extraordinary memory, especially for dates, so that he knew when he was very old the day of the birth, marriage & death of a multitude of persons in Shropshire; & he once told me that this power annoyed him, for if he once heard a date he could not forget it; & thus the deaths of many friends were often recalled to his mind. Owing to his strong memory he knew an extraordinary number of curious stories, which he liked to tell, as he was a great talker. He was generally in high spirits, & laughed & joked with everyone, often with his servants with the utmost freedom; yet he had the art of making everyone obey him to the letter. Many persons were much afraid of him. I remember my father telling us one day with a laugh that several persons had asked him whether Miss Pigott (a grand old lady in Shropshire) had called on him, so that at last he enquired why they asked him, & was told that Miss Pigott, whom my father had somehow mortally offended, was telling everybody that she would call & tell "that fat old doctor very plainly what she thought of him." She had already called, but her courage had failed, & no one could have been more courteous & friendly.— As a boy I went to stay at the house of Major B^y, whose wife was insane;* & the poor creature, as soon as she saw me, was in the most abject state of terror that I ever saw, weeping bitterly & asking me over & over again "is your father coming?"; but was soon pacified. On my return home I asked my father why she was so frightened, & he answered he [was] very glad to hear it, as he had frightened her on purpose, feeling sure that she could be kept in safety & much happier without any restraint, if her husband could influence her, whenever she became at all violent, by proposing to send for Dr Darwin; & these words succeeded perfectly during the rest of her long life.

My father was very sensitive so that many small events annoyed or pained him much. I once asked him when he was old & could not walk, why he did not drive out for exercise; & he answered "every road out of Shrewsbury is associated in my mind with some painful event". Yet he was generally in high spirits. He was easily made very angry, but as his kindness was unbounded, he was widely & deeply loved.

He was a cautious & good man of business, so that he hardly ever lost money by any investment, & left to his children a very large property. I remember a story, showing how easily utterly false beliefs originate & spread. Mr E. a squire of one of the oldest families in

Shropshire & head partner in a Bank,* committed suicide. My father was sent for as a matter of form, & found him dead. I may mention by the way to show how matters were managed in those old days, that because M^r· E was a rather great man & universally respected, no inquest was held over his body. My father in returning home thought it proper to call at the Bank (where he had an account) to tell the managing partner of the event, as it was not improbable it would cause a run on the bank. Well the story was spread far & wide that my father went into the bank, drew out all his money, left the bank, came back again & said "I may just tell you that M^r· E. has killed himself": & then departed. It seems that it was then a common belief that money withdrawn from a bank was not safe, until the person had passed out through the door of the bank. My father did not hear this story till some little time afterwards, when the managing partner said that he had departed from his invariable rule of never allowing anyone to see the account of another man, by having shown the ledger with my fathers account to several persons, as this proved that my father had not drawn out a penny on that day. It would have been dishonourable in my father to have used his professional knowledge for his private advantage. Nevertheless the supposed act was greatly admired by some persons; & many years afterwards a gentleman remarked, "ah Doctor what a splendid man of business you were in so cleverly getting all your money safe out of that bank."—

My fathers mind was not scientific, & he did not try to generalise his knowledge under general laws; yet he formed a theory for almost everything which occurred. I do not think that I gained much from him intellectually; but his example ought to have been of much moral service to all his children. One of his golden rules (a hard one to follow) was "never become the friend of anyone whom you cannot respect."

With respect to my father's father, the author of the Botanic Garden &c,* I have put together all the facts which I could collect in his published Life.

Having said this much about my father, I will add a few words about my brother & sisters. My brother Erasmus possessed a remarkably clear mind, with extensive & diversified tastes & knowledge in literature, art & even in science. For a short time he collected & dried plants, & during a somewhat longer time experimented in chemistry. He was extremely agreeable, & his wit often reminded me of that in the letters & works of Charles Lamb. He was very kind-hearted; but his health from his boyhood had been weak, & as a consequence he failed in energy. His spirits were not high, sometimes low, more especially during early & middle man-hood. He read much, even whilst a boy, & at school encouraged me to read, lending me books. Our minds & tastes were, however, so different, that I do not think that I owe much to him intellectually, nor to my four sisters who possessed very different characters, & some of them had strongly marked characters. All were extremely kind & affectionate towards me during their whole lives. I am inclined to agree with Francis Galton in believing that education & environment produce only a small effect on the mind of anyone, & that most of our qualities are innate. The above sketch of my brother's character was written before that which was published in Carlyle's Remembrances, & which appears to me to have little truth & no merit.

Looking back as well as I can at my character during my school-life, the only qualities which at this period promised well for the future, were, that I had strong & diversified tastes, much zeal for whatever interested me, & a keen pleasure in understanding any complex subject or thing. I was taught Euclid by a private tutor, & I distinctly remember the intense satisfaction which the clear geometrical proofs gave me. I remember with equal distinctness the delight which my uncle gave me (the father of Francis Galton) by explaining the principle of the vernier to a barometer. With respect to diversified tastes, independently of science, I was fond of reading various books, & I used to sit for hours reading the historical plays of Shakespeare, generally in an old window in the thick walls of the school. I read also other poetry, such as the recently published poems of Byron, Scott, Thompson's Seasons. I mention this because later in life I wholly lost, to my great regret, all pleasure from poetry of any kind, including Shakespeare. In connection with pleasure from poetry I may add that in 1822 a vivid delight in scenery was first awakened in my mind, during a riding tour on the borders of Wales, & which has lasted longer than any other æsthetic pleasure. Early in my school days a boy had a copy of the Wonders of the World,* which I often read & disputed with other boys about the veracity of some of the statements; & I believe this book first gave me a wish to travel in remote countries which was ultimately fulfilled by the voyage of the Beagle. In the latter part of my school-life I became passionately fond of shooting, & I do not believe that any one could have shown more zeal for the most holy cause than I did for shooting birds. How well I remember killing my first snipe, & my excitement was so great that I had much difficulty in reloading my gun from the trembling of my hands. This taste long continued & I became a very good shot. When at Cambridge I used to practice throwing up my gun to my shoulder before a looking glass to see that I threw it up straight. Another & better plan was to get a friend to wave about a lighted candle & then to fire at it with a cap on the nipple, & if the aim was accurate the little puff of air would blow out the candle. The explosion of the cap caused a sharp crack, & I was told that the Tutor of the College* remarked, "what an extraordinary thing it is Mr Darwin seems to spend hours in cracking a horsewhip in his room, for I often hear the crack when I pass under his windows."

I had many friends amongst the school-boys, whom I loved dearly, & I think that my disposition was then very affectionate. Some of these boys were rather clever, but I may add on the principle of 'noscitur a socio'* that not one of them ever became in the least distinguished.

With respect to science, I continued collecting minerals with much zeal, but quite unscientifically,—all that I cared for was a new *named* mineral, & I hardly attempted to classify them. I must have observed insects with some little care, for when 10 years old (1819) I went for 3 weeks to Plas Edwards on the sea-coast in Wales, I was very much interested & surprised at seeing a large black & scarlet Hemipterous insect, many moths (Zygæna), & a Cicindela which are not found in Shropshire.* I almost made up my mind to begin collecting all the insects which I could find dead, for on consulting my sister I concluded that it was not right to kill insects for the sake of making a collection. From reading White's Selbourne I took much pleasure in watching the habits of birds & even made notes on this subject. In my simplicity I remember wondering why every gentleman did not become an ornithologist.

Towards the close of my school-life, my brother worked hard at chemistry & made a fair laboratory with proper apparatus in the Tool-House in the garden, & I was allowed to aid him as a servant in most of his experiments. He made all the gases & many compounds, & I read with care several books on chemistry, such as Henry & Parkes Ch. Catechism. The subject interested me greatly, & we often used to go on working till rather late at night. This was the best part of my education at school, for it showed me practically the meaning of experimental science. The fact that we worked at chemistry somehow got known at school, & as it was an unprecedented fact, I was nicknamed 'gas'. I was also once publickly rebuked by the headmaster, D[r.] Butler, for thus wasting my time over such useless subjects; & he called me very unjustly a 'poco curante',* & as I did not understand what he meant it seemed to me a fearful reproach.

As I was doing no good at school, my father wisely took me away at a rather earlier age than usual, & sent me (Oct. 1825) to to[sic] Edinburgh University with my brother, where I staid for two years or sessions. My brother was completing his medical studies, though I do not believe he ever really intended to practice, & I was sent there to commence them. But soon after this period I became convinced from various small circumstances that my Father would leave me property enough to subsist on with some comfort, though I never imagined that I sh[d.] be so rich a man as I am; but my belief was sufficient to check any strenuous effort to learn medicine. The instruction at Edinburgh was altogether by Lectures, & these were intolerably dull, with the exception of those in chemistry by Hope; but to my mind there are no advantages & many disadvantages in lectures compared with reading. D[r] Duncan's lectures on materia medica* at 8 oclock on a winter's morning are something fearful to remember. D[r.] Munro made his lectures on human anatomy as dull, as he was himself, & the subject disgusted me. It has proved one of the greatest evils in my life that I was not urged to practice dissection, for I sh[d.] soon have got over my disgust; & the practice would have been invaluable for all my future work. This has been an irremediable evil, as well as my incapacity to draw. I also attended regularly the clinical wards in the Hospital. Some of the cases distressed me a good deal, & I still have vivid pictures before me of some of them; but I was not so foolish as to allow this to lessen my attendance. I cannot understand why this part of my medical course did not interest me in a greater degree; for during the summer before coming to Edinburgh I began attending some of the poor people, chiefly children & women, in Shrewsbury: I wrote down as full an account as I could of the cases with all the symptoms, & read them aloud to my Father, who suggested further enquiries, & advised me what medicines to give, which I made up myself. At one time I had at least a dozen patients, & I felt a keen interest in the work. My Father, who was by far the best judge of character whom I ever knew, declared that I sh^d. make a successful physician,—meaning by this one who got many patients. He maintained that the chief element of success was exciting confidence; but what he saw in me which convinced him that I should create confidence I know not. I also attended on two occasions the operating theatre in the Hospital at Edinburgh, & saw two very bad operations, one on a child, but I rushed away before they were completed. Nor did I ever attend again, for hardly any inducement would have been strong enough to make me do so; this being long before the blessed days of chloroform. The two cases fairly haunted me for many a long year.

My Brother staid only one year at the University, so that during the second year I was left to my own resources; & this was an advantage, for I became well acquainted with several young men fond of natural science. One of these was Ainsworth, who afterwards published his travels in Assyria; he was a Wernerian geologist* & knew a little about many subjects, but was superficial & very glib with his tongue. D^r. Coldstream was a very different young man, prim, formal, highly religious & most kind-hearted; he afterwards published some good zoological articles. A third young man was Hardie, who w^d. I think have made a good botanist, but died early in India. Lastly Dr Grant, my senior by several years, but how I became acquainted with him I cannot remember: he published some first-rate zoological papers, but after coming to London as Professor in University College he did nothing more in science,—a fact which has always been inexplicable to me. I knew him well; he was dry & formal in manner, but with much enthusiasm beneath the outer crust. He one day when we were walking together burst forth in high admiration of Lamarck & his views on evolution. I listened in silent astonishment, & as far as I can judge without any effect on my mind. I had previously read the Zoonomia by my grandfather, in which similar views are maintained, but without producing any effect on me. Nevertheless it is probable that the hearing rather early in life such views maintained & praised may have favoured my upholding them under a different form in my Origin of Species. At this time I admired greatly the Zoonomia; but on reading it a second time after an interval of 10 or 15 years, I was much disappointed; the proportion of speculation being so large to the facts given.

D^r. Grant & Coldstream attended much to marine zoology, & I often accompanied the former to collect animals in the tidal pools which I dissected as well as I could. I also became friends with some of the Newhaven fishermen, & sometimes accompanied them when they trawled for oysters, & thus got many specimens. But from not having had any regular practice in dissection & from possessing only a wretched microscope my attempts were very poor. Nevertheless I made one interesting little discovery, & read about the beginning of the year 1826 a short paper on the subject before the Plinian Soc^y. This was that the so-called ova of Flustra had the power of independent movement by means of cilia, & were in fact larvæ. In another short paper I showed that little globular bodies, which had been supposed to be the young state of Fucus loreus were the egg-cases of the worm-like Pontobdella muricata.*

The Plinian Society was encouraged & I believe founded by Prof. Jameson: it consisted of students & met in an underground room in the University for the sake of reading papers on natural science & discussing them. I used regularly to attend, & the meetings had a good effect on me in stimulating my zeal & giving me new congenial acquaintances. One evening a poor young man got up & after stammering for a prodigious length of time, blushing crimson, he at last slowly got out the words "Mr President I have forgotten what I was going to say". The poor fellow looked quite overwhelmed & all the members were so surprised that no one could think of a word to say to cover his confusion. The papers which were read to our little Society, were not printed, so that I had not the satisfaction of seeing my paper in print; but I believe D^r. Grant noticed my small discovery in his excellent memoir on Flustra. I was also a member of the R. Medical Soc^y & attended pretty regularly, but as the subjects were exclusively medical, I did not much care about them. Much rubbish was talked there, but there were some good speakers, of whom the best was the present Sir J. Kay Shuttleworth. D^r. Grant took me occasionally to the meetings of the Wernerian Soc^y, where various papers on natural history were read, discussed, & afterwards published in the Transactions. I heard Audubon deliver there some interesting discourses on the habits of N. American birds, sneering somewhat unjustly at Waterton. By the way a negro lived in Edinburgh, who had travelled with Waterton, & gained his livelihood by stuffing birds, which he did excellently: he gave me lessons for payment, & I used often to sit with him, for he was a very pleasant & intelligent man.*

Mr Leonard Horner also took me once to a meeting of the Royal Soc. of Edinburgh,* where I saw Sir W. Scott in the chair as President, & he apologised to the meeting as not feeling fitted for such a position. I looked at him & at the whole scene with some awe & reverence; & I think it was owing to this visit during my youth & to my having attended the R. Medical Soc^y, that I felt the honour of being elected a few years ago an Honorary Member of both these Societies, more than any other similar honour. If I had been told at that time that I should one day have been thus honoured, I declare that I sh^d have thought it as ridiculous & improbable, as if I had been told that I sh^d be elected King of England.

During my second year in Edinburgh I attended Jameson's lectures on Geology & Zoology, but they were incredibly dull. The sole effect they produced on me was the determination never as long as I lived to read a book on geology or in any way to study the science. Yet I feel sure that I was prepared for a philosophical treatment of the subject; for an old M^r. Cotton in Shropshire who knew a good deal about rocks had pointed out to me, 2 or 3 years previously, a well-known large erratic boulder* in the town of Shrewsbury called the bell-stone; he told me that there was no rock of the same kind nearer than Cumberland or Scotland, & he solemnly assured me that the world would come to an end before anyone would be able to explain how this stone came where it now lay. This produced a deep impression on me & I meditated over this wonderful stone. So that I felt the keenest delight when I first read of the action of icebergs in transporting boulders, & I gloried in the progress of geology. Equally striking is the fact that I though now only 67 years old heard Prof Jameson in a field lecture at Salisbury Craigs discoursing on a trap-dyke, with amygdaloidal margins & the strata indurated on each side,* with volcanic rocks all around us, & say that it was a fissure filled with sediment from above, adding with a sneer that there were men who maintained that it had been injected from beneath in a molten condition. When I think of this lecture, I do not wonder that I determined never to attend to Geology.

From attending Jameson's lectures, I became acquainted with the curator of the Museum, Mr Macgillivray, who afterwards published a large & excellent book on the birds of Scotland. He had not much of the appearance or manners of the gentleman, but I had much interesting natural history talk with him, & he was very kind to me. He gave me some rare shells, for I at that time collected marine mollusca, but with no great zeal.

My summer vacations during these two years were wholly given up to amusements, though I always had some book in hand which I read with interest. During the summer of 1826 I took a long walking tour with two friends with knapsacks on our backs through North Wales. We walked 30 miles most days, including one day the ascent of Snowdon. I also went with my sister Caroline a riding tour in N. Wales, a servant with saddle-bags carrying our clothes. The autumns were devoted to shooting, chiefly at Mr Owens at Woodhouse, & at my Uncle Jos' at Maer.* My zeal was so great that I used to place my shooting boots open by my bed side, when I went to bed so as not to lose half-a-minute in putting them on in the morning; & on one occasion I reached a distant part of the Maer Estate on the 20th of August for black-game shooting before I could see: I then toiled on with the gamekeeper the whole day through thick heath & young Scotch-firs. I kept an exact record of every bird which I shot throughout the whole season. One day when shooting at Woodhouse with Capt. Owen, the eldest son & Major Hill, his cousin, afterwards Ld Berwick, both of whom I liked very much, I thought myself shamefully used, for every time after I had fired & thought that I had killed a bird, one of the two acted as if loading his gun & cried out you must not count that bird, for I fired at the same time; & the gamekeeper perceiving the joke backed them up. After some hours they told me the joke, but it was no joke to me for I had shot a large number of birds, but did not know how many & could not add them to my list, which I used to do by making a knot to a piece of string tied to a button-hole. This my wicked friends had perceived. How I did enjoy shooting, but I think that I must have been half-consciously ashamed of my zeal, for I tried to persuade myself that shooting was almost an intellectual employment; it required so much skill to judge where to find most game & to hunt the dogs well.

One of my autumnal visits to Maer in 1827 was memorable from meeting there Sir J. Mackintosh, who was the best converser I ever listened to. I heard afterward with a glow of pride that he had said "there is something in that young man which interests me". This must have been chiefly due to his perceiving that I listened with much interest to everything which he said, for I was as ignorant as a pig about his subjects of history, politicks & moral philosophy. To hear of praise from an eminent person, though no doubt apt or certain to excite vanity, is, I think, good for a young man, as it helps to keep him in the right course.

My visits to Maer during these two & the three succeeding years were quite delightful, independently of the autumnal shooting. Life there was perfectly free; the country was very pleasant for walking or riding; & in the evening there was much very agreeable conversation, not so personal as it generally is in large family parties, together with music. In the summer the whole family used often to sit on the steps of the old portico, with the flower-garden in front, & with the steep wooded bank, opposite to the house, reflected in the lake, with here & there a fish rising or a water-bird paddling about. Nothing has left a more vivid picture on my mind than those evenings at Maer. I was also attached to & greatly revered my Uncle Jos: he was silent & reserved so as to be a rather awful man; but he sometimes talked openly with me. He was the very type of an upright man with the clearest judgment. I do not believe that any power on earth could have made him swerve an inch from what he considered the right course. I used to apply to him in my mind, the well-known ode of Horace, now forgotten by me, in which the words 'nec vultus tyranni &c'* come in.

Cambridge 1828–1831.—

After having spent two sessions in Edinburgh, my Father perceived or he heard from my sisters that I did not like the thought of being a physician, so he proposed that I sh^d. become a clergyman. He was very properly vehement against my turning an idle sporting man, which then seemed my probable destination. I asked for some time to consider, as from what little I had heard & thought on the subject I has scruples about declaring my belief in all the dogmas of the Church of England; though otherwise I liked the thought of being a country clergyman. Accordingly I read with care Pearson on the Creeds & a few other books on divinity; & as I did not then in the least doubt the strict & literal truth of every word in the Bible, I soon persuaded myself that our Creed must be fully accepted. It never struck me how illogical it was to say that I believed in what I could not understand & what is in fact unintelligible. I might have said with entire truth that I had no wish to dispute any dogma; but I never was such a fool as to feel & say 'credo quia incredibile'.* Considering how fiercely I have been attacked by the orthodox, it seems ludicrous that I once intended to be a clergyman. Nor was this intention & my Father's wish ever formally given up, but died a natural death when on leaving Cambridge I joined the Beagle as Naturalist. If the phrenologists* are to be trusted, I was well fitted in one respect to be a clergyman. A few years ago the Secretaries of a German psychological Soc^y asked me earnestly by letter for a photograph of myself; & some time afterwards I received the Proceedings of one of the meetings, in which it seemed that the shape of my head had been the subject of a public discussion, & one of the speakers declared that I had the bump of Reverence developed enough for ten Priests.

As it was decided that I sh^d be a clergyman, it was necessary that I sh^d go to one of the English Universities & take a degree; but as I had never opened a classical book since leaving school, I found to my dismay that in the two intervening years I had actually forgotten, incredible as it may appear, almost everything which I had learnt even to some few of the Greek letters. I did not therefore proceed to Cambridge at the usual time in October, but worked with a private tutor in Shrewsbury & went to Cambridge after the Christmas vacation, early in 1828. I soon recovered my school standard of knowledge, & could translate easy Greek books, such as Homer & the Greek Testament with moderate facility.

During the three years which I spent at Cambridge my time was wasted, as far as the academical studies were concerned, as completely as at Edinburgh & at school. I attempted mathematicks, & even went during the summer of 1828 with a private tutor* (a very dull man) to Barmouth, but I got on very slowly. The work was repugnant to me, chiefly from not my not being able to see any meaning in the early steps in algebra. This impatience was very foolish, & in after years I have deeply regretted that I did not proceed far enough at least to understand something of the great leading principles of mathematicks; for men thus endowed seem to have an extra sense. But I do not believe that I sh^d ever have succeeded beyond a very low grade. With respect to classics I did nothing except attend a few compulsory college lectures, & the attendance was almost nominal. In my second year I had to work for a month or two to pass the Little Go,* which

I did easily. Again in my last year I worked with some earnestness for my final degree of B.A., & brushed up my Classics together with a little algebra & Euclid, which latter gave me much pleasure as it did whilst at school. In order to pass the B.A. examination it was, also, necessary to get up Paleys Evidences of Christianity & his Moral Philosophy. This was done in a thorough manner, & I am convinced that I could have written out the whole of the Evidences with perfect correctness, but not of course in the clear language of Paley. The logic of this book, & as I may add of his Natural Theology* gave me as much delight as did Euclid. The careful study of these works, without attempting to learn any part by rote, was the only part of the Academical Course, which as I then felt & as I still believe, was of the least use to me in the education of my mind. I did not at that time trouble myself about Paley's premises; & taking these on trust I was charmed & convinced by the long line of argumentation. By answering well the examination questions in Paley, by doing Euclid well, & by not failing miserably in Classics, I gained a good place amongst the οι πολλοι, or crowd of men who do not go in for honours. Oddly enough I cannot remember how high I stood, & my memory fluctuates between the fifth, tenth or twelfth name on the list.*

Public lectures on several branches were given in the University, attendance being quite voluntary; but I was so sickened with lectures at Edinburgh that I did not even attend Sedgwicks eloquent & interesting lectures. Had I done so I sh^d. probably have become a geologist earlier than I did. I attended, however, Henslow's lectures on Botany, & liked them much for their extreme clearness & the admirable illustrations; but I did not study botany. Henslow used to take his pupils, including several of the older members of the University, field excursions, on foot, or in coaches to distant places, or in a barge down the river, & lectured on the rarer plants or animals which were observed. These excursions were delightful.

Although as we shall presently see there were some redeeming features in my life at Cambridge, my time was sadly wasted there & worse than wasted. From my passion for shooting & for hunting & when this failed for riding across country I got into a sporting set, including some dissipated, low-minded young men. We used often to dine together in the evening, though these dinners often included men of a higher stamp, & we sometimes drank too much, with jolly singing & playing at cards afterwards. I know that I ought to feel ashamed of days & evenings thus spent, but as some of my friends were very pleasant, & we were all in the highest spirits, I cannot help looking back to those times with much pleasure.

But I am glad to think that I had many other friends of a widely different nature. I was very intimate with Whitley, who was afterwards Senior Wrangler,* & we used continually to take long walks together. He inoculated me with a taste for pictures & good engravings, of which I bought some. I frequently went to the Fitzwilliam Gallery,* & my taste must have been fairly good, for I certainly admired the best pictures, which I discussed with the old curator. I read also with much interest Sir J. Reynolds book. This taste, though not natural to me, lasted for several years, & many of the pictures in the National Gallery in London gave me much pleasure; that of Sebastian del Piombo exciting in me a sense of sublimity.

I also got into a musical set, I believe by means of my warm-hearted friend, Herbert, who took a high wrangler's degree. From associating with these men & hearing them play, I acquired a strong taste for music & used very often to time my walks so as to hear on week-days the anthem in King's College Chapel. This gave me intense pleasure, so that my back-bone would sometimes shiver. I am sure that there was no affectation or mere imitation in this taste, for I used generally to go by myself to Kings College, & I sometimes hired the chorister boys to sing in my rooms. Nevertheless I am so utterly destitute of an ear, that I cannot perceive a discord or keep time & hum a tune correctly; & it is a mystery how I could possibly have derived pleasure from music. My musical friends soon perceived my state, & sometimes amused themselves by making me pass an examination, which consisted in ascertaining how many tunes I could recognise, when they were played rather more quickly or slowly than usual. 'God save the King' when thus played was a sore puzzle. There was another man with almost as bad an ear as I had, & strange to say he played a little on the flute. Once I had the triumph of beating him in one of our musical examinations.

But no pursuit at Cambridge was followed with nearly so much eagerness or gave me so much pleasure as collecting beetles. It was the mere passion for collecting, for I did not dissect them & rarely compared their external characters with published descriptions, but got them named anyhow. I will give a proof of my zeal: one day on tearing off some old bark, I saw two rare beetles & seized one in each hand; then I saw a third & new kind, which I could not bear to lose, so that I popped the one which I held in my right hand into my mouth. Alas it ejected some intensely acrid fluid, which burnt my tongue so that I was forced to spit the beetle out, which was lost, as well as the third one. I was very successful in collecting & invented two new methods; I employed a labourer to scrape during the winter moss off old trees & place in a large bag, & likewise to collect the rubbish at the bottom of the barges in which reeds are brought from the fens, & thus I got some very rare species. No poet ever felt more delight at seeing his first poem published than I did at seeing in 'Stephen's Illustrations of British Insects' the magic words "captured by C. Darwin E^sq." I was introduced to Entomology by my second-cousin, W. Darwin Fox, a clever & most pleasant man, who was then at Christ College & with whom I became extremely intimate. Afterwards I became well acquainted with & went out collecting with Albert Way of Trinity, who in after years became a well-known archæologist; also with H. Thompson of the same college, afterwards a leading agriculturist, chairman of a great Railway & member of parliament. It seems therefore that a taste for collecting beetles is some indication of future success in life.!

I am surprised what an indelible impression many of the beetles which I caught at Cambridge have left on my mind. I can remember the exact appearance of certain posts, old trees & banks where I made a good capture. The pretty Panagæus crux-major* was a treasure in those days, & here at Down I saw a beetle running across a walk, & on picking it up instantly perceived that it differed slightly from P. crux major, & it turned out to be P. quadripunctatus, which is only a variety or closely allied species differing from it very slightly in outline. I had never seen in those old days Licinus alive, which to an uneducated eye hardly differs from many other black Carabidous* beetles; but my sons found here a specimen & I instantly recognised that it was new to me; yet I had not looked at a British beetle for the last twenty years.

I have not as yet mentioned a circumstance which influenced my whole career more than any other. This was my friendship with Prof. Henslow. Before coming up to Cambridge, I had heard of him from my brother as a man who knew every branch of science, & I was accordingly prepared to reverence him. He kept open house once every week, where all undergraduates & several older members of the Universitea, who were attached to science, used to meet in the evening. I soon got through Fox an invitation & went there regularly. Before long I became well acquainted with Henslow, & during the latter half of my time at Cambridge took long walks with him on most days; so that I was called by some of the dons "the man who walks with Henslow"; & in the evening I was very often asked to join his family dinner. His knowledge was great in botany, entomology, chemistry, mineralogy & geology. His strongest taste was to draw conclusions from long-continued minute observations. His judgment was excellent, & his whole mind well balanced; but I do not suppose that anyone would say that he possessed much original genius. He was deeply religious, & so orthodox that he told me one day, he sh^d be grieved if a single word in the 39 Articles* were altered. His moral qualities were in every way admirable. He was free from every tinge of vanity or other petty feeling; & I never saw a man who thought so little about himself or his own concerns. His temper was imperturbably good, with the most winning & courteous manners; yet, as I have seen, he could be roused by any bad action to the warmest indignation & prompt action. I once saw in his company in the streets of Cambridge, almost as horrid a scene, as could have been witnessed during the French Revolution. Two body-snatchers had been arrested & whilst being taken to prison had been torn from the constable by a crowd of the roughest men, who dragged them by their legs along the muddy & stony road. They were covered from head to foot with mud & their faces were bleeding either from having been kicked or from the stones; they looked like corpses, but the crowd was so dense that I got only a few momentary glimpses of the wretched creatures. Never in my life have I seen such wrath painted on a man's face, as was shown by Henslow at this horrid scene. He tried repeatedly to penetrate the mob; but it was simply impossible. He then rushed away to the mayor, telling me not to follow him, to get more policemen. I forget the issue, except that the two were got into the prison before being killed.

Henslow's benevolence was unbounded, as he proved by his many excellent schemes for his poor parishioners, when in after years he held the living of Hitcham.* My intimacy with such a man ought to have been & I hope was an inestimable benefit. I cannot resist mentioning a trifling incident, which showed his kind consideration.

Whilst examining some pollen-grains on a damp surface I saw the tubes exserted,* & instantly rushed off to communicate my surprising discovery to him. Now I do not suppose any other Professor of Botany could have helped laughing at my coming in a hurry to make such a communication. But he agreed how interesting the phenomenon was & explained its meaning, but made me clearly understand how well it was known; so I left him not in the least mortified, but well pleased at having discovered for myself so remarkable a fact, but determined not to be in such a hurry again to communicate my discoveries.

D^r. Whewell was one of the older & distinguished men who sometimes visited Henslow, & on several occasions I walked home with him at night. Next to Sir J. Mackintosh he was the best converser on grave subjects to whom I ever listened. Leonard Jenyns, (grandson of the famous Soames Jenyns) who afterwards published some good essays in Natural History, often staid with Henslow, who was his brother-in-law. At first I disliked him from his somewhat grim & sarcastic expression; & it is not often that a first impression is lost; but I was completely mistaken & found him very kind-hearted, pleasant & with a good stock of humour. I visited him at his parsonage on the borders of the Fens, & had many a good walk & talk with him about Natural History. I became also acquainted with several other men older than me, who did not care much about science, but were friends of Henslow. One was a scotchman, brother of Sir Alexander Ramsay, & tutor of Jesus College;* he was a delightful man, but did not live for many years. Another was M^r. Dawes, afterwards Dean of Hereford & famous for his success in the education of the poor. These men & others of the same standing, together with Henslow used sometimes to take distant excursions into the country, which I was allowed to join & they were most agreeable. Looking back, I infer that there must have been something in me a little superior to the common run of youths, otherwise the above-mentioned men, so much older than me & higher in academical position, would never have allowed me to associate with them. Certainly I was not aware of any such superiority, & I remember one of my sporting friends, Turner, who saw me at work on my beetles, saying that I should some day be a fellow of the Royal Society, & the notion seemed to me preposterous.

During my last year at Cambridge I read with care & profound interest Humboldts Personal Narrative. This work & Sir J. Herschels

Introduction to the Study of Natural Philosophy stirred up in me a burning zeal to add even the most humble contribution to the noble structure of Natural Science. No one or a dozen other books influenced me nearly so much as these two. I copied out from Humboldt long passages about Teneriffe, & read them aloud on one of the above-mentioned excursions, to (I think) Henslow, Ramsay & Dawes; for on a previous occasion I had talked about the glories of Teneriffe & some of the party declared they would endeavour to go there; but I think that they were only half in earnest. I was, however, quite in earnest, & got an introduction to a merchant in London to enquire about ships; but the scheme was of course knocked on the head by the voyage of the Beagle.

My summer vacations were given up to collecting beetles, to some reading & short tours. In the autumn my whole time was devoted to shooting, chiefly at Woodhouse & Maer, & sometimes with young Eyton of Eyton. Upon the whole the three years which I spent at Cambridge were the most joyful in my happy life; for I was then in excellent health, & almost always in high spirits.

As I had at first come up to Cambridge at Christmas, I was forced to keep two terms after passing my final examination, at the commencement of 1831; & Henslow then persuaded me to begin the study of geology. Therefore on my return to Shropshire I examined sections & coloured a map of parts round Shrewsbury. Prof. Sedgwick intended to visit N. Wales in the beginning of August to pursue his famous geological investigation amongst the older rocks, & Henslow asked him to allow me to accompany him. Accordingly he came & slept at my Father's house. A short conversation with him during this evening produced a strong impression on my mind. Whilst examining an old gravel pit near Shrewsbury a labourer told me that he had found in it, a large worn tropical Volute shell,* such as may be seen on the chimney-pieces of cottages; & as he would not sell the shell I was convinced that he had really found it in the pit. I told Sedgwick of the fact, & he at once said (no doubt truly) that it must have been thrown away by some one into the pit; but then added, if really embedded there it would be the greatest misfortune to geology, as it would overthrow all that we know about the superficial deposits of the midland counties. These gravel-beds belonged in fact to the glacial period, & in after years I found in them broken arctic shells. But I was then utterly astonished at Sedgwick not being delighted at so wonderful a fact as a tropical shell being found near the surface in the middle of England. Nothing before had ever made me thoroughly realise, though I had read various scientific books, that science consists in grouping facts so that general laws or conclusions may be drawn from them.

Next morning we started for Llangollen, Conway Bangor & Capel Curig. This tour was of decided use in teaching me a little how to make out the geology of a country. Sedgwick often sent me on a line parallel to his, telling me to bring back specimens of the rocks & to mark the stratification on a map. I have little doubt that he did this for my good, as I was too ignorant to have aided him. On this tour I had a striking instance how easy it is to overlook phenomena, however conspicuous, before they have been observed by any one. We spent many hours in Cwm Idwall, examining all the rocks with extreme care, as Sedgwick was anxious to find fossils in them; but neither of us saw a trace of the wonderful glacial phenomena all around us: we did not notice the plainly scored rocks, the perched boulders, the lateral & terminal moraines.* Yet these phenomena are so conspicuous, that as I declared in a paper published many years afterwards in the Philosophical magazine, a house burnt down by fire did not tell its story more plainly than did this valley. If it had still been filled by a glacier, the phenomena would have been less distinct than they now are.

At Capel-Curig I left Sedgwick & went in a straight line by compass & map across the mountains to Barmouth, never following any track unless it coincided with my course. I thus came on some strange wild places & enjoyed much this manner of travelling. I visited Barmouth to see some Cambridge friends who were reading there, & thence returned to Shrewsbury & to Maer for shooting; for at that time I sh^d. have thought myself mad to give up the first days of partridge-shooting for geology or any other science.

Voyage of the Beagle from Decr 27th, 1831 to Oct 2^d. 1836.—

On returning home from my short geological tour in N. Wales, I found a letter from Henslow, informing me that Captain FitzRoy was willing to give up part of his own cabin to any young man who would volunteer to go with him, without pay as naturalist to the voyage of the Beagle. I have given as I believe in my M.S. Journal* an account of all the circumstances which then occurred; & will here only say that I was instantly eager to accept the offer, but my Father strongly objected, adding the words fortunate for me,—"if you can find any man of common sense, who advises you to go, I will give my consent." So I wrote that evening & refused the offer. On the next morning I went to Maer to be ready for Sept. 1ˢᵗ, & whilst out shooting, my Uncle sent for me offering to drive me over to Shrewsbury & talk with my Father. As my Uncle thought it would be wise in me to accept the offer, & as my Father always maintained that he was one of the most sensible men in the world, he at once consented in the kindest manner. I had been rather extravagant at Cambridge & to console my Father said "that I shᵈ· be deuced clever to spend more than my allowance whilst on board the Beagle"; but he answered with a smile, "but they all tell me you are very clever". Next day I started for Cambridge to see Henslow, & thence to London to see FitzRoy, & all was soon arranged. Afterwards on becoming very intimate with FitzRoy, I heard that I had run a very narrow risk of being rejected, on account of the shape of my nose! He was an ardent disciple of Lavater, & was convinced that he could judge of a man's character by the outline of his features; & he doubted whether anyone with my nose could possess sufficient energy & determination for the voyage. But I think he was afterwards well-satisfied that my nose had spoken falsely.

FitzRoy's character was a singular one, with many very noble features: he was devoted to his duty, generous to a fault, bold, determined, indomitably energetic, & an ardent friend to all under his sway. He would undertake any amount of trouble to assist those whom he thought deserved assistance. He was a handsome man, strikingly like a gentleman, with highly courteous manners, which resembled those of his maternal uncle, the famous Lᵈ· Castlereagh, as I was told by the Minister at Rio. Nevertheless he must have inherited much in his appearance from Charles II., for Dʳ· Wallich gave me a collection of photographs, which he had made, & I was struck with the resemblance of one to FitzRoy; on looking at the name & found it Ch. E. Sobieski Stuart, Count d'Albanie, an illegitimate descent of this same monarch. FitzRoy's temper was a most unfortunate one, & this was shown not only by passion, but by fits of long-continued moroseness against those who had offended him. His temper was usually worst in the early morning, & with his eagle eye he could generally detect something amiss about the ship, & was then unsparing in his blame. The junior officers, when they relieved each other in the forenoon used to ask "whether much hot coffee had been served out this morning,"—which meant how was the captain's temper? He was also somewhat suspicious & occasionally in very low spirits, on one occasion bordering on insanity. He seemed to me often to fail in sound judgment or common sense. He was extremely kind to me, but was a man very difficult to live with on the intimate terms which necessarily followed from our messing by ourselves in the same cabin. We had several quarrels, for when out of temper he was utterly unreasonable. For instance, early in the voyage at Bahia in Brazil he defended & praised slavery which I abominated, & told me that he had just visited a great slave-owner, who had called up many of his slaves & asked them whether they were happy, & whether they wished to be free, & all answered no. I then asked him, perhaps with a sneer, whether he thought that the answers of slaves in the presence of their master was worth anything. This made him excessively angry, & he said, that as I doubted his word, we could not live any longer together. I thought that I sh^d have been compelled to leave the ship; but as soon as the news spread, which it did quickly, as the Captain sent for the first Lieutenant* to assuage his anger by abusing me, I was deeply gratified by receiving an invitation from all the gun-room officers to mess with them. But after a few hours FitzRoy showed his usual magnanimity, by sending an officer to me with an apology & a request that I would continue to live with him.

I remember another instance of his candour. At Plymouth, before we sailed, he was extremely angry with a dealer in crockery who refused to exchange some article purchased in his shop: the Captain asked the man the price of a very expensive set of china & said "I should have purchased this, if you had not been so disobliging". As I knew that the cabin was amply stocked with crockery, I doubted whether he had any such intention; & I must have shown my doubts in my face, for I said not a word. After leaving the shop, he looked at me, saying you do not believe what I have said, & I was forced to own that it was so. He was silent for a few minutes, & then said you are right, & I acted wrongly in my anger at the blackguard.

At Conception in Chile, poor FitzRoy was sadly overworked & in very low spirits; he complained bitterly to me that he must give a great party to all the inhabitants of the place. I remonstrated & said that I could see no such necessity on his part under the circumstances. He then burst out into a fury, declaring that I was the sort of man who would receive any favours & make no return. I got up & left the cabin without saying a word, & returned to Conception, where I was then lodging. After a few days I came back to the ship, & was received by the Captain, as cordially as ever, for the storm had by that time quite blown over. The first Lieutenant, however, said to me: "Confound you, philosopher, I wish you would not quarrel with the skipper; the day you left the ship I was dead-tired (the ship was refitting) & he kept me walking the deck till midnight, abusing you all the time."

The difficulty of living on good terms with a Captain of a Man of War is much increased by its being almost mutinous to answer him, as one would answer anyone else; & by the awe in which he is held or was held in my time, by all on board. I remember hearing a curious instance of this in the case of the Purser of the Adventure,*—the ship which sailed with the Beagle during the first Voyage. The Purser was in a store in Rio de Janeiro, purchasing rum for the ship's company, & a little gentleman in plain clothes walked in. The Purser said to him, "Now, Sir, be so kind as to taste this rum, & give me your opinion of it". The gentleman did as he was asked, & soon left the store. The store-keeper then asked the Purser, whether he knew that he had been speaking to the Captain of a Line of Battle Ship which had just come into the harbour. The poor purser was struck dumb with horror; he let the glass of spirits drop from his hand on the floor, & immediately went on board, & no persuasion, as an officer in the Adventure assured me, could make him go on shore again for fear of meeting the Captain after his dreadful act of familiarity.

I saw FitzRoy only occasionally after our return home, for I was always afraid of unintentionally offending him, & did do so once, almost beyond mutual reconciliation. He was afterward very indignant with me for having published so unorthodox a book (for he became very religious) as the Origin of Species. Towards the close of his life he was, as I fear, much impoverished, & this was largely due to his generosity. Anyhow after his death a subscription was raised to pay his debts. His end was a melancholy one, namely suicide exactly like that of his uncle L^d. Castlereagh, whom he resembled closely, in manners & appearance. His character was in several respects one of the most noble which I have ever known, though tarnished by grave blemishes.

The voyage of the Beagle has been by far the most important event in my life & has determined my whole career; yet it depended on so small a circumstance as my Uncle offering to drive me 30 miles to Shrewsbury, which few uncles would have done, & on such a trifle as the shape of my nose. I have always felt that I owe to the Voyage the first real training or education of my mind. I was led to attend closely to several branches of natural history, & thus my powers of observation were improved, though they were already fairly developed. The investigation of the geology of all the places visited was far more important, as reasoning here comes into play. On first examining a new district nothing can appear more hopeless than the chaos of rocks; but by recording the stratification & nature of the rocks & fossils at many points, always reasoning & predicting what will be found elsewhere, light soon begins to dawn on the district, & the structure of the whole becomes more or less intelligible. I had brought with me the first volume of Lyell's Principles of Geology, which I studied attentively; & this book was of the highest service to me in many ways. The very first place which I examined, namely St. Jago in the Cape Verde islands showed me clearly the wonderful superiority of Lyell's manner of treating geology, compared with that of any other author, whose works I had with me or ever afterwards read. Another of my occupations was collecting animals of all classes, briefly describing & roughly dissecting many of the marine ones; but from not being able to draw & from not having sufficient anatomical knowledge a great pile of M.S which I made during the voyage has proved almost useless. I thus lost much time, with the exception of that spent in acquiring some knowledge of the Crustaceans, as this was of service when in after years I undertook a monograph of the Cirripedia.* During some part of the day I wrote my Journal, & took much pains in describing carefully & vividly all that I had seen; & this was good practice. My Journal served, also, in part as letters to my home, & portions were sent to England, whenever there was an opportunity.

The above various special studies were, however, of no importance compared with the habit of energetic industry & of concentrated attention to whatever I was engaged in, which I then acquired. Everything about which I thought or read was made to bear directly on what I had seen & was likely to see; & this habit of mind was continued during the five years of the voyage. I feel sure that it was this training which has enabled me to do whatever I have done in science.

Looking backwards I can now perceive how my love for science gradually preponderated over every other taste. During the first two years my old passion for shooting survived in nearly full force, & I shot myself all the birds & animals for my collection; but gradually I gave up my gun more & more, & finally altogether to my servant, as shooting interfered with my work, more especially with making out the geological structure of a country. I discovered, though unconsciously & insensibly that the pleasure of observing & reasoning was a much higher one than that of skill & sport. The primeval instincts of the barbarian slowly yielded to the acquired tastes of the civilised man. That my mind became developed through my pursuits during the voyage, is rendered probable by a remark made by my Father, who was the most acute observer whom I ever saw, of a sceptical disposition, & far from being a believer in phrenology; for on first seeing me after the Voyage, he turned round to my sisters, & exclaimed, "why the shape of his head is quite altered."

To return to the voyage. On Sept 11th (1831) I paid a flying visit with FitzRoy to the Beagle at Plymouth. Thence to Shrewsbury to wish my Father & sisters a long farewell. On Oct 24th, I took up my residence at Plymouth & remained there, until December, 27th, when the Beagle finally left the shores of England for her circumnavigation of the world. We made two earlier attempts to sail, but were driven back each time by heavy gales. These two months at Plymouth were the most miserable which I ever spent, though I exerted myself in various ways. I was out of spirits at the thought of leaving all my family & friends for so long a time, & the weather seemed to me inexpressibly gloomy. I was also troubled with palpitations & pain about the heart, & like many a young ignorant man, especially one with a smattering of medical knowledge, was convinced that I had heart-disease. I did not consult any doctor, as I fully expected to hear the verdict that I was not fit for the voyage, & I was resolved to go at all hazards.

I need not here refer to the events of the Voyage,—where we went & what we did—as I have given a sufficiently full account in my published Journal. The glories of the vegetation of the Tropics rise before my mind at the present time more vividly than anything else.

Though the sense of sublimity, which the great deserts of Patagonia & the forest-clad mountains of Tierra del Fuego excited in me, has left an indelible impression on my mind. The sight of a naked savage in his native land is an event which can never be forgotten. Many of my excursions on horseback through wild countries, or in the boats, some of which lasted several weeks, were deeply interesting: their discomfort & some degree of danger were at the time hardly a draw-back & none at all afterwards. I also reflect with high satisfaction on some of my scientific work, such as solving the problem of coral-islands & making out the geological structure of certain islands, for instance St. Helena. Nor must I pass over the discovery of the singular relations of the animals & plants inhabiting the several islands of the Galapagos archipelago, & of all of them to the inhabitants of South America.

As far as I can judge of myself, I worked to the utmost during the Voyage from the mere pleasure of investigation, & from my strong desire to add a few facts to the great mass of facts in natural science. But I was also ambitious to take a fair place amongst scientific men,—whether more ambitious or less so than most of my fellow-workers I can form no opinion. The geology of St. Jago is very strik-ing yet simple: a stream of lava formerly flowed over the bed of the sea, formed of triturated* recent shells & corals, which it has baked into a hard white rock. Since then the whole island has been upheaved. But the line of white rock revealed to me a new & import-ant fact, namely that there had been afterwards subsidence round the craters which had since been in action & had poured forth lava. It then first dawned on me that I might perhaps write a book on the geology of the various countries visited, & this made me thrill with delight. That was a memorable hour to me, & how distinctly I can call to mind the low cliff of lava beneath which I rested, with the sun glaring hot, a few strange desert plants growing near, & with living corals in the tidal pools at my feet. Later in the voyage FitzRoy asked to read some of my Journal, & declared it would be worth publishing; so here was a second book in prospect! Towards the close of our voyage I received a letter whilst at Ascension, in which my sisters told me that Sedgwick had called on my Father & said that I sh^d take a place amongst the leading scientific men. I could not at the time understand how he could have learnt anything of my proceed-ing, but I heard (I believe afterwards) that Henslow had read some of the letters which I wrote to him, before the Philosophical Soc. of Cambridge & had printed them for private distribution. My collection of fossil bones, which had been sent to Henslow, also excited considerable attention amongst palæontologists. After reading this letter I clambered over the mountains of Ascension with a bounding step, & made the volcanic rocks resound under my geological hammer! All this shows how ambitious I was; but I think that I can say with truth that in after years, though I cared in the highest degree for the approbation of such men as Lyell & Hooker, who were my friends, I did not care much about the general public. I do not mean to say that a favourable review or a large sale of my books did not please me greatly; but the pleasure was a fleeting one, & I am sure that I have never turned one inch out of my course to gain fame.

From my return to England (Oct 2 1836) to my marriage (Jany 29th 1839)

These two years & three months were the most active ones which I ever spent, though I was occasionally unwell & so lost some time. After going backwards & forwards several times between Shrewsbury, Maer, Cambridge & London, I settled in lodgings at Cambridge on Dec. 13th, where all my collections were under the care of Henslow. I staid here three months & got my minerals & rocks examined by the aid of Prof. Miller. I began preparing my Journal of Travels, which was not hard work, as my M.S Journal had been written with care, & my chief labour was making an abstract of my more interesting scientific results. I sent also at the request of Lyell a short account of my observations on the elevation of the coast of Chile to the Geological Society. On March 7th 1837 I took lodgings in Great Marlborough St in London & remained there for nearly two years until I was married. During these two years I finished my Journal, read several papers before the Geological Society, began preparing the M.S for my Geological Observations & arranged for the publication of the Zoology of the Voyage of the Beagle. In July I opened my first note-book for facts in relation to the Origin of Species, about which I had long reflected, & never ceased working on for the next 20 years. During these two years I also went a little into Society, & acted as one of the Hon. Secretaries of the Geological Society. I saw a great deal of Lyell. One of his chief characteristics was his sympathy with the work of others; & I was as much astonished as delighted at the interest which he showed when on my return to England I explained to him my views on coral-reefs. This encouraged me greatly, & his advice & example had much influence on me. During this time I saw also a good deal of Robert Brown, "facile princeps botanicorum"*: I used often to call & sit with him during his breakfast on Sunday mornings, & he poured forth a rich treasure of curious observations & acute remarks, but they almost always related to minute points, & he never with me discussed large & general questions in science.

During these two years I took several short excursions as a relaxation, & one longer one to the Parallel roads of Glen Roy,* an account of which was published in the Philosophical Transactions. This paper was a great failure, & I am ashamed of it. Having been deeply impressed with what I had seen of the elevation of the land in S. America, I attributed the parallel lines to the action of the sea; but I had to give up this view when Agassiz propounded his glacier-lake theory. Because no other explanation was possible under our then state of knowledge, I argued in favour of sea-action; & my error has been a good lesson to me never to trust in science to the principle of exclusion.

As I was not able to work all day at science I read a good deal during these two years on various subjects, including some metaphysical books, but I was not at [all] well fitted for such studies. About this time I took much delight in Wordsworths & Coleridge's poetry, & can boast that I read the Excursion twice through. Formerly Milton's Paradise Lost had been my chief favourite, & in my excursions during the Voyage of the Beagle, when I could take only a single small volume, I always chose Milton.

Religious belief. —

During these two years I was led to think much about religion. Whilst on board the Beagle I was quite orthodox, & I remember being heartily laughed at by several of the officers (though themselves orthodox) for quoting the Bible as an unanswerable authority on some point of morality. I suppose it was the novelty of the argument that amused them. But I had gradually come by this time (i.e. 1836 to 1839) to see that the Old Testament, from its manifestly false history of the world, with the Tower of Babel, the rain-bow as a sign &c &c, & from its attributing to God the feelings of a revengeful tyrant, was no more to be trusted than the sacred books of the Hindoos, or the beliefs of any barbarian. The question then continually rose before my mind & would not be banished,—is it credible that if God were now to make a revelation to the Hindoos, would he permit it to be connected with the belief in Vishnu, Siva &c, as Christianity is connected with the Old Testament. This appeared to me utterly incredible. By further reflecting that the clearest evidence would be requisite to make any sane man believe in the miracles by which Christianity is supported,—that the more we know of the fixed laws of nature the more incredible do miracles become,—that the men at that time were ignorant & credulous to a degree almost incomprehensible by us—that the Gospels can not be proved to have been written simultaneously with the events,—that they differ in many important details, far too important as it seemed to me to be admitted as the usual inaccuracies of eye-witnesses—by such reflections as these which I give not as having the least novelty or value, but as they influenced me, I gradually came to disbelieve in Christianity as a divine revelation.

The fact that many false religions have spread over large portions of the earth like wild-fire had some weight with me. Beautiful as is the morality of the New Testament, it can hardly be denied that its perfection depends in part on the interpretation which we now put on metaphors & allegories. But I was very unwilling to give up my belief.—I feel sure of this for I can well remember often & often inventing day-dreams of old letters between distinguished Romans & manuscripts being discovered at Pompeii or elsewhere which confirmed in the most striking manner all that was written in the Gospels. But I found it more & more difficult, with free scope given to my imagination to invent evidence which would suffice to convince me. Thus disbelief crept over me at a very slow rate, but was at last complete. The rate was so slow that I felt no distress, & have never since doubted* even for a single second that my conclusion was correct. I can indeed hardly see how anyone ought to wish Christianity to be true; for if so, the plain language of the text seems to show that the men who do not believe, & this would include my Father, Brother & almost all my best friends, will be everlastingly punished.

And this is a damnable doctrine.

Although I did not think much about the existence of a personal God until a considerably later period of my life, I will here give the vague conclusions to which I have been driven. The old argument from design in nature, as given by Paley, which formerly seemed to me so conclusive, fails now that the law of natural selection has been discovered. We can no longer argue that, for instance, the beautiful hinge of a bivalve shell must have been made by an intelligent being, like the hinge of a door by man. There seems to be no more design in the variability of organic beings & in the action of natural selection, than in the course which the wind blows. Everything in nature is the result of fixed laws. But I have discussed this subject at the end of my book on the Variation of Domesticated Animals & plants, & the argument there given has never, as far as I can see, been answered.

But passing over the endless beautiful adaptations which we everywhere meet with, it may be asked how can the generally beneficent arrangement of the world be accounted for? Some writers indeed are so much impressed with the amount of suffering in the world, that they doubt if we look to all sentient beings, whether there is more of misery or of happiness,—whether the world as a whole is a good or bad one. According to my judgment happiness decidedly prevails, though this would be very difficult to prove. If the truth of this conclusion be granted it harmonises well with the effects which we might expect from natural selection. If all the individuals of any species were habitually to suffer to an extreme degree they would neglect to propagate their kind; but we have no reason to believe that this has ever or at least often occurred. Some other considerations, moreover, lead to the belief that all sentient beings have been formed so as to enjoy as a general rule happiness. Everyone who believes, as I do, that all the corporeal & mental organs (excepting those which are neither advantageous or disadvantageous to the possessor) of all beings have been developed through natural selection or the survival of the fittest, together with use or habit, will admit that these organs have been formed so that their possessors may compete successfully with other beings, & thus increase in number. Now an animal may be led to pursue that course of action which is the most beneficial to the species by suffering, such as pain, hunger, thirst & fear,—or by pleasure, as in eating & drinking & in the propagation of the species &c, or by both means combined as in the search for food. But pain or suffering of any kind, if long continued, causes depression & lessens the power of action; yet is well adapted to make a creature guard itself against any great or sudden evil. Pleasurable sensations, on the other hand may be long continued without any depressing effect; on the contrary they stimulate the whole system to increased action. Hence it has come to pass that most or all sentient beings have been developed in such a manner through natural selection that pleasurable sensations serve as their habitual guides. We see this in the pleasure from exertion, even occasionally from great exertion of the body or mind,—in the pleasure of our daily meals, & especially in the pleasure derived from sociability & from loving our families. The sum of such pleasures as these, which are habitual or frequently recurrent give, as I can hardly doubt, to most sentient beings an excess of happiness over misery, although many occasionally suffer much. Such suffering, is quite compatible with the belief in natural selection, which is not perfect in its action, but tends only to render each species as successful as possible in the battle for life with other species under wonderfully complex & changing circumstances.

That there is much suffering in the world no one disputes. Some have attempted to explain this in reference to man; by imagining that it serves for his moral improvement. But the number of men in the world is as nothing compared with that of all other sentient beings, & these often suffer greatly without any moral improvement. A being so powerful & so full of knowledge as a God who could create the universe, is to our finite minds omnipotent & omniscient, & it revolts our understanding to suppose that his benevolence is not unbounded, for what advantage can there be in the sufferings of millions of the lower animals throughout almost endless time? This very old argument from the existence of suffering against the existence of an intelligent first cause seems to me a strong one; whereas, as just remarked, the presence of much suffering agrees well with the view that all organic beings have been developed through variation & natural selection.

At the present day the most usual argument for the existence of an intelligent God is drawn from deep inward conviction & feelings which are experienced by most persons. But it cannot be doubted that Hindoos, Mahomedans & others might argue in the same manner & with equal force in favour of the existence of one God, or of many Gods, or as with the Buddists of no God. There are also many barbarian tribes who cannot be said with any truth to believe in what we call God: they believe indeed in spirits or ghosts, & it can be explained, as Tyler* & Herbert Spencer have shown, how such a belief would be likely to arise. Formerly I was led by feelings such as those just referred to (although I do not think that the religious sentiment was ever strongly developed in me) to the firm conviction of the existence of God, & of the immortality of the soul. In my Journal I wrote that whilst standing in the midst of the grandeur of a Brazilian forest, "it is not possible to give an adequate idea of the higher feelings of wonder, admiration & devotion, which fill & elevate the mind." I well remember my conviction that there is more in man than the mere breath of his body. But now the grandest scenes would not cause any such convictions & feelings to rise in my mind. It may be truly said that I am like a man who has become colour-blind, & the universal belief by men of the existence of red-ness makes my present loss of perception of not the least value as evidence. This argument would be a valid one, if all men of all races had the same inward conviction of the existence of one God; but we know that this is very far from being the case. Therefore I cannot see that such inward convictions & feelings are of any weight as evidence of what really exists. The state of mind which grand scenes formerly excited in me, & which was intimately connected with a belief in God, did not essentially differ from that which is often called the sense of sublimity; & however difficult it may be to explain the genesis of this sense, it can hardly be advanced as an argument for the existence of God, any more than the powerful though vague & similar feelings excited by music.

With respect to immortality, nothing shows me how strong & almost instinctive a belief it is, as the consideration of the view now held by most physicists, namely that the sun with all the planets will in time grow too cold for life, unless indeed some great body dashes into the sun & thus gives it fresh life.—Believing as I do that man in the distant future will be a far more perfect creature than he now is, it is an intolerable thought that he & all other sentient beings are doomed to complete annihilation after such long-continued slow progress. To those who fully admit the immortality of the human soul, the destruction of our world will not appear so dreadful.

Another source of conviction in the existence of God, connected with the reason & not with the feelings, impresses me as having much more weight. This follows from the extreme difficulty or rather impossibility of conceiving this immense & wonderful universe, including man with his capacity of looking far backward & far into futurity, as the result of blind chance or necessity. When thus reflecting I feel compelled to look to a First Cause having an intelligent mind in some degree analogous to that of man; & I deserve to be called a Theist. This conclusion was strong in my mind about the time, as far as I can remember, when I wrote the Origin of Species; and it is since that time that it has very gradually with many fluctuations become weaker.* But then arises the doubt can the mind of man, which has, as I fully believe, been developed from a mind as low as that possessed by the lowest animal, be trusted when it draws such grand conclusions? May not these be the result of that connection between cause & effect which strikes us as a necessary one, but probably depends merely on inherited experience? Nor must we overlook the probability of the constant inculcation of a belief in God on the minds of children producing so strong & perhaps an inherited effect on their brains, not as yet fully developed, that it would be as difficult for them to throw off their belief in God, as for a monkey to throw off its instinctive fear & hatred of a snake.* I cannot pretend to throw the least light on such abstruse problems. The mystery of the beginning of all things is insoluble by us; & I for one must be content to remain an Agnostic.*

A man who has no assured & ever present belief in the existence of a personal God or of a future existence with retribution & reward, can have for his rule of life, as far as I can see, only to follow those impulses & instincts which are the strongest or which seem to him the best ones. A dog acts in this manner, but he does so blindly. A man, on the other hand, looks forward & backwards, & compares his various feelings, desires & recollections. He then finds, in accordance with the verdict of all the wisest men, that the highest satisfaction is derived from following certain impulses, namely the social instincts. If he acts for the good of others, he will receive the approbation of his fellow-men & gain the love of those with whom he lives; & this latter gain undoubtedly is the highest pleasure on this earth. By degrees it will become intolerable to him to obey his sensuous passions rather than his higher impulses, which when rendered habitual may be almost called instincts. His reason may occasionally tell him to act in opposition to the opinion of others, whose approbation he will then not receive; but he will still have the solid satisfaction of knowing that he has followed his innermost guide or conscience.—As for myself I believe that I have acted rightly in steadily following & devoting my life to science. I feel no remorse from having committed any great sin, but have often & often regretted that I have not done more direct good to my fellow-creatures. My sole & poor excuse is much ill-health, & my mental constitution which makes it extremely difficult for me to turn from one subject or occupation to another. I can imagine with high satisfaction giving up my whole time to philanthropy, but not a portion of it; though this would have been a far better line of conduct.

Nothing is more remarkable than the spread of scepticism or rationalism during the latter half of my life. Before I was engaged to be married, my father advised me to conceal carefully my doubts, for he said that he had known extreme misery thus caused with married persons. Things went on pretty well until the wife or husband became out of health, & then some women suffered miserably by doubting about the salvation of their husbands, thus making them likewise to suffer. My father added that he had known during his whole long life only three women who were sceptics; & it sh^d. be remembered that he knew well a multitude of persons & possessed extraordinary powers of winning confidence. When I asked him who the three women were, he had to own with respect to one of them, his sister-in-law, Kitty Wedgwood, that he had no good evidence, only the vaguest hints, aided by the conviction that so clear-sighted a woman could not be a believer. At the present time, with my small acquaintance I know (or have known) several married ladies, who believe very little more than their husbands. My father used to quote an unanswerable argument, by which an old lady, a M^rs. Barlow,* who suspected him of unorthodoxy, hoped to convert him: "Doctor I know that sugar is sweet in my mouth, & I know that my Redeemer liveth."*

From my marriage, Jan 29th. 1839, & residence in Upper Gower St. to our leaving London & settling at Down, Sept 14th 1842. —

You all know well your mother,* & what a good mother she has ever been to all of you. She has been my greatest blessing, & I can declare that in my whole life I have never heard her utter one word which I would rather have been unsaid. She has never failed in the kindest sympathy towards me, & has borne with the utmost patience my frequent complaints from ill-health & discomfort. I do not believe she has ever missed an opportunity of doing a kind action to anyone near her. I marvel at my good fortune that she so infinitely my superior in every single moral quality consented to be my wife. She has been my wise adviser & cheerful comforter throughout life, which without her would have been during a very long period a miserable one from ill-health. She has earned the love & admiration of every soul near her.

(Mem: her beautiful letter to myself preserved, shortly after our marriage.)*

I have indeed been most happy in my family, & I must say to you my children that not one of you has ever given me one minutes anxiety, except on the score of health. There are, I suspect very few fathers of five sons who could say this with entire truth. When you were very young it was my delight to play with you all, & I think with a sigh that such days can never return. From your earliest days to now that you are grown up, you have all, sons & daughters, ever been most pleasant, sympathetic & affectionate to us & to one another. When all or most of you are at home (as thank Heavens happens pretty frequently) no party can be, according to my taste, more agreeable, & I wish for no other society. We have suffered only one very severe grief in the death of Annie at Malvern on April 24th 1851, when she was just over ten years old. She was a most sweet & affectionate child, & I feel sure would have grown into a delightful woman. But I need say nothing here of her character, as I wrote a short sketch of it shortly after her death. Tears still sometimes come into my eyes, when I think of her sweet ways.

During the three years & eight months whilst we resided in London, I did less scientific work, though I worked as hard as I possibly could, than during any other equal length of time in my life. This was owing to frequently recurrent unwellness & to one long & serious illness. The greater part of my time, when I could do anything, was devoted to my work on Coral Reefs, which I had begun before my marriage, & of which the last proof sheet was corrected on May 6th 1842. This book, though a small one, cost me 20 months of hard work, as I had to read every work on the islands of the Pacific & to consult many charts. It was thought highly of by scientific men, & the theory therein given, is, I think now well established. No other work of mine was begun in so deductive a spirit as this; for the whole theory was thought out on the west coast of S. America before I had seen a true coral-reef. I had therefore only to verify & extend my views by a careful examination of living reefs. But it should be observed that I had during the two previous years been incessantly attending to the effects on the shores of S. America of the intermittent elevation of the land, together with denudation* & the deposition of sediment. This necessarily led me to reflect much on the effects of subsidence, & it was easy to replace in imagination the continued deposition of sediment by the upward growth of coral. To do this was to form my theory of the formation of barrier-reefs & atolls.

Besides my work on coral-reefs, during my residence in London, I read before the Geological Society papers on the Erratic boulders of S. America, on Earthquakes, & on the formation by the agency of earth-worms of mould.* I also continued to superintend the publication of the Zoology of the Voyage of the Beagle. Nor did I ever intermit collecting facts bearing on the origin of species; & I could sometimes do this when I could do nothing else from illness. In the summer of 1842 I was stronger than I had been for some time & took a little tour by myself in N. Wales for the sake of observing the effects of the old glaciers which formerly filled all the larger valleys. I published a short account of what I saw in the Philosophical Magazine. This excursion interested me greatly, & it was the last time I was ever strong enough to climb mountains or to take long walks, such as are necessary for geological work.

During the early part of our life in London,* I was strong enough to go into general society & saw a good deal of several scientific men, & other more or less distinguished men. I will give my impressions with respect to some of them, though I have little to say worth saying.

I saw more of Lyell than of any other man both before & after my marriage. His mind was characterised, as it appeared to me, by clearness, caution, sound judgment & a good deal of originality. When I made any remark to him on geology, he never rested until he saw the whole case clearly, & often made me see it more clearly than I had done before. He would advance all possible objections to my suggestion, & even after these were exhausted would long remain dubious. A second characteristic was his hearty sympathy with the work of other scientific men. On my return from the voyage of the Beagle, I explained to him my views on coral-reefs, which differed from his, & I was greatly surprised & encouraged by the vivid interest which he showed. On such occasions, while absorbed in thought, he would throw himself into the strangest attitudes, often resting his head on the seat of a chair, while standing up. His delight in science was ardent, & he felt the keenest interest in the future progress of mankind. He was very kind-hearted, & thoroughly liberal in his religious beliefs or rather disbeliefs; but he was a strong theist. His candour was highly remarkable. He exhibited this by becoming a convert to the Descent-theory, though he had gained much fame by opposing Lamarck's views, & this after he had grown old. He reminded me that I had many years before said to him, when discussing the opposition of the old school of geologists to his new views, "what a good thing it would be, if every scientific man were to die when 60 years old, as afterwards he would be sure to oppose all new doctrines." But he hoped that now he might be allowed to live.

He had a strong sense of humour & often told amusing anecdotes. He was very fond of society, especially of eminent men & of persons high in rank; & this over estimation of a man's position in the world seemed to me his chief foible. He used to discuss with Lady Lyell as a most serious question, whether or not they should accept some particular invitation. But as he would not dine out more than three times a week on account of the loss of time, he was justified in weighing his invitations with some care. He looked forward to going out oftener in the evening with advancing years, as to a great reward; but the good time never came, as his strength failed.

The science of geology is enormously indebted to Lyell, more so, as I believe, than to any other man who ever lived. When starting on the voyage of the Beagle, the sagacious Henslow, who like all other geologists believed at that time in successive cataclysms, advised me to get & study the first volume of the Principles, which had then just been published, but on no account to accept the views therein advocated. How differently would any one now speak of the Principles! I am proud to remember that the first place, namely St. Jago in the Cape Verd archipelago, which I geologised, convinced me of the infinite superiority of Lyell's views over those advocated in any other work known to me. The powerful effects of Lyells works could formerly be plainly seen in the different progress of the science in France & England. The present total oblivion of Elie de Beaumonts wild hypotheses, such as his 'Craters of Elevation' & 'Lines of Elevation'* (which latter hypothesis I heard Sedgwick at the Geolog. Soc. lauding to the skies) may be largely attributed to Lyell.

All the leading geologists were more or less known by me at the time when geology was advancing with triumphant steps. I liked most of them, with the exception of Buckland, who though very good-humoured & good-natured seemed to me a vulgar & almost coarse man. He was incited more by a craving for notoriety, which sometimes made him act like a buffoon, than by a love of science. He was not, however, selfish in his desire for notoriety; for Lyell, when a very young man, consulted him about communicating a poor paper to the Geolog. Soc. which had been sent him by a stranger, & Buckland answered "you had better do so, for it will be headed, 'Communicated by Charles Lyell', & thus your name will be brought before the public."

The services rendered to geology by Murchison by his classification of the older formations cannot be over-estimated; but he was very far from possessing a philosophical mind. He was very kind-hearted & would exert himself to the utmost to oblige anyone. The degree to which he valued rank was ludicrous, & he displayed this feeling & his vanity with the simplicity of a child. He related with the utmost glee to a large circle, including many mere acquaintances, in the room of the Geolog. Soc^y how the Czar, Nicholas, when in London, had patted him on the shoulder & had said, alluding to his geological work "Mon ami, Russia is grateful to you"; & then Murchison added, rubbing his hands together, "the best of it was that Prince Albert heard it all." He announced one day to the Council of the Geolog. Soc that his great work on the Silurian system was at last published; & he then looked at all who were present & said "you will every one of you find your name in the Index," as if this was the height of glory.—

I saw a good deal of R. Brown "facile Princeps Botanicorum"* as he was called by Humboldt; & before I was married I used to go & sit with him almost every Sunday morning. He seemed to me to be chiefly remarkable for the minuteness of his observations & their perfect accuracy. He never propounded to me any large scientific views in biology. His knowledge was extraordinarily great, & much died with him, owing to his excessive fear of ever making a mistake. He poured out his knowledge to me in the most unreserved manner, yet was strangely jealous on some points. I called on him two or three times before the voyage of the Beagle, & on one occasion he asked me to look through a microscope & describe what I saw. This I did, & believe now that it was the marvellous currents of protoplasm in some vegetable cell. I then asked him what I had seen; but he answered me, who was then hardly more than a boy & on the point of leaving England for five years, "that is my little secret." I suppose that he was afraid that I might steal his discovery. Hooker told me that he was a complete miser, & knew himself to be a miser, about his dried plants; & he would not lend specimens to Hooker, who was describing the plants of Tierra del Fuego, although well knowing that he himself would never make any use of the collections from this country. On the other hand he was capable of the most generous actions. When old, much out of health & quite unfit for any exertion, he daily visited (as Hooker told me) an old man-servant, who lived at a distance & whom he supported, & read aloud to him. This is enough to make up for any degree of scientific penuriousness or jealousy. He was rather given to sneering at anyone who wrote about what he did not fully understand: I remember praising Whewell's History of the Inductive Sciences to him, & he answered "yes I suppose that he has read the prefaces of very many books."

I often saw Owen, whilst living in London & admired him greatly, but was never able to understand his character & never became intimate with him. After the publication of the Origin of Species he became my bitter enemy, not owing to any quarrel between us, but as far as I could judge out of jealousy at its success. Poor dear Falconer, who was a charming man, had a very bad opinion of him, being convinced that he was not only ambitious, very envious & arrogant, but untruthful & dishonest. His power of hatred was certainly unsurpassed. When in former days I used to defend Owen, Falconer often said "you will find him out some day", & so it has proved.

At a somewhat later period I became very intimate with Hooker, who has been one of my best friends throughout life. He is a delightfully pleasant companion & most kind-hearted. One can see at once that he is honourable to the back-bone. His intellect is very acute, & he has great power of generalisation. He is the most untirable worker that I have ever seen, & will sit the whole day working with the microscope, & be in the evening as fresh & pleasant as ever. He is in all ways very impulsive & somewhat peppery in temper; but the clouds pass away almost immediately. He once sent me an almost savage letter from a cause which will appear ludicrously small to an outsider, viz. because I maintained for a time the silly notion that our coal-plants had lived in shallow water in the sea. His indignation was all the greater because he could not pretend that he sh^d ever have suspected that the mangrove (& a few other marine plants which I named) had lived in the sea, if they had been found only in a fossil state. On another occasion he was almost equally indignant because I rejected with scorn the notion that a continent had formerly extended between Australia & S. America. I have known hardly any man more loveable than Hooker.—

A little later I became intimate with Huxley. His mind is as quick as a flash of lightning & as sharp as a razor. He is the best talker whom I have known. He never says & never writes anything flat. From his conversation no one would suppose that he could cut up his opponents in so trenchant a manner as he can do & does do. He has been a most kind friend to me & would always take any trouble for me. He has been the mainstay in England of the principle of the gradual evolution of organic beings. Much splendid work as he has done in Zoology, he would have done far more, if his time had not been so largely consumed by official & literary work, & by his efforts to improve the education of the country. He would allow me to say anything to him: many years ago I thought that it was a pity that he attacked so many scientific men, although I believe that he was right in each particular case, & I said so to him. He denied the charge indignantly, & I answered that I was very glad to hear that I was mistaken. We had been talking about his well-deserved attacks on Owen, so I said after a time, "how well you have exposed Ehrenberg's blunders"; he agreed & added that it was necessary for science that such mistakes should be exposed. Again after a time, I added "poor Agassiz has fared ill under your hands." Again I added another name, & now his bright eyes flashed on me, & he burst out laughing, anathematising me in some manner. He is a splendid man & has worked well for the good of mankind.

I may here mention a few other eminent men, whom I have occasionally seen, but I have little to say about them worth saying. I felt a high reverence for Sir J. Herschel, & was delighted to dine with him at his charming house at the C. of Good Hope, & afterwards at his London house. I saw him, also, on a few other occasions.

He never talked much, but every word which he uttered was worth listening to. He was very shy & he often had a distressed expression. Lady Caroline (?) Bell,* at whose house I dined at the C. of Good Hope, admired Herschel much, but said that he always came into a room, as if knew that his hands were dirty, & that he knew that his wife knew that they were dirty.

I once met at breakfast at Sir R. Murchison's house the illustrious Humboldt, who honoured me by expressing a wish to see me. I was a little disappointed with the great man, but my anticipations probably were too high. I can remember nothing distinctly about our interview, except that Humboldt was very cheerful & talked much.

I used to call pretty often on Babbage & regularly attended his famous evening parties. He was always worth listening to, but he was a disappointed & discontented man; & his expression was often or generally morose. I do not believe that he was half as sullen as he pretended to be. One day he told me that he had invented a plan by which all fires could be effectually stopped, but added, "I shan't publish it—damn them all, let all their houses be burnt". The all were the inhabitants of London. Another day he told me that he had seen a pump on a road-side in Italy, with a pious inscription on it to the effect that the owner had erected the pump for the love of God & his country, that the tired way-farer might drink. This led Babbage to examine the pump closely & he soon discovered that every time that a way-farer pumped some water for himself, he pumped a larger quantity into the owner's house. Babbage then added "there is only one thing which I hate more than piety, & that is patriotism." But I believe that his bark was much worse than his bite.

Herbert Spencers conversation seemed to me very interesting, but I did not like him particularly & did not feel that I could easily have become intimate with him. I think that he was extremely egotistical. After reading any of his books I generally feel enthusiastic admiration of his transcendent talents, & have often wondered whether in the distant future he would rank with such great men, as Descartes, Leibnitz &c, about whom, however, I know very little. Nevertheless I am not conscious of having profited in my own work by Spencer's writings. His deductive manner of treating every subject is wholly opposed to my frame of mind. His conclusions never convince me; & over & over again I have said to myself, after reading one of his discussions, "here would be a fine subject for half-a-dozen years work."—His fundamental generalisations, (which have been compared in importance by some persons with Newton's laws!) which I daresay may be very valuable under a philosophical point of view, are of such a nature that they do not seem to me to be of any strictly scientific use. They partake more of the nature of definitions than of laws of nature. They do not aid one in predicting what will happen in any particular case. Anyhow they have not been of any use to me.

Speaking of H. Spencer reminds me of Buckle, whom I once met at Hensleigh Wedgwoods. I was very glad to learn from him his system of collecting facts. He told me that he bought all the books which he read, & made a full index to each of the facts which he thought might prove serviceable to him, & that he could always remember in what book he had read anything, for his memory was wonderful. I then asked him how at first he could judge what facts would be serviceable, & he answered that he did not know, but that a sort of instinct guided him. From this habit of making indices he was enabled to give the astonishing number of references on all sorts of subjects, which may be found in his History of Civilisation. This book I thought most interesting & read it twice; but I doubt whether his generalisations are worth anything. H. Spencer told me that he had never read a line of it! Buckle was a great talker, & I listened to him without saying hardly a word; nor indeed could I have done so, for he left no gaps. When Effie* began to sing, I jumped up & said that I must listen to her. This, I suppose, offended him, for after I had moved away, he turned round to a friend, & said (as was overheard by my brother) "well Mr Darwin's books are much better than his conversation". What he really meant, was that I did not properly appreciate his conversation.

Of other great literary men, I once met Sidney Smith at Dean Milman's house. There was something inexplicably amusing in every word which he uttered. Perhaps this was partly due to the expectation of being amused. He was talking about Lady Cork, who was then extremely old. This was the lady, who, as he said, was once so much affected by one of his charity sermons, that she *borrowed* a guinea from a friend to put into the Plate. He now said "it is generally believed that my dear old friend Lady Cork has been overlooked"; & he said this in such a manner that no one could for a

moment doubt that he meant that his dear old friend had been overlooked by the Devil. How he managed to express this, I know not.

I likewise once met Macaulay at L$^{d.}$ Stanhope's (the Historian)
house, & as there was only one other man at dinner, I had a grand
opportunity of hearing him converse, & he was very agreeable. He did
not talk at all too much; nor indeed could such a man talk too much
as long as he allowed others to turn the stream of his conversation,
& this he did allow. L$^{d.}$ Stanhope once gave me a curious little proof
of the accuracy & fulness of Macaulay's memory: many historians
used often to meet at L$^{d.}$ Stanhope's house, & in discussing various
subjects they would sometimes differ from Macaulay, & formerly
they often referred to some book to see who was right; but latterly,
as L$^{d.}$ Stanhope noticed, no historian ever took this trouble, &
whatever Macaulay said was final.

On another occasion I met at Ld Stanhope's house one of his
parties of historians & other literary men, & amongst them were
Motley & Grote. After luncheon I walked about Chevening Park for
nearly an hour with Grote, & was much interested by his conversation & pleased by the simplicity & absence of all pretension in his
manners. I met another set of great men at breakfast at L$^{d.}$ Stanhope's
house in London. After breakfast was quite over, Monckton Milnes
(L$^{d.}$ Houghton now) walked in, & after looking round exclaimed
(justifying Sidney Smith's nickname of "the cool of the evening")
"well I declare you are all very premature."

Long ago I dined occasionally with the old Earl,* the father of the
historian. I have heard that his father, the democratic earl well-known
at the time of the French Revolution, had his son educated as a
blacksmith, as he declared that every man ought to know some trade.
The old Earl, whom I knew, was a strange man, but what little I saw
of him, I liked much. He was frank, genial & pleasant. He had
strongly marked features with a brown complexion & his clothes,
when I saw him, were all brown. He seemed to believe in everything
which was to others utterly incredible. He said one day to me "why
don't you give up your fiddle-faddle of geology & zoology, & turn to
the occult sciences?" The historian (then Ld Mahon) seemed shocked
at such a speech to me, & his charming wife much amused.

The last man whom I will mention is Carlyle, seen by me several
times at my brother's house & 2 or 3 times at my own house. His talk
was very racy & interesting, just like his writings, but he sometimes

went on too long on the same subject. I remember a funny dinner at my brother's, where amongst a few others were Babbage & Lyell, both of whom liked to talk. Carlyle, however, silenced everyone by haranguing during the whole dinner on the advantages of silence. After dinner Babbage in his grimmest manner thanked Carlyle for his very interesting Lecture on Silence. Carlyle sneered at almost everyone: one day in my house, he called Grote's History "a fetid quagmire, with nothing spiritual about it." I always thought, until his Reminiscences appeared, that his sneers were partly jokes, but this now seems rather doubtful. His expression was that of a depressed, almost despondent, yet benevolent man; & it is notorious how heartily he laughed. I believe that his benevolence was real, though stained by not a little jealousy. No one can doubt about his extraordinary power of drawing vivid pictures of things & men,—far more vivid, as it appears to me, than any drawn by Macaulay. Whether his pictures of men were true ones is another question. He has been all powerful in impressing some grand moral truths on the minds of men. On the other hand his views about slavery were revolting. In his eyes might was right. His mind seemed to me a very narrow one; even if all branches of science, which he despised, are excluded. It is astonishing to me that Kingsley should have spoken of him, as a man well fitted to advance science. He laughed to scorn the idea that a mathematician, such as Whewell, could judge, as I maintained he could, of Goethes views on light.* He thought it a most ridiculous thing that anyone should care whether a glacier moved a little quicker or a little slower, or moved at all. As far as I could judge, I never met a man with a mind so ill adapted for scientific research.—

Whilst living in London, I attended as regularly as I could the meetings of several Scientific societies, & acted as Secretary to the Geological Soc^y· But such attendance & ordinary society suited my health so badly that we resolved to live in the country, which we both preferred & have never repented of.—

Residence at Down from Sept 14th 1842 to the present time 1876.

After several fruitless searches in Surrey & elsewhere, we found this house & purchased it. I was pleased with the diversified appearance of the vegetation, proper to a chalk district, & so unlike what I had been accustomed to in the midland·counties; & still more pleased with the extreme quietness & rusticity of the place. It is not, however, quite so retired a place as a writer in a German periodical makes it, who says that my house can be approached only by a mule track! Our fixing ourselves here has answered admirably in one way, which we did not anticipate, namely by being very convenient for frequent visits from our children, who never miss an opportunity of doing so when they can.

Few persons can have lived a more retired life than we have done. Besides short visits to the houses of relations, & occasionally to the sea-side or elsewhere, we have gone nowhere. During the first part of our residence we went a little into society & received a few friends here; but my health almost always suffered from the excitement, violent shivering & vomiting attacks being thus brought on. I have therefore been compelled for many years to give up all dinner-parties; & this has been somewhat of a deprivation to me, as such parties always put me into high spirits. From the same cause I have been able to invite here very few scientific acquaintances. Whilst I was young & strong I was capable of very warm attachments, but of late years, though I still have very friendly feelings towards many persons, I have lost the power of becoming deeply attached to anyone, not even so deeply to my good & dear friends Hooker & Huxley, as I should formerly have been. As far as I can judge this grievous loss of feeling has gradually crept over me, from the expectation of much distress afterwards from exhaustion having become firmly associated in my mind with seeing & talking with anyone for an hour, except my wife & children.

My chief enjoyment & sole employment throughout life has been scientific work; & the excitement from such work makes me for the time forget or drives quite away my daily discomfort. I have therefore nothing to record during the rest of my life, except the publication of my several books. Perhaps a few details how they arose may be worth giving.

My several publications

In the early part of 1844, my observations on the Volcanic Islands visited during the voyage of the Beagle were published. In 1845 I took much pains in correcting a new edition of my Journal of Researches, which was originally published in 1839, as part of FitzRoys work. The success of this my first literary child always tickles my vanity more than that of any of my other books. Even to this day it sells steadily in England & the United States, & has been translated for the second time into German, & into French & other languages. This success of a book of travels, especially of a scientific one, so many years after its first publication, is surprising. Ten thousand copies have now been sold in England of the second edition. In 1846 my geological observations on S. America were published. I record in a little diary which I have always kept, that my three geological books (Coral-reefs included) consumed four & a half years steady work; "& now it is ten years since my return to England. How much time have I lost by illness." I have nothing to say about these three books, except that to my surprise new editions have lately been called for.

In October 1846 I began to work on Cirripedia. When on the coast of Chile I found a most curious form, which burrowed into the shells of Concholepas,* & which differed so much from all other cirripedes that I had to form a new sub-order for its sole reception. Lately an allied burrowing genus has been found on the shores of Portugal. To understand the structure of my new cirripede I had to examine & dissect many of the common forms; & this gradually led me on to take up the whole group. I worked steadily on the subject for the next eight years, & ultimately published two thick volumes describing all the known living species & two thin quartos on the extinct species. I do not doubt that Sir E. Lytton Bulwer had me in his mind, when he introduces in one of his novels, a Professor Long who had written two huge volumes on Limpets. Although I was employed during eight years on this work, yet I record in my diary that about two years out of this time was lost by illness. On this account I went in 1848 for some months to Malvern for hydropathic treatment,* which did me much good so that on my return home I was able to resume work. So much was I out of health that when my dear Father died on November 13th 1847 I was unable to attend his funeral or to act as one of his executors.

My work on the Cirripedia possessed, I think, considerable value, as besides describing several new & remarkable forms, I made out the homologies of the various parts—I discovered the cementing apparatus, though I blundered dreadfully about the cement-glands —& lastly I proved the existence in certain genera of minute males complemental to & parasitic on the hermaphrodites.* This latter discovery has at last been fully confirmed; though at one time a German writer* was pleased to attribute the whole account to my fertile imagination. The cirripedes form a highly varying & difficult group of species to class; & my work was of considerable use to me, when I had to discuss in the Origin of Species the principles of a natural classification. Nevertheless I doubt whether the work was worth the consumption of so much time.

From September 1854 onwards I devoted all my time to arranging my huge pile of notes, to observing & experimenting, in relation to the transmutation of species. During the voyage of the Beagle I had been deeply impressed by discovering in the Pampean formation great fossil animals covered with armour like that on the existing armadillos; secondly by the manner in which closely allied animals replace one another in proceeding southwards over the continent; & thirdly by the S. American character of most of the productions of the Galapagos archipelago, & more especially by the manner in which they differ slightly on each island of the group; none of these islands appearing to be very ancient in a geological sense. It was evident that such facts as these as well as many others could be explained on the supposition that species gradually become modified; & the subject haunted me. But it was equally evident that neither the action of the surrounding conditions, nor the will of the organisms (especially in the case of plants) could account for the innumerable cases in which organisms of every kind are beautifully adapted to their habits of life,—for instance a woodpecker or tree-frog to climb trees, or a seed for dispersal by hooks or plumes. I had always been much struck by such adaptations, & until these could be explained it seemed to me almost useless to endeavour to prove by indirect evidence that species have been modified.

After my return to England it appeared to me that by following the example of Lyell in Geology, & by collecting all facts which bore in any way on the variation of animals & plants under domestication & nature, some light might perhaps be thrown on the whole subject. My first note-book was opened in July 1837. I worked on true Baconian principles & without any theory collected facts on a whole-sale scale, more especially with respect to domesticated productions, by printed enquiries, by conversation with skilful breeders & garden-ers, & by extensive reading. When I see the list of books of all kinds which I read & abstracted, including whole series of Journals & Transactions, I am surprised at my industry. I soon perceived that Selection was the key-stone of man's success in making useful races of animals & plants. But how selection could be applied to organisms living in a state of nature remained for some time a mystery to me. In October 1838, that is fifteen months after I had begun my systematic enquiry, I happened to read for amusement 'Malthus on Population', & being well prepared to appreciate the struggle for existence which everywhere goes on from long-continued observation of the habits of animals & plants, it at once struck me that under these circumstances favourable variations would tend to be preserved & unfavourable ones to be destroyed. The result of this would be the formation of new species.

Here then I had at last got a theory by which to work; but I was so anxious to avoid prejudice, that I determined not for some time to write even the briefest sketch of it. In June 1842 I first allowed myself the satisfaction of writing a very brief abstract of my theory in pencil in 35 pages; & this was enlarged, during the summer of 1844 into one of 230 pages, which I had fairly copied out & still possess. But at that time I overlooked one problem of great importance; & it is astonishing to me, except on the principle of Columbus & his egg,* how I could have overlooked it & its solution. This problem is the tendency in organic beings descended from the same stock to diverge in character as they become modified. That they have diverged greatly is obvious from the manner in which species of all kinds can be classed under genera, genera under families, families under sub-orders & so forth; & I can remember the very spot in the road whilst in my carriage, when to my joy the solution occurred to me; & this was long after I had come to Down. The solution, as I believe is that the modified offspring of all dominant & increasing forms tend to become adapted to many & highly diversified places in the economy of nature.

Early in 1856 Lyell advised me to write out my views pretty fully, & I began at once to do so on a scale three or four times as extensive as that which was afterwards followed in my Origin of Species; yet it was only an abstract of the materials which I had collected, & I got through about half the work on this scale. But my plans were overthrown, for early in the summer of 1858 Mr Wallace, who was then in the Malay Archipelago, sent me an essay "on the tendency of varieties to depart indefinitely from the original type"; & this essay contained exactly the same theory as mine. Mr Wallace expressed the wish that if I thought well of his essay, I sh^d. send it to Lyell for perusal. The circumstances under which I consented at the request of Lyell & Hooker to allow of an extract from my M.S., together with a letter to Asa Gray dated Sept 5 1857, to be published at the same time with Wallace's Essay, are given in the Journal of the Proceedings of the Linn: Soc. 1858 p. 45.* I was at first very unwilling to consent, as I thought M^r· Wallace might consider my doing so unjustifiable, for I did not then know how generous & noble was his disposition. The extract from my M.S & the letter to Asa Gray had neither been intended for publication & were badly written. Mr Wallace's essay, on the other hand was admirably expressed & quite clear. Nevertheless our joint productions excited very little attention, & the only published notice of them which I can remember was by Prof. Haughton of Dublin, whose verdict was that all that was new in them was false, & what was true was old. This shows how necessary it is that any new view should be explained at considerable length in order to arouse public attention.

In September 1858 I set to work by the strong advice of Lyell & Hooker to prepare a volume on the transmutation of species, but was often interrupted by ill-health, & short visits to Dr Lanes delightful hydropathic establishment at Moor Park. I abstracted the M.S begun on a much larger scale in 1856, & completed the volume on the same reduced scale. It cost me 13 months & ten days hard labour. It was published under the title of the 'Origin of Species' in November 1859. Though considerably added to & corrected in the later editions it has remained substantially the same book.

It is no doubt the chief work of my life. It was from the first highly successful. The first small edition of 1250 copies was sold on the day of publication, & a second edition of 3000 copies soon afterwards. Sixteen thousand copies have now (1876) been sold in England, & considering how stiff a book it is this is large sale. It has been translated into almost every European tongue, even into such languages as Spanish, Bohemian, Polish & Russian. (1880) It has also, according to Miss Bird, been translated in Japanese & is there much studied. Even an essay in Hebrew has appeared on it, showing that the theory is contained in the Old Testament! The Reviews were very numerous; for a time I collected all that appeared on the Origin & on my related books; & these amount (excluding newspaper reviews) to 265; but after a time I gave up the attempt in despair. Many separate essays & books on the subject have appeared; & in Germany a catalogue or Bibliography on "Darwinismus" has appeared every year or two.

The success of the Origin may, I think, be attributed in large part to my having long before written two condensed sketches, & to my having finally abstracted a much larger manuscript, which was itself an abstract. By this means I was enabled to select the more striking facts & conclusions. I had, also, during many years followed a golden rule, namely that whenever a published fact, new observation or thought came across me, which was opposed to my general results, to make a memorandum of it without fail & at once; for I had found by experience that such facts & thoughts were far more apt to escape from the memory, than favourable ones. Owing to this habit, very few objections were raised against my views which I had not at least noticed & attempted to answer. It has sometimes been said that the success of Origin proved "that the subject was in the air" or "that men's minds were prepared for it." I do not think that this is strictly true, for I occasionally sounded out a few naturalists, & never happened to come across a single one who seemed to doubt about the permanence of species. Even Lyell & Hooker, though they would listen with interest to me never seemed to agree. I tried once or twice to explain to able men what I meant by natural selection, but signally failed. What I believe was strictly true is that innumerable well-observed facts were stored in the minds of naturalists ready to take their proper places, as soon as any theory which would receive them was sufficiently explained. Another element in the success of the book was its moderate size, & this I owe to the appearance of Wallace's essay; had I published on the scale in which I began to write in 1856, the book would have been four or five times as large as the Origin, & very few would have had the patience to read it.

I gained much by my delay in publishing from about 1839, when the theory was clearly conceived, to 1859; & I lost nothing by it, for I cared very little whether men attributed most originality to me or Wallace, & his essay no doubt aided in the reception of the theory. I was forestalled in only one important point, which my vanity has always made me regret, namely the explanation by means of the Glacial period of the presence of the same species of plants & of some few animals on distant mountain-summits & in the Arctic regions. This view pleased me so much that I wrote it out in extenso & it was read by Hooker, some years before E. Forbes published his celebrated memoir on the subject. In the very few points in which we differed, I still think that I was in the right. I have never of course alluded in print to my having independently worked out this view.

Hardly any point gave me so much satisfaction when I was at work on the Origin, as the explanation of the wide difference in many classes between the embryo & the adult animal, & of the close resemblance of the embryos within the same class. No notice of this point was taken, as far as I remember, in the early reviews of the Origin, & I recollect expressing my surprise on this head in a letter to Asa Gray. Within late years several reviewers have given the whole credit of the idea to Fritz Müller & Häckel, who undoubtedly have worked it out much more fully & in some respects more correctly than I did. I had materials for a whole chapter on the subject, & I ought to have made the discussion longer; for it is clear that I failed to impress my readers; & he who succeeds in doing so deserves in my opinion all the credit. This leads me to remark that I have almost always been treated honestly by my reviewers, passing over those without scientific knowledge as not worthy of notice. My views have often been grossly misrepresented, bitterly opposed & ridiculed, but this has been generally done as I believe in good faith. I must however except Mr Mivart, who as an American expressed it in a letter has acted towards me "like a pettifogger", or as Huxley has said "like an Old Bailey lawyer." On the whole I do not doubt that my work has been over & over again greatly overpraised. I rejoice that I have avoided controversies, & this I owe to Lyell who many years ago in reference to my geological works strongly advised me never to get entangled in a controversy, as it rarely did any good & caused a miserable loss of time & temper.

Whenever I have found out that I have blundered or that my work has been imperfect, & when I have been contemptuously criticised & even when I have been overpraised, so that I have felt mortified, it has been my greatest comfort to say hundreds of times to myself that "I have worked as hard & as well as I could, & no man can do more than this." I remember when in Good Success Bay in Tierra del Fuego, thinking, (& I believe that I wrote home to this effect) that I could not employ my life better than in adding a little to natural science. This I have done to the best of my abilities, & critics may say what they like, but they cannot destroy this conviction.

During the two last months of the year 1859 I was fully occupied in preparing a second edition of the Origin, & by an enormous correspondence. On Jan 9th 1860 I began arranging my notes for my work on the Variation of Animals & Plants under Domestication; but it was not published until the beginning of 1868; the delay having been caused partly by frequent illnesses, one of which lasted seven months, & partly by having been tempted to publish on other subjects which at the time interested me more.

On May 15 1862, my little book on the Fertilisation of Orchids, which cost me ten months work, was published: most of the facts had been slowly accumulated during several previous years. During the summer of 1839 & I believe during the previous summer, I was led to attend to the cross-fertilisation of flowers by the aid of insects, from having come to the conclusion in my speculations on the origin of species, that crossing played an important part in keeping specific forms constant. I attended to the subject more or less during every subsequent summer; & my interest in it was greatly enhanced by having procured & read in November 1841 through the advice of Robert Brown, a copy of C. K. Sprengels wonderful book "Das Entdeckte Geheimniss der Natur". For some years before 1862 I had specially attended to the fertilisation of our British orchids; & it seemed to me the best plan to prepare as complete a treatise on this group of plants as well as I could, rather than to utilise the great mass of matter which I had slowly collected with respect to other plants. My resolve proved a wise one; for since the appearance of my book, a surprising number of papers & separate works on the fertilisation of all kinds of flowers have appeared; & these are far better done than I could possibly have effected. The merits of poor old Sprengel, so long overlooked, are now fully recognised many years after his death.

During this same year I published in the Journal of the Linnean Society a paper "On the two forms or dimorphic condition of Primula", & during the next five years, five other papers on dimorphic & trimorphic plants.* I do not think anything in my scientific life has given me so much satisfaction as making out the meaning of the structure of these plants. I had noticed in 1838 or 1839 the dimorphism of *Linum flavum*, & had at first thought that it was merely a case of unmeaning variability. But on examining the common species of Primula I found that the two forms were much too regular & constant to be thus viewed. I therefore became almost convinced that the common cowslip & primrose were on the high road to become diœcious;—that the short pistil in the one form, & the short stamens in the other form were tending towards abortion.* The plants were therefore subjected under this point of view to trial; but as soon as the flowers with short pistils fertilised with pollen from the short stamens, were found to yield more seeds than any other of the four possible unions, the abortion-theory was knocked on the head. After some additional experiments, it became evident that the two forms, though both were perfect hermaphrodites, bore almost the same relation to one another as do the two sexes of an ordinary animal. With Lythrum we have the still more wonderful case of three forms standing in a similar relation to one another. I afterwards found that the offspring from the union of two plants belonging to the same form presented a close & curious analogy with hybrids from the union of two distinct species.

In the autumn of 1864 I finished a long paper on Climbing Plants & sent it to the Linnean Society. The writing of this paper cost me four months; but I was so unwell when I received the proof-sheets that I was forced to leave them very badly & often obscurely expressed. The paper was little noticed, but when in 1875 it was corrected & published as a separate book it sold well. I was led to take up this subject by reading a short paper by Asa Gray, published in 1858, on the movements of the tendrils of a Cucurbitacean plant.* He sent me seeds, & on raising some plants I was so much fascinated & perplexed by the revolving movements of the tendrils & stems, which movements are really very simple though appearing at first very complex, that I procured various other kinds of Climbing Plants, & studied the whole subject. I was all the more attracted to it, from not being at all satisfied with the explanation which Henslow gave us in his Lectures, about Twining plants, namely that they had a natural tendency to grow up in a spire. This explanation proved quite erroneous. Some of the adaptations displayed by Climbing Plants are as beautiful as those by Orchids for ensuring cross-fertilisation.

My Variation of Animals & Plants under Domestication was begun, as already stated in the beginning of 1860, but was not published until the beginning of 1868. It is a big book & cost me four years & two months hard labour. It gives all my observations & an immense number of facts collected from various sources about our domestic productions. In the second volume the causes & laws of variation, inheritance, &c are discussed, as far as our present state of knowledge permits. Towards the end of the work I give my well-abused hypothesis of Pangenesis.* An unverified hypothesis is of little or no value; but if anyone should hereafter be led to make observations by which some such hypothesis could be established, I shall have done good service, as an astonishing number of isolated facts can thus be connected together & rendered intelligible. In 1875 a second & largely corrected edition, which cost me a good deal of labour, was brought out.

My Descent of Man was published in Feb. 1871. As soon as I had become in the year 1837 or 1838 convinced that species were mutable productions, I could not avoid the belief that man must come under the same law. Accordingly I collected notes on the subject for my own satisfaction, & not for a long time with any intention of publishing. Although in the Origin of Species, the derivation of any particular species is never discussed, yet I thought it best, in order that no honourable man should accuse me of concealing my views, to add that by the work in question "light would be thrown on the origin of man & his history." It would have been useless & injurious to the success of the book to have paraded without giving any evidence my conviction with respect to his origin. But when I found that many naturalists fully accepted the doctrine of the evolution of species, it seemed to me advisable to work up such notes as I possessed & to publish a special treatise on the origin of man. I was the more glad to do so, as it gave me an opportunity of fully discussing Sexual Selection,—a subject which had always greatly interested me. This subject, & that of the variation of our domestic productions together with the causes & laws of variation, inheritance &c, & the intercrossing of Plants are the sole subjects which I have been able to write about in full, so as to use all the materials which I had collected. The Descent of Man took me three years to write, but then as usual some of this time was lost by ill-health, & some was consumed by preparing new editions & other minor works. A second & largely corrected Edition of the Descent appeared in 1874.

My book on the Expression of the Emotions in Men & Animals was published in the autumn of 1872. I had intended to give only a chapter on the subject in the Descent of Man, but as soon as I began to put my notes together, I saw that it would require a separate Treatise.

My first child* was born on Dec 27th 1839, & I at once commenced to make notes on the first dawn of the various expressions which he exhibited, for I felt convinced even at that early period that the most complex & fine shades of expression must all have had a gradual & natural origin. During the summer of the following year, 1840, I read Sir C. Bell's admirable work on Expression, & this greatly increased the interest which I felt in the subject, though I could not at all agree with his belief that various muscles had been specially created for the sake of expression. From this time forward I occasionally attended to the subject, both with respect to man & our domestic animals. My book sold largely; 5267 copies having been disposed of on the day of publication.

In the summer of 1860 I was idling & resting near Hartfield, where two species of Drosera abound*; & I noticed that numerous insects had been entrapped by the leaves. I carried home some plants & on giving them insects saw the movements of the tentacles, & this made me think it probable that the insects were caught for some special purpose. Fortunately a crucial test occurred to me, that of placing a large number of leaves in various nitrogenous* & non-nitrogenous fluids of equal density; & as soon as I found that the former alone excited energetic movements, it was obvious that here was a fine new field for investigation. During subsequent years, whenever I had leisure I pursued my experiments, & my book on "Insectivorous Plants" was published July 1875,—that is 16 years after my first observations. The delay in this case, as with all my other books has been a great advantage to me; for a man after a long interval can criticise his own work, almost as well as if it were that of another person. The fact that a plant should secrete when properly excited a fluid containing an acid & ferment, closely analogous to the digestive fluid of an animal, was certainly a remarkable discovery.

During this autumn of 1876 I shall publish on the "Effects of Cross & Self Fertilisation in the Vegetable Kingdom". This book will form a complement to that on the Fertilisation of Orchids, in which I showed how perfect were the means for cross-fertilisation, & here I shall show how important are the results. I was led to make during eleven years the numerous experiments recorded in this volume by a mere accidental observation; & indeed it required the accident to be repeated before my attention was thoroughly aroused to the remarkable fact that seedlings of self-fertilised parentage are inferior even in the first generation in height & vigour to seedlings of cross-fertilised parentage. I hope also to republish a revised edition of my book on Orchids, & hereafter my papers on dimorphic & trimorphic plants, together with some additional observations on allied points which I never have had time to arrange. My strength will then probably be exhausted, & I shall be ready to exclaim "nunc dimittis."*

["]The Effects of Cross & Self-fertilisation"* was published in the autumn of 1876; & the results there arrived at explain, as I believe, the endless & wonderful contrivances for the transportal of pollen from one plant to another of the same species. I now believe, however, chiefly from the observations of Hermann Müller, that I ought to have insisted more strongly than I did on the many adaptations for self-fertilisation; though I was well aware of many such adaptations. A much enlarged Edit. of my Fertilisation of Orchids was published in 1877. In this same year "The Different Forms of Flowers &c." appeared, & in 1880 a 2ᵈ Edition. This book consists chiefly of the several papers on Heterostyled flowers,* originally published by the Linnean Socʸ, corrected with much new matter added, together with observations on some other cases in which the same plant bears two kinds of flowers. As before remarked no little discovery of mine ever gave me so much pleasure as the making out the meaning of heterostyled flowers. The results of crossing such flowers in an illegitimate manner, I believe to be very important as bearing on the sterility of hybrids; although these results have been noticed by only a few persons.

In 1879, I had a translation of Dʳ· Ernst Krause's life of Erasmus Darwin published, & I added a sketch of his character & habits from materials in my possession. Many persons have been much interested by this little life, & I am surprised that only 800 or 900 copies were sold. Owing to my having accidentally omitted to mention that Dʳ Krause had enlarged & corrected his article in German before it was translated, Mʳ Samuel Butler abused me with almost insane virulence. How I offended him so bitterly, I have never been able to understand. The subject gave rise to some controversy in the Athenæum newspaper & Nature. I laid all the documents before some good judges, viz Huxley, Leslie Stephen, Litchfield &c, & they were all unanimous that the attack was so baseless that it did not deserve any public answer; for I had already expressed privately my regret to Mr Butler for my accidental omission. Huxley consoled me by quoting some German lines from Goethe, who had been attacked by some one, to the effect "that every Whale has its Louse".*

In 1880 I published with Frank's assistance,* our "Power of Movement in Plants". This was a tough piece of work. The book bears somewhat the same relation to my little book on Climbing Plants, which "Cross-Fertilisation" did to the "Fertilisation of Orchids"; for in accordance with the principles of evolution it was impossible to account for climbing plants having been developed in so many widely different groups, unless all kinds of plants possess some slight power of movement of an analogous kind. This I proved to be the case, & I was further led to a rather wide generalisation, viz that the great & important classes of movement, excited by light, the attraction of gravity &c are all modified forms of the fundamental movement of circumnutation. It has always pleased me to exalt plants in the scale of organised beings; & I therefore felt an especial pleasure in showing how many & what admirably well adapted movements the tip of a root possesses.

I have now (May 1st 1881) sent to the Printers the M.S. of a little book on "The Formation of vegetable mould, through the action of worms." This is a subject of but small importance; & I know not whether it will interest any readers, but it has interested me. It is the completion of a short paper read before the Geological Society more than 40 years ago, & has revived old geological thoughts.

I have now mentioned all the books which I have published, & these have been the mile-stones in my life, so that little remains to be said. I am not conscious of any change in my mind during the last 30 years, excepting in one point presently to be mentioned; nor indeed could any change have been expected unless one of general deterioration. But my Father lived to his 83^d year with his mind as lively as ever it was, & all his faculties undimmed; & I hope that I may die before mine fails to a sensible extent. I think that I have become a little more skilful in guessing right explanations & in devising experimental tests; but this may probably be the result of mere practice & of a larger store of knowledge. I have as much difficulty as ever in expressing myself clearly & concisely; & this difficulty has caused me a very great loss of time; but it has had the compensating advantage of forcing me to think long & intently about every sentence, & thus I have been often led to see errors in reasoning & in my own observations or those of others. There seems to be a sort of fatality in my mind leading me to put at first my statement & proposition in a wrong or awkward form. Formerly I used to think about my sentences before writing them down; but for several years I have found that it saves time to scribble in a vile hand whole pages as quickly as I possibly can, contracting half the words; & then correct deliberately. Sentences thus scribbled down are often better ones than I could have written deliberately.

Having said this much about my manner of writing, I will add that with my larger books I spend a good deal of time over the general arrangement of the matter. I first make the rudest outline in two or three pages, & then a larger one in several pages, a few words or one word standing for a whole discussion or series of facts. Each one of these headings is again enlarged & often transposed before I begin to write in extenso. As in several of my books facts observed by others have been very extensively used, & as I have always had several quite distinct subjects in hand at the same time, I may mention that I keep from 30 to 40 large portfolios, in cabinets with labelled shelves, into which I can at once put a detached reference or memorandum. I have bought many books & at their ends I make an index of all the facts which concern my work; or if the book is not my own write out a separate abstract, & of such abstracts I have a large drawer full. Before beginning on any subject I look to all the short indexes & make a general & classified index, & by taking the one or more proper portfolios I have all the information collected during my life ready for use.

I have said that in one respect my mind has changed during the last 20 or 30 years. Up to the age of thirty, or beyond it, poetry of many kinds, such as the works of Milton, Gray, Byron, Wordsworth Coleridge & Shelley, gave me great pleasure, & even as a school-boy I took intense delight in Shakspeare especially in the historical plays. I have also said that formerly Pictures gave me considerable, & music very great delight. But now for many years I cannot endure to read a line of poetry: I have tried lately to read Shakespeare & found it so intolerably dull that it nauseated me. I have also almost lost any taste for pictures or music.—Music generally sets me thinking too energetically on what I have been at work on, instead of giving me pleasure. I retain some taste for fine scenery, but it does not cause me the exquisite delight which it formerly did. On the other hand, novels which are works of the imagination, though not of a very high order, have been for years a wonderful relief & pleasure to me, & I often bless all novelists. A surprising number have been read aloud to me, & I like all if moderately good, & if they do not end unhappily,—against which a law ought to be passed. A novel, according to my taste, does not come into the first class, unless it contains some person whom one can thoroughly love, & if it be a pretty woman all the better.

This curious & lamentable loss of the higher æsthetic tastes is all the odder, as books on history, biographies & travels (independently of any scientific facts which they may contain) & essays on all sorts of subjects interest me as much as ever they did. My mind seems to have become a kind of machine for grinding general laws out of large collections of facts, but why this should have caused the atrophy of that part of the brain alone, on which the higher tastes depend, I cannot conceive. A man with a mind more highly organised or better constituted than mine, would not I suppose have thus suffered; & if I had to live my life again I would have made a rule to read some poetry & to listen to some music at least once every week; for perhaps the parts of my brain now atrophied could thus have been kept active through use. The loss of these tastes is a loss of happiness, & may possibly be injurious to the intellect & more probably to the moral character by enfeebling the emotional part of our nature.

My books have sold largely in England, have been translated into many languages & passed through several editions in foreign countries. I have heard it said that the success of a work abroad is the best test of its enduring value. I doubt whether this is at all trustworthy, but judged by this standard my name ought to last for a few years. Therefore it may be worth while for me to try to analyse the mental qualities & the conditions on which my success has depended; though I am aware that no man can do this correctly. I have no great quickness of apprehension or wit which is so remarkable in some clever men, for instance Huxley. I am therefore a poor critic: a paper or book, when first read, generally excites my admiration, & it is only after considerable reflection that I perceive the weak points. My power to follow a long & purely abstract train of thought is very limited; I should, therefore, never have succeeded with metaphysics or mathematics. My memory is extensive, yet hazy: it suffices to make me cautious by vaguely telling me that I have observed or read something opposed to the conclusion which I am drawing, or on the other hand in favour of it; & after a time I can generally recollect where to search for my authority. So poor in one sense is my memory, that I have never been able to remember for more than a few days a single date or a line of poetry. Some of my critics have said "Oh he is a good observer but has no power of reasoning". I do not think that this can be true, for the Origin of Species is one long argument from the beginning to the end, & it has convinced not a few able men. No one could have written it without having some power of reasoning. I have a fair share of invention & of common sense or judgment, such as every fairly successful lawyer or doctor must have, but not I believe in any higher degree.

On the favourable side of the balance I think that I am superior to the common run of men in noticing things which easily escape attention, & in observing them carefully. My industry has been nearly as great as it could have been in the observation & collection of facts. What is far more important my love of natural science has been steady & ardent. This pure love has however been much aided by the ambition to be esteemed by my fellow naturalists. From my early youth I have had the strongest desire to understand or explain whatever I observed,—that is to group all facts under some general laws. These causes combined have given me the patience to reflect or ponder for any number of years over any unexplained problem. As far as I can judge I am not apt to follow blindly the lead of other men. I have steadily endeavoured to keep my mind free, so as to give up any hypothesis, however much beloved (& I cannot resist forming one on every subject) as soon as facts are shown to be opposed to it. Indeed I have had no choice but to act in this manner, for with the exception of the Coral Reefs I cannot remember a single first-formed hypothesis which had not after a time to be given up or greatly modified. This has naturally led me to distrust greatly deductive reasoning in the mixed sciences. On the other hand I am not very sceptical,—a frame of mind which I believe to be injurious to the progress of science; for I have met with not a few men, who I feel sure have often thus been deterred from experiment or observations, which would have proved directly or indirectly serviceable.

A good deal of scepticism in a scientific man is advisable to avoid much loss of time. In illustration, I will give the oddest case which I have known. A gentleman (who as I afterwards heard was a good local botanist) wrote to me from the Eastern counties that the seeds or beans in the common field-bean had this year everywhere grown on the wrong side of the pod. I wrote back, asking for further information, as I did not understand what was meant; but I did not receive any answer for a long time. I then saw in two newspapers, one published in Kent & the other in Yorkshire, paragraphs stating that it was a most remarkable fact that the beans this year had all grown on the wrong side. So I thought that there must be some foundation for so general a statement. Accordingly I went to my gardener, an old Kentish man,* & asked him whether he had heard anything about it; & he answered "oh no Sir, it must be a mistake, for the beans grow on the wrong side only on Leap-year, & this is not Leap-year." I then asked him how they grew on common years & how on leap-years, but soon found out that he knew absolutely nothing of how they grew at any time; but he stuck to his belief. After a time I heard from my first informant, who with many apologies said that he should not have written to me had he not heard the statement from several intelligent farmers; but that he had since spoken again to every one of them, & not one knew in the least what he had himself meant. So that here a belief—if indeed a statement with no definite idea attached to it can be called a belief—had spread over almost the whole of England without any vestige of evidence.—*

I have known in the course of my life only three intentionally falsified statements, & one of these may have been a hoax (and there have been several scientific hoaxes) which, however took in an American Agricultural Journal. It related to the formation in Holland of a new breed of oxen by the crossing of distinct species of Bos* (some of which I happen to know are sterile together), & the author had the impudence to state that he had corresponded with me & that I had been deeply impressed with the importance of his results. The article was sent to me by the editor of an English Agricult. Journal, asking for my opinion before republishing it.—

A second case was an account of several varieties raised by the author from several species of Primula, which had spontaneously yielded a full complement of seed, although the parent-plants had been carefully protected from the access of insects. This account was published before I had discovered the meaning of heterostylism, & the whole statement must have been fraudulent, or there was neglect in excluding insects so gross as to be scarcely credible.

The third case was more curious: Mr Huth published in his book on consanguineous* marriage some long extracts from a Belgian author, who stated that he had interbred rabbits in the closest manner for very many generations without the least injurious effects. The account was published in a most respectable Journal, that of the R. Medical Soc. of Belgium; but I could not avoid feeling doubts,— I hardly know why, except that there were no accidents of any kind, & my experience in breeding animals made me think this improbable. So with much hesitation I wrote to Prof. Van Beneden asking him whether the author was a trustworthy man. I soon heard in answer that the Society had been greatly shocked by discovering that the whole account was a fraud. The writer had been publicly challenged in the Journal to say where he had resided & kept his large stock of rabbits while carrying on his experiments, which must have consumed several years, & no answer could be extracted from him. I informed poor Mr Huth, that the account which formed the cornerstone of his argument was fraudulent; & he in the most honourable manner immediately had a slip printed to this effect to be inserted in all future copies of his book which might be sold.

My habits are methodical, & this has been of not a little use for my particular line of work. Lastly I have had ample leisure from not having to earn my own bread. Even ill-health, though it has annihilated several years of my life, has saved me from the distractions of society & amusement.

Therefore my success as a man of science, whatever this may have amounted to, has been determined, as far as I can judge, by complex & diversified mental qualities & conditions. Of these the most important have been—the love of science—unbounded patience in long reflecting over any subject—industry in observing & collecting facts—& a fair share of invention as well as of common sense. With such moderate abilities as I possess, it is truly surprising that thus I sh^d have influenced to a considerable extent the belief of scientific men on some important points.

Aug 3^d. 1876

This sketch of my life was begun about May 28th at Hopedene,* & since then I have written for nearly an hour on most afternoons.

THE MAKING OF A CELEBRITY

Darwin is dead. Dying in his seventy-third year, he had earned a great reputation before he was thirty, but it was in 1859, on the publication of his "Origin of Species," that his name became so familiarly known all the world over.

> Announcement in *The Times of India* on 22 Apr. 1860, carried by electrical telegraph and published in Bombay (Mumbai) within twenty-four hours of its appearance in the London press

The Abbey has its orators and Ministers who have convinced senates and swayed nations. Not one of them all has wielded a power over men and their intelligences more complete than that which for the last twenty-three years has emanated from a simple country house in Kent. . . . The moment the thought arose, not, apparently, in any single mind, but spontaneously and everywhere, that the body of the great naturalist ought to be buried at Westminster, it was felt that the Abbey needed it more than it needed the Abbey.

> *The Times of London* (26 Apr. 1882) on Darwin's burial in Westminster Abbey

What is less well known perhaps to the general public, but which his death will bring to light and which we cannot mention without blushing, is that in France—in France alone!—scientific officials, rentiers, pensioners, prize-winners, after having struggled helplessly to resist and having nearly proscribed him, remain adamant. In 1878, the greatest naturalist of the century was not judged worthy of the dignity of becoming a Corresponding Member of our Institute. He died without the Academy of Sciences having elevated him from the grade of Foreign Associate to the First Class!

Will the Academy of Sciences imitate the example, not reparation, but the humiliation given already by her sister of letters? Will she admit a bust of Darwin into its meeting hall?

> French journalist and socialist Charles Longuet, from the Parisian radical newspaper *La Justice* (23 Apr. 1882)

Mr. Darwin was a Shrewsbury man: what will his native town, to which he has lent an imperishable renown as the birthplace of one of the world's greatest thinkers, do to honour his memory, or rather to record her own honour? . . . The people of Shrewsbury, we are afraid, may be slow to appreciate a man like Darwin, the greatest of their townsmen; but it is scarcely possible they will fail in some way to show themselves not altogether unworthy of kindred with the great Englishman who to-day, amid the affectionate regrets of the civilized world, is buried in Westminster Abbey, where, of all the worthies who rest beside him, so few can be reckoned as his peers.

Oswestry Advertizer and Montgomeryshire Mercury (26 Apr. 1882)

In all fields of human activity, there is a trait held in common by truly superior individuals. That is, never to lose sight of particular facts when dealing with theories or actions of major significance. Thus, a great general at the same time takes care of the food or even the footwear of his men, as well as strategic plans; a great judge can argue a case about a party wall, and draw up a legal code. Among naturalists, Darwin had this exceptional ability. . . . He always took care to give both sides of the case, arguments both strong and weak, for the reader to compare, sort, and conclude. The method is not didactic. It is purely scientific.

Swiss botanist Alphonse de Candolle, 'Darwin considéré au point de vue des causes de son succès et de l'importance de ses travaux', *Bibliothèque universelle* (*Archives de sciences physiques et naturelles*), ser. 3, vol. 7 (May 1882), 481–95, at 487–8

Gentlemen, I can truthfully claim that the name of Darwin has been known to me for forty years, since the Beagle, under Fitz Roy's command, visited the extreme south of our continent. I knew the ship, and her crew, and soon the *Journal of Researches*, which I consulted frequently in discussions of the Strait [of Magellan]. As you recall, I was never particularly zealous in defending our southern-most possessions, because notwithstanding the fantastic descriptions issuing from the imaginations of credible folk who still hoped to find El Dorado, I never thought them worth a single barrel of powder. Our fathers had scoured those same regions too long in vain for us to contemplate yet another war in pursuit of some mythological Holy Sepulcher.

And there are many among us whose own reason obliges them to believe, practice, and test the doctrines of the illustrious sage.

They, in turn, are enriched by their beliefs, a benefit which not all those who believe in human progress are privileged to enjoy. The clever breeders of sheep are positively consumed by Darwinism, and without equals in the art of *varying species*.

Here in our own pastures they gave Darwin his first inklings, notions he later perfected in the breeding of pigeons. In Europe this pastime consists in fashioning varieties after the whim of the breeder. . . . Our country is home to hundreds of landowners devoted to the breeding of sheep and other livestock. Among them the Pereiras, Duportals, Chás, Ocampos, Casares, Kemmis, and Dowrys have particularly distinguished themselves, and they read with exquisite care Darwin's discussions of variation under natural selection, for they accomplish it artificially, by choosing the reproducers. Given the slightest scrap of change, they have enough material for the next cross, and the next selection.

With our Argentine fossils and breeds, we gave Darwin science, and fame; our landowners, in turn, enrich themselves by following his directions.

It seems to me that we Argentines have sufficient motive for subscribing to the transformist doctrine, given how we transmute one variety of sheep in another. We have constituted a new species, the *argentiferous sheep*, so-called both because of its Argentine origins, and because it brings in the silver.

Former president of Argentina, Domingo Sarmiento, 'Lecture on Darwin', presented at a public meeting of the Círculo Médico [Medical Circle] in the Teatro Nacional [National Theatre], 30 May 1882, in *Obras de D. F. Sarmiento*, xxii (Buenos Aires: Imprenta Mariano Moreno, 1899), 105–33, at 106–10, trans. in A. Novoa and A. Levine, *¡Darwinistas!* (forthcoming)

Consciously or unconsciously, Charles Darwin, so recently gone from us, was the most formidable foe religion ever had. . . .

It will be said that to-day that the better class of clergy accept Darwinism. But not a captain in the Salvation Army accepts it. Your bishops and archbishops, most of them Freethinkers in their heart of hearts, smile approval of the great truth. But the truly religious people, with whom the fashionable religionists are contemplating amalgamation, reject it with shrieks. The clergy of 1859 rejected it to a man. When the "Origin of Species" first appeared the clergy were true to their creed for once. They denounced the truth as they have denounced every new great truth yet given to the world. Darwinism was Atheism then. It is Atheism now, thank man. . . .

Only the Roman Catholic Church is consistent. That Church is the most truly religious, and therefore the most wicked and danger-ous, of all. Its Bishop of Salford but yesterday declared that Darwin was at this moment in hell, suffering the tortures of the damned. The position of the Bishop of Salford is the only truly religious position, and the religious people who try to reconcile the irreconcilable, who would accept modern truths while they cling to ancient fables, who pretend to believe in religion and in the irreligious truths of science, are palterers with their own souls.

English socialist Edward B. Aveling, 'Darwin and Religion', *The Freethinker* (25 June 1882), 202–3

Looking up I saw the devil, who made a sign for me to follow. We crossed the hall, and passing through a doorway, entered a still larger room. It contained every scientific book, instrument, and piece of apparatus the world had ever produced. The walls were covered with diagrams of all descriptions, dealing with all the sciences. Tables were scattered over the floor. At these sat students each engaged on his favorite subject; and I saw a man passing among them, helping all when any difficulty arose. I looked him in the face and saw he was Charles Darwin.

C. Herbert Pring, 'Hell', *The Freethinker* (3 June 1882), 171

If ever a man's ancestors transmitted to him ability to succeed in a particular field, Charles Darwin's did. If ever early surroundings were calculated to call out inherited ability, Charles Darwin's were. If ever a man grew up when a ferment of thought was disturbing old convictions in the domain of knowledge for which he was adapted, Charles Darwin did. If ever a man was fitted by worldly position to undertake unbiassed and long-continued investigations, Charles Darwin was such a man. And he indisputably found realms waiting for a conqueror. Yet Darwin's achievements far transcend his advantages of ancestry, surroundings, previous suggestion, position. He stands magnificently conspicuous as a genius of rare simplicity of soul, of unwearied patience of observation, of striking fertility and ingenuity of method, of unflinching devotion to and belief in the efficacy of truth. He revolutionised not merely half-a-dozen sciences, but the whole current of thinking men's mental life.

English botanist George Thomas Bettany, *Life of Charles Darwin* (London, 1887), 11

One who is deeply convinced that the mind of man has developed out of the lowest movements of instinct, and owns no higher origin than the blind struggle for food and safety by which a medusa wriggles into solidity and gets itself bones and muscles, must naturally be nauseated by Shakespeare—for Shakespeare is a tremendous nut for an evolutionist to crack. He may doubt or reject the existence of God, but he cannot reject the fact of that supreme and sovran poet. . . . It shows, however, the perfect honesty and good faith of Darwin, that he admits this damaging result—or, perhaps, a certain comfortable satisfaction with himself which did not see how damaging it was.

Scottish novelist Margaret Oliphant, reviewing anonymously the *Life and Letters of Charles Darwin*, *Blackwood's Edinburgh Magazine*, 143 (Jan. 1888), 104–15, at 112–13

Religious-minded men most justly opposed Darwin, for we feel persuaded that his influence has been more hostile and fatal to Theism than the influence of any other writer for centuries past—perhaps since the foundation of Christianity itself—and his many excellent qualities made his influence all the more disastrous. Not that he was himself an anti-religious man, or that he desired or rejoiced at the spread of irreligion. His mind (like that of many another man who in his day has gained the esteem and love of his fellows) was simply non-religious as well as non-philosophical; and though, within narrow limits, so intelligent and so excellent in his human sentiments and actions, he had no conception of intellect, morality, or religion as such. He attempted to ideally construct a world without them, and it was a world of insects and pigeons, apes and curious plants; but man, as he exists, had no place within it.

The English anatomist St George Jackson Mivart, in an anonymous article on 'The Life and Letters of Charles Darwin', *Edinburgh Review*, 167 (Apr. 1888), 407–47, at 434–5

Why, after all, want to make Darwin a theologian? Let us rather admire him as a great man of learning, modest and affable, the tireless and honest researcher. . . .

French Catholic modernist theologian Marcel Hébert, 'La Vie et la correspondence de Charles Darwin', *Bulletin critique*, 9 (15 Apr. 1888), 152–6, at 156

Abū Bakr ibn Bashrūn said in his epistle on alchemy . . . that minerals transform into plants, and plants into animals, and that the last of these three transformations and the highest link in the chain is man. . . . If this is what the theory of evolution is based on, then the Arab scientists preceded Darwin. We should however acknowledge his efforts and his dogged standing by these principles and his service to most fields of natural history, even as I disagree with him and his followers regarding the issue of the 'breath of life' which is created by the glorious and exalted Creator and not by means of evolution from apes to men or from stormy water! Or with the idea that a flea can become an elephant because they see something similar in a flea with the trunk of an elephant and claim that it will change into it after thousands of years, and so on. This I refuted in my *Refutation of the Materialists*, which was an answer to Darwin and his claims. . . .

And the leader [*imām*] of the school of evolution and progress without a doubt is Darwin, and when this sage reached the essential point, and that is the existence of 'the breath of life', he could only stop and say that it was the Creator who gave life to the living, writing as follows: 'I see that all the living creatures which lived on this earth are from one primitive form, which the Creator breathed life into!'

That Darwin said this means that he rejects the emergence of life by naturalistic means and that he does not go as far as the materialistic natural scientists, and that they deny that Darwin said this and accused him of fear of the people of his religion. They said that this claim would make the theory incomplete and undermine its foundation because the aim, as we mentioned, of the school of the naturalist is 'the denial of the Creator' and to attribute [His] works to nature [alone]. . . .

In short, all that the school of the naturalists do by way of limiting living beings into a few groups, from which larger numbers branch off, none of this is harmful to believe in. And it does not benefit them to claim that life and the appearance of the living is the result of a natural power. Yes, if they had proof for spontaneous generation, then their claims would have had meaning and their theoretical principles would have been supported.

Islamic modernizer Jamāl al-Dīn al-Afghānī, *'madhab al-nushu' wa-l-irtiqat (The School of Evolution and Progress), Khātirāt Jamāl al-Dīn al-Afghānī* [Works of Jamal al-Din al-Afghani] (Cairo, 2002; originally published *c.*1892–6), vi, 154–8, trans. Marwa Elshakry

Mr. Darwin was not purely and simply in religious matters a simpleton. He was a hypocrite and he knew it; he tried to smother his qualms; he volunteered a few extreme expressions and one attempt at sarcasm as sets-off to his fictitious pieties. . . . Into his remark that there is grandeur in this view of life having been originally breathed into a few forms or into one—a remark with which the concluding, most conspicuous, and most commented upon sentence of the *Origin of Species* commences—Darwin inserted words originally absent therefrom; and these words were—"by the Creator". . . . For he saw that the black beasts [priests] and their wood [for burning heretics] were no mere relics of the past. He feared, he propitiated, he won. And his body was placed within the four walls of Westminster Abbey in London—not scattered to the four winds of the Campo di Fiori in Rome. And the black beasts sang his praise.

English freethinker Oswald Dawson, *An Indictment of Darwin* (London, 1888), 35–6

A glance at any photograph of Darwin is sufficient to convince any one that his brain was so imperfectly developed that he was not naturally capable of exhibiting any higher functions of mind, and could only be a keen observer of facts and a steady plodder in experiments. Even his experiments on the influence of worms were due to the suggestions of another, and he originated nothing. Darwin may occupy an honourable place in heaven, because he made the most of his abilities, and acted well according to his light; but a man with a brain and intellect like that of Shakespeare could not believe as Darwin did without a violation of his consciousness of truth, and so would be guilty of a grievous sin. Although the evolution theory was contrary to reason and to scientific principles, the imperfection of his brain and the deficiency of his education in the knowledge of perfect archetypes made Darwin incapable of feeling the full force of the absurdity of his notion that poverty and ignorance are scientific indications of natural inferiority.

Scottish physician Samuel B. G. McKinney, *The Science and Art of Religion* (London, 1888), 35–6

Without the recent biographical and autobiographical notes published by his son, no one could have imagined that Darwin, a model father and citizen, so self-controlled and even so free from vanity, was a neuropath. His son tells us that for forty years he never enjoyed twenty-four hours of health like other men. Of the eight years devoted to the study of the cirripedes, two, as he himself writes, were lost through illness. Like all neuropaths he could bear neither heat nor cold; half an hour of conversation beyond his habitual time was sufficient to cause insomnia and hinder his work on the following day. He suffered also from dyspepsia, from spinal anæmia and giddiness (which last is known to be frequently the equivalent of epilepsy; and he could not work more than three hours a day. He had curious crochets. Finding that eating sweets made him ill, he resolved not to touch them again, but was unable to keep his resolution, unless he had repeated it aloud. . . . He frequently, says his daughter, inverted his sentences, both in speaking and writing, and had a difficulty in pronouncing some letters, especially *w*. Like . . . Socrates, he had a short snub nose, and his ears were large and long. Nor were degenerative characteristics wanting among his ancestors. . . . Between the physiology of the man of genius, therefore, and the pathology of the insane, there are many points of coincidence; there is even actual continuity.

Italian criminologist Cesare Lombroso, *The Man of Genius* (London, 1891), 356–7, 359

No matter what may be opinions as to evolution, the story of Charles Darwin's life alone is one of the greatest and grandest that can be studied. In no other biography can there be found such an expression of will or determination, of patience, and of the triumph of a man's intellect over physical weakness. Think of the arduous labors of Charles Darwin, who for forty years hardly enjoyed a day free from suffering! His autobiography here printed, which bears this title: "Recollections of the Development of My Mind and Character," is the most fascinating piece of writing we know of. There never were words written that had a truer ring.

'Darwin's Life Once More', *New York Times* (29 Jan. 1893)

Darwin was an English scholar of animals and plants, who as a young man, carrying on the scholarship of his family, sailed round the globe, collecting in great profusion the most rare and exotic of flora and fauna. Then, after infinitely careful research, extending over many decades, he wrote a book entitled *In Search of the Origin of Species*. After it came out, it was soon to be found in almost every home in Europe and America, and Western scholarship, government, and philosophy all drastically changed. Someone has said, and it is no idle boast, that Darwin's theories have given men new eyes and ears and have changed their thinking more even than did the science and astronomy of Newton.

Scholar and translator Yan Fu challenging the Chinese authorities in 'Yuan qiang' (Whence strength?), in the Tianjin newspaper *Zhibao* (1895), trans. in J. R. Pusey, *China and Charles Darwin* (Cambridge, Mass., 1983), 59

It has been somewhat of a fashion since the publication of the 'Life and Letters' to speak with gentle pity of Darwin. "Poor fellow, quite crushed in the end by science; mind quite atrophied, sense of religion dead. When a young man he never moved without a Milton in his pocket but could not read a line of poetry later on." But such critics will find little evidence of an "atrophied" mind in his letters, and may if they be candid with themselves, realise that Darwin, with his ill-health and the enormous mass of work that he was carrying, was forced to confine his whole mind to one thing if he was to live at all.

Review of *More Letters of Charles Darwin* in the English *Saturday Review*, 95 (16 May 1903), 619

We refer to Darwin in his final aspect as the great imaginative writer of his century. For already it has been recognized that amid all the richness and variety of genius in the Victorian era, not the Arthurial, nor the Pantheon of Carlyle, nor the vast portrait gallery of Dickens, but "The Descent of Man" of Darwin, is the one dominant work of imagination. The age's true epic. More and more it is being felt that the true position of Darwin with respect to the longer range of thought and feeling beyond his own generation is at the side of Dante and Goethe. . . . a potent organizer and assembler of the age's most vital and controlling dreams, the finished exponent of the peculiar imaginative energy of his time.

Poet Revd Obadiah Cyrus Auringer, writing from Forestport, New York in the *New York Times Saturday Review of Books* (1 May 1909)

Charles Darwin, the great English man of Science of the nineteenth century, has not only revolutionized the aims and methods of biological sciences by the profound researches of his own, but has given a great impetus to the progress of human thought in general all over the world. . . .

It is fortunate that when Japan's task of national renovation was just beginning by the introduction of western sciences and institutions, our men of science came face to face with such a new and far-reaching theory as advanced by Darwin which represented the highest tide mark of European thought of the nineteenth century.

Furthermore the true spirit of investigation which characterized Darwin, his constant and untiring love for truth, have set a noble ideal for the nation's youth and all students of science.

Imperial University of Tokyo, hand-painted scroll in lacquer box, presented at the centenary celebration of Darwin's birth at Cambridge. CUL: UA Conf. I.34.116*

Prof. Hertwig referred to the influence of Darwin's work upon German biology, particularly at Jena. It was through Haeckel, who hailed Darwinism with delight, and said that evolution was the key of man's destiny, that the theory became predominant in German science. It had been the starting point for all the researches of the younger men, and had entered into the life of the German people. Earlier this year festivals in commemoration of Darwin's work were held in Hamburg, Munich, Frankfurt, and other towns in Germany. The celebration at Cambridge was the acme of these festivals, and would give an immense stimulus to the scientific work of the delegates privileged to be present at it. . . .

Prof. Metchnikoff in his address referred to the debt which medical science owes to the theory of organic evolution founded by Darwin. Diseases undergo evolution in accordance with the Darwinian law, and the recognition of this fact led to the science of comparative pathology. . . .

Prof. Arrhenius [the Swedish chemist] then spoke as follows:— . . . All of us are profoundly sensible that the great intellectual revolution which is due to the introduction of evolutionism is the most important event in the development of the human mind, since the mighty political movement which began with the storming of the Bastille 120 years ago. There is, however, this significant difference between that time and this, that whereas in such a period every mighty change in the social, political, and intellectual development of mankind was only effected by strife and horrors of war, to-day, thanks to the civilising progress, this change has been accomplished by reason and persuasion. "The pen has been mightier than the sword." How much may we not congratulate ourselves that we have lived in such a period? In reality, the doctrine of evolution is inconsistent with violence, and we may hope, therefore, that it will give a mighty impetus to the maintenance of peace and a good understanding between civilized nations.

'The Darwin Celebrations at Cambridge', *Nature*, 81 (1 July 1909), 7–14, at 8 and 11

EXPLANATORY NOTES

JOURNAL OF RESEARCHES

3 *chronometrical measurements round the World*: the voyage's aim was to use chronometers (extremely accurate timepieces for measuring longitude at sea) to improve navigational charts.

4 *Peninsular war*: the war of 1808–14 in the Iberian peninsula fought between the allied forces of Spain, Great Britain, and Portugal against France during the Napoleonic Wars.

5 *Vênda*: the Portuguese name for an inn.

6 *some vintéms*: Portuguese copper coins of relatively low value.

hygrometer: an instrument for measuring atmospheric humidity.

infusoria with siliceous shields: infusoria (originally so named because found in decaying plant infusions) are simple microscopic organisms. Those with hard shells made of silica include the diatoms.

harmattan: a dusty North African trade wind that blows from the end of November to the middle of March.

7 *cryptogamic plants*: the class of plants, including mosses, liverworts, and ferns, which lack flowers and reproduce by spores.

scoriaceous: having the appearance of rough, slag-like masses formed when molten lava is exposed to the air.

arragonite: a carbonate of lime.

8 *galvanism*: current electricity from an electrical battery.

9 *tertiary formations*: in early nineteenth-century geology, the Tertiary rocks were sedimentary rocks above, and more recently formed than, the underlying Secondary and Primary strata. Approximately equivalent to the modern Cenozoic.

10 *pumiceous*: like pumice, a glassy frothy lava.

porphyry: an igneous rock with large detached crystals of feldspar in a fine-grained ground mass.

11 *the guanaco and llama*: the guanaco is a member of the camel family common in South America, and the llama is its domesticated form. The *Palaeotherium* is an extinct hoofed animal with a tapir-like snout. On the complex affinities of *Macrauchenia*, which had been identified from Darwin's specimens by Richard Owen, see S. P. Rachootin, 'Owen and Darwin Reading a Fossil: *Macrauchenia* in a Boney Light', in D. Kohn (ed.), *The Darwinian Heritage* (Princeton, 1985), 155–83.

11 *Toxodon and the Capybara*: the Toxodon, identified for Darwin by Richard Owen, was an extinct rodent the size of a hippopotamus; the capybara, also found in South America, is the largest living rodent.

Edentata: an order of mammals having no front teeth or teeth at all, such as the sloths, anteaters, and armadillos.

Ctenomys and Hydrochærus: commonly known as tuco-tucos, *Ctenomys* is a genus of several dozen species of South American burrowing rodents; *Hydrochaerus* is the genus of the giant water-dwelling rodent, the capybara.

13 *the little tucutuco at Bahia Blanca*: Owen had described Darwin's fossils as belonging to a hitherto-unknown species of extinct rodent, *Ctenomys priscus*.

14 *Megalonyx*: a genus of extinct giant ground sloth.

16 *These Fuegians . . . Strait of Magellen*: the native peoples of Tierra del Fuego included four major tribal groups with distinct customs and languages. The first group the *Beagle* met were the Haush, a hunting people; the 'miserable wretches further westward' were the Yamana, who built canoes and lived by fishing. The tall, tent-dwelling Tehuelches (commonly called Patagonians by Europeans in this period) lived further north and had had more contacts through trade than any of the Fuegian groups.

Der Freischutz: a three-act opera by Carl Maria von Weber, famous for the uncanny supernatural effects of the Wolf's Glen scene. Facial and body painting was a common practice among the tribal groups, with specific designs employed for hunting, competitive sports, ceremonies, and peace meetings.

scarcely deserves to be called articulate: in fact these languages were highly complex. Yamana had more inflections than Greek and an extensive vocabulary.

17 *Caffres*: Kaffir, the name given by Europeans to indigenous peoples of southern Africa, particularly the Xhosa.

23 *the skin of her naked baby!*: the diet of the Yamana, rich in meat and blubber, afforded considerable protection against the cold.

24 *the slaughter-house at their own fire-sides!*: the anthropological literature is unequivocal in denying that the inhabitants of Tierra del Fuego ate shipwrecked sailors, tribal enemies, or old women. Human life was sacred and any death had to be avenged by relatives.

a wizard or conjuring doctor: on the role of shamans, see A. Chapman, *Drama and Power in a Hunting Society: The Selk'nam of Tierra del Fuego* (Cambridge, 1982), esp. 44–7.

28 *"yammerschooner," which means "give me"*: the phrase actually meant 'be kind to me' or 'be kind to us'.

29 *Oens men*: the Ona, or Selknam tribe, who lived further east.

30 *the Tekenika*: another name (used by Robert Fitzroy in his extensive ethological study of these groups) for the Yamana.

32 *several huge whales*: Darwin's note reads: 'One day, off the east coast of Tierra del Fuego, we saw a grand sight in several spermaceti whales jumping upright quite out of the water, with the exception of their tail-fins. As they fell down sideways, they splashed the water high up, and the sound reverberated like a distant broadside.'

36 *"chef-d'oeuvres . . . et ses phénomènes"*: 'masterpieces of human industry, as they treat the laws of nature and their phenomena' (Fr.), quoted from Louis-Antoine de Bougainville, *Voyage autour du monde par la frégate du roi La Boudeuse, et la flûte L'Étoile, en 1766, 1767, 1768 et 1769* (Paris, 1771).

37 *nice-looking wife*: lassaweea, met also by the *Allen Gardiner* in 1858. The crew gave her the name Jamesina Button.

his wife Fuegia: Darwin's note reads: 'Captain Sulivan, who, since his voyage in the Beagle, has been employed on the survey of the Falkland Islands, heard from a sealer in (1842?), that when in the western part of the Strait of Magellan, he was astonished by a native woman coming on board, who could talk some English. Without doubt this was Fuegia Basket. She lived (I fear the term probably bears a double interpretation) some days on board.'

38 *Otaheite*: Tahiti.

41 *Euphorbiaceæ*: the spurge genus, with a milky, usually poisonous juice.

antediluvian animals: literally, before the flood; used to refer to any of the large prehistoric creatures from early geological ages.

45 *Polybori*: John Gould had described this unusual species of hawk; for more on the identifications of the birds mentioned in this paragraph, see F. Sulloway, 'Darwin's Conversion: The *Beagle* Voyage and its Aftermath', *Journal of the History of Biology*, 15 (1982), 325–96.

Progne purpurea: the common Purple Martin.

47 *Totanus*: a wader, in a new species determined by John Gould.

49 *guayavita*: a species of the guava tree, later found to be unique to the Galapagos.

pericardium: the sac round the heart.

51 *Wood and Rogers*: a reference to Woodes Rogers (*c.*1679–1732).

56 *Secondary epochs*: in early nineteenth-century geology, the rocks between the Primary and the Tertiary; approximately equivalent to the modern Mesozoic. At the time Darwin left on the *Beagle*, the recently discovered herbivorous iguanodon and the carnivorous megalosaurus were the best known of the extinct reptiles.

Prionotus . . . on the eastern side of America: Darwin's specimens of these searobins, after identification by Leonard Jenyns, extended the range of species hitherto only known from the Atlantic.

57 *Diptera and Hymenoptera*: diptera are insects with two wings; hymenoptera (such as ants, bees, and wasps) have four transparent wings.

61 *Leguminosæ*: flowering plants of the pea or bean family.

arborescent genus of the Compositæ: endemic to the Galapagos, *Scalesia* is an unusual tree-like genus of the aster family.

64 *Opetiorhynchus*: the tussock bird of the Falklands is identified as *Opetiorhynchus antarcticus* in *Zoology of the Voyage of H.M.S. Beagle* (1839), pt. 3, 67–8. Like others in its genus (now reclassified as *Cinclodes*) it is in the ovenbird family.

69 *chama*: a family of marine bivalve clams.

71 *"C'est une meruille de voir . . . point d'artifice humain"*: it is a marvel to see each of these atolls completely surrounded by a great shelf of stone, at no point involving human artifice (Fr.).

88 *Serpulæ, together with some few barnacles and nulliporæ*: a group of marine worms that secret a contorted or spiral calcareous tube. Nulliporae are algae that secrete carbonate of lime on their surface.

ORIGIN OF SPECIES

107 *that mystery of mysteries . . . by one of our greatest philosophers*: John Herschel, in a letter to Charles Lyell printed in C. Babbage, *The Ninth Bridgewater Treatise* (1837), 203, had written of 'that mystery of mysteries, the replacement of extinct species by others'.

108 *coadaptation*: the mutual adaptation of species or parts of an organism, so that they may come to depend upon one another.

109 *The author of the 'Vestiges of Creation'*: Darwin had guessed that the author of this widely read evolutionary survey, published anonymously in 1844, was Robert Chambers, but the secret was not revealed until 1884.

113 *carunculated*: having the character of a fleshy excrescence.

114 *crop*: a pouch-like enlargement of the oesophagus in many birds.

ramus of the lower jaw: the ascending part of the lower jaw, which joins to meet the skull.

caudal and sacral vertebræ: caudal vertebrae are those of the tail; sacral vertebrae belong to the sacrum, or the bone composed of fused vertebrae and connecting to the pelvis.

furcula: the forked bone below the neck of a bird; the wishbone.

scutellæ: the horny plates which cover the feet of many birds.

120 *fuller's teazle . . . the wild Dipsacus*: the heads of fuller's teasel have hooked prickles between the flowers, and are used for teasing cloth; wild teasel (*D. sylvestris*) has straight instead of hooked prickles.

the turnspit dog . . . the ancon sheep: turnspit dogs (now extinct) had long bodies and short legs, suitable for their role in running within a tread-wheel to turn a roasting-spit; ancon sheep were bred from a single lamb born in 1791 with a long body and very short legs.

121 *Collins*: Darwin's mistaken reference (retained in all editions of *Origin*) to the stockbreeder Charles Colling.

128 *Rubus, Rosa, and Hieracium amongst plants*: *Rubus* are the genus of brambles, including raspberries and blueberries; *Rosa* is the genus of roses; and *Hieracium* are hawkweed, a large genus of the sunflower family.

Brachiopod shells: brachiopods are marine animals with two-valved shells, and fringed arms for carrying food to the mouth.

138 *charlock*: wild mustard, or other plants of the same order.

148 *stipules in some Leguminosæ*: stipules are small leafy organs placed at the base of the stalks of many flowering plants, in this case those of the pea or bean family.

149 *larger anthers, would be selected*: in flowering plants, anthers are the summits of the stamens, in which the pollen—the male element—is produced; it is then received by the stigma, part of the pistil (the female organ of the plant), which also contains the style (or ovary).

150 *"physiological division of labour"*: just as contemporary economists argued for the effectiveness of employing specialized workers in different steps of a manufacturing process, so the naturalist Henri Milne Edwards suggested that the efficiency of the organism was enhanced by having different organs of the body employed in specific tasks.

corollas: the corolla is the inner envelope of a flower, usually composed of coloured leaves (petals).

155 *tertiary*: the Tertiary rocks were strata above, and more recently formed than, the underlying Secondary and Primary.

Ganoid fishes: a group of fishes, now mostly extinct, covered with enamelled bony scales.

Ornithorhynchus and Lepidosiren: the Australian platypus (an egg-laying mammal) and the South American lungfish (a fish with both lungs and gills).

158 *"as if by some murderous pestilence:"* a quotation adapted from W. Youatt, *Cattle* (1834), 200.

159 *"fanciers do not . . . but like extremes:"* from J. M. Eaton, *A Treatise on the Art of Breeding and Managing Tame, Domestic, Foreign, and Fancy Pigeons* (1858), 86.

177 *Balanus*: the acorn-shell genus of barnacles.

179 *the Mustela vison*: the American mink.

181 *monstrous as a whale*: Darwin removed the details of this much-criticized hypothetical genealogy from all further editions.

182 *the Puffinuria berardi . . . profoundly modified*: Darwin based this conclusion on John Gould's studies of the *Beagle* specimens, which showed this

seagoing bird was 'a complete auk in its habits, although from its structure it must be classed with the Petrels' (*Zoology of the Voyage of H.M.S. Beagle* (London, 1841), pt. 3, no. 5, p. 138).

182 *grallatores*: a former order of wading-birds (e.g. storks, cranes, snipes), generally with long legs featherless above the heel, and without membranes between the toes.

183 *spherical and chromatic aberration*: spherical aberration occurs when rays of light passing through different parts of a convex lens are brought into focus at slightly different distances; chromatic aberration takes place when the coloured rays are separated by the prismatic action of the lens and are also brought into focus at different distances.

184 *Articulata*: in the taxonomic system of Cuvier, a division of the animal kingdom generally characterized by a segmented body plan.

188 *"Natura non facit saltum"*: 'Nature does not make a leap' (Lat.).

189 *I have attempted to show*: in Chapter VII, not included here.

192 *the Silurian system*: in the classification used by Darwin, these were the oldest rocks in which fossils had been discovered.

198 *ichneumonidæ*: a major family of parasitic wasps, which lay their eggs in the bodies or eggs of other insects.

the burrowing tucutucu: Darwin collected the rodent *Ctenomys* in Brazil, noting that they had eyes but were commonly blind. As he wrote in *Journal of Researches* (2nd edn. [1845], 51–2), 'Lamarck would have been delighted with this fact, had he known it, when speculating (probably with more truth than usual with him) on the gradually-*acquired* blindness of the Aspalax, a Gnawer living under ground, and of the Proteus, a reptile living in dark caverns filled with water; in both of which animals the eye is in an almost rudimentary state, and is covered by a tendinous membrane and skin.'.

200 *I have attempted to show*: in Chapter VII, not included here.

206 *vera causa*: 'true cause' (Lat.). For Darwin, as for his mentor Lyell and for Isaac Newton, a *vera causa* had to be a cause in operation, sufficient to explain the phenomena, that could be verified by the senses. This notion had been developed in the Scottish philosophy of common sense during the eighteenth century, and in John Stuart Mill's *System of Logic* (1843). The degree to which natural selection met this criterion was much debated.

germinal vesicles: minute vesicles in the eggs of animals, from which development of the embryo proceeds.

207 *life was first breathed*: in the second edition of 1860, Darwin added the phrase 'by the Creator', but removed it in the third and subsequent editions.

211 *having been originally breathed*: in the second and subsequent editions, Darwin added the phrase 'by the Creator'.

DESCENT OF MAN

233 *"light would be . . . and his history"*: this edn. p. 210.

"personne, en Europe . . . des espèces": no one, in Europe at least, still dares to support the separate and independent creation of every species (Fr.).

234 *homological structure*: structures are said to be homological when they result from the development of corresponding embryonic parts, so that they are similar in position and structure but not necessarily in function. An example would be the wing of a bird and foreleg of a horse.

rudimentary organs: vestigial organs, often of little or no apparent use.

anthropomorphous apes: apes resembling man.

235 *"Natürliche Schöpfungsgeschichte"*: translated from German into English in 1876 as *History of Creation*.

first edition, p. 199: this edn. p. 235.

in full detail: Darwin's note reads: 'Prof. Häckel is the sole author who, since the publication of the "Origin," has discussed, in his various works, in a very able manner, the subject of sexual selection, and has seen its full importance.'

236 *a lamprey or lancelet*: lampreys look superficially like eels, but lack scales and have a skeleton made of cartilage. Lancelets, also known as amphioxus, live in warm seas, and possess a nerve cord running along the back which is not protected (as in vertebrates) by bone. In Darwin's time they were thought to be the most primitive fish.

238 *philology:* the scientific study of language.

239 *microcephalous*: abnormally small-headed.

242 *a Crinoid*: a sea-lily of the class Crinoidea, with a radially symmetrical cup-shaped body and arms, usually attached by a jointed stalk.

243 *"I abide. . . consequently the hypothesis is a false one"*: Darwin puts the words of the Revd Dr J. M'Cann, *Anti-Darwinism* (1869), 13 into the mouth of the dog.

244 *"as one of the most remarkable . . . savages and brutes"*: Darwin cites an anonymous review of Lubbock's *Prehistoric Times* in the weekly *Spectator* (4 Dec. 1869), 1428–30, at 1430.

246 *an intermediate condition*: the rich traditions of belief and ritual among these tribes were recorded in the late nineteenth century before their complete extermination; see J. Wilbert (ed.), *Folk Literature of the Yamana Indians: Martin Gusinde's Collection of Yamana Narratives* (Berkeley and Los Angeles, 1977).

247 *"Duty! Wondrous thought . . . whence thy original?"*: Darwin cites the English translation of Kant, *Metaphysics of Ethics* (1838), 136. His subsequent footnotes indicate that his answer is targeted not only against

certain theological writers, but also liberal utilitarian philosophers John Stuart Mill and Alexander Bain.

249 *derivative school of morals . . . "Greatest Happiness principle"*: on the 'derivative school of morals', Darwin praises 'an able article' on 'The Natural History of Morals', *Westminster Review*, NS 36 (Oct. 1869), 494–531, at 498. On the greatest happiness principle Darwin's reference is to J. S. Mill, *Utilitarianism* (see *On Liberty and other Essays*, ed. J. Gray (Oxford, 1991), 137).

252 *"not even . . . past so pleasant to us"*: Darwin cites Tennyson, *Idylls of the King*, 'Guinevere', ll. 372–3 (1859), 244–5.

253 *intuitionists*: those who believe in the perception of truth by immediate apprehension rather than logical reasoning.

255 *"As ye would . . . do ye to them likewise"*: a reference to Luke 6: 31.

Mr. Wallace, in an admirable paper before referred to: in the fourth chapter, not included here, Darwin had discussed A. R. Wallace's 'The Origin of Human Races and the Antiquity of Man Deduced from the Theory of "Natural Selection"', *Anthropological Review*, 2 (May 1864), pp. clviii–clxx.

257 *Bronze period*: the era between the Stone Age and the Iron Age, in the three-age system widely adopted among evolutionists during the 1860s.

262 *we institute poor-laws*: laws relating to the support of the poor, such as the system of workhouses introduced by the New Poor Law in 1834.

263 *Primogeniture with entailed estates*: a common practice under British law, primogeniture settled an estate exclusively on the eldest son; if the estates were entailed he could not dispose of them during his lifetime.

267 *twice as many . . . excessively high*: for the statistics in this and the following paragraph, Darwin cites 'our highest authority on such questions', William Farr, in 'The Influence of Marriage on the Mortality of the French People', *Transactions of the National Association for the Promotion of Social Sciences* (1858), 504–13.

268 *"they were . . . corrupt to the very core"*: Darwin cites W. R. Greg, 'On the Failure of "Natural Selection" in the Case of Man', *Fraser's Magazine*, 78 (Sept. 1868), 353–62, at 357.

269 *"daring and persistent energy"*: an unattributed quotation from Greg, 'On the Failure', *Fraser's Magazine*, 78 (Sept. 1868), 358.

271 *Botocudos*: Portuguese name for a tribe in eastern South America, known as the Aimorés or Aimborés.

vigesimal: based on the number twenty.

272 *paleolithic and neolithic*: Old Stone Age and New Stone Age.

276 *a highly developed mandrill*: Semnopithecus is the long-tailed entellus monkey of India; the monkeys of the *Macacus* genus include the rhesus monkey and barbary ape; the Mandrill is a large West African baboon.

analogical resemblances: structures with similar functions.

277 *a full-blooded negro*: John Edmondston, who taught Darwin taxidermy in Edinburgh.

280 *principle of reversion*: Darwin had defined this in Chapter IV as 'Whenever a structure is arrested in its development, but still continues growing until it closely resembles a corresponding structure in some lower and adult member of the same group, we may in one sense consider it as a case of reversion'. Reversion offered vital keys in suggesting the characteristics of the common progenitor of a group.

Les Eyzies: a town in the Dordogne region of France.

281 *the deadly influence of the Terai*: a belt of marshy grasslands, savannas, and forests, rich in wildlife but notorious among Europeans for breeding disease.

283 *the same Aryan stock*: in nineteenth-century ethnography, the Indo-Europeans or Aryans were a racial group found from India to the Atlantic Ocean, characterized by a symmetrical, oval skull.

the Semitic stock: speakers of the Semitic languages (including Hebrew, Arabic, Phoenician, and Assyrian) were thought to belong to a single race.

286 *the dolichocephalic type*: long-headed.

289 *ovipositor*: an egg-laying organ.

mandibles: in insects, the first or uppermost pair of jaws.

Complemental males of certain cirripedes: as Darwin explained in *Origin* (p. 357): 'in some genera the larvæ become developed either into herm-aphrodites having the ordinary structure, or into what I have called com-plemental males: and in the latter, the development has assuredly been retrograde; for the male is a mere sack, which lives for a short time, and is destitute of mouth, stomach, or other organ of importance, excepting for reproduction'.

epiphytic: a plant growing on another plant, but not parasitic.

prehensile: capable of grasping.

290 *natatory*: adapted for swimming.

cæteris paribus: all things being equal (Lat.).

tarsi: the five-jointed feet of insects.

291 *prehension*: the process of grasping.

293 *blackcap*: a distinctive grayish warbler; the male has a black cap.

297 *spurs*: the claw-like projection at the back of the leg, usually found in cocks.

the hypothesis of pangenesis: developed from his early transmutation theor-izing, pangenesis was Darwin's theory of heredity, and was outlined in full in *Variation of Animals and Plants under Domestication* (1868), ii. 357–404. Although widely criticized, its great advantage was flexibility in explaining the vast range of phenomena he had uncovered during his years of research.

300 *the superciliary ridge*: the bony ridge on the skull above the eye.

301 *the Guaranys of Paraguay*: the Guaraní, a major tribal group in South America between the Uruguay and lower Paraguay Rivers.

302 *Cebus azaræ*: the capuchin monkey of South America, so-called because their hair gives them the appearance of wearing a friar's cowl..

sagittal crest: a ridge of bone running lengthwise along the top of the skull.

304 *a well-known passage in Mungo Park's Travels*: 'I do not recollect a single instance of hard-heartedness towards me in the women. In all my wanderings and wretchedness I found them uniformly kind and compassionate', M. Park, *Travels in the Interior Districts of Africa* (1799), 263.

305 *gain the victory*: Darwin's note reads: 'J. Stuart Mill remarks ("the subjection of Women", 1869, p. 122), "the things in which man most excels women are those which require most plodding, and long hammering at single thoughts." What is this but energy and perseverance?' This is a slight misquotation, see *On Liberty and other Essays*, ed. Gray, 542.

306 *Reindeer period*: that part of the Palaeolithic era when reindeer were common across central Europe.

307 *"to be great personal attractions"*: S. Baker, *The Nile Tributaries of Abyssinia* (1867); Kordofan and Durfur are regions of central and western Sudan.

"temples have been gashed": S. Baker, *The Albert N'yanza* (1866), i. 218.

"but simply vie . . . of their own style": S. Baker, *The Nile Tributaries* (1867), 121.

314 *Charruas*: the Charrúa were a nomadic people inhabiting Uruguay, southern Brazil, and north-eastern Argentina.

"it frequently happens. . . very mention of marriage": the quotation, as a later note by Darwin makes clear, is from M. Dobrizhoffer, *An Account of the Abipones* (1822), ii. 207. The Abipones were a tribal group in the region of Paraguay, who after the Spanish conquest became known as fierce warriors and skilled horsemen.

"on reaching the home . . . the matter is settled forthwith": Darwin cites the Revd John Williams, as quoted in J. Lubbock, *Origin of Civilisation* (1870), 79.

"but if she is unwilling . . . this seldom happens": quoted from R. Fitzroy, *Narrative of the Surveying Voyages of His Majesty's Ships Adventure and Beagle* (1839), ii. 182.

315 *M. Burien's account*: a reference to the translator of a work written by Pierre Henri Dumoulin Borie (1820–91).

"when a girl . . . as that of the parents": W. Burchell, *Travels in the Interior of Southern Africa* (1824), ii. 59.

316 *Pithecia satanas*: the black saki monkey of South America.

321 *arboreal*: tree-dwelling.

322 *branchiæ*: gills or organs for respiration in water.

marine Ascidians: sea-squirts or tunicates, the larvae of which possess a simple nerve cord and certain other features of the basic chordate body plan.

326 *Arthropoda*: animals with segmented bodies and jointed appendages, including crustaceans, insects, and spiders.

AUTOBIOGRAPHIES

351 *Carolines*: a reference to Caroline Sarah Darwin, Charles Darwin's sister.

Catherine: Emily Catherine Darwin, Charles Darwin's younger sister.

damascenes: damsons, a small oval dark purple variety of plum.

Parkfields with poor Betty Harvey: Parkfield was the home of Sarah Wedgwood, Darwin's grandmother, and Betty Harvey was almost certainly a servant.

352 *old Peter Hailes the bricklayer*: Peter Hales, who lived near the Darwins in Shrewsbury. Information on Hales, Bage, and other Shrewsbury figures mentioned in the autobiographical writings has generously been provided by Donald F. Harris of Shrewsbury, and is detailed in his *Notes on Dr. Darwin* (copy on deposit with the Darwin Correspondence Project at the University of Cambridge).

Sarah & Kitty: Darwin's aunts, Sarah and Catherine (Kitty) Wedgwood.

Susan: Susan Elizabeth Darwin, Charles Darwin's sister.

353 *Erasmus*: Erasmus Alvey Darwin, Charles Darwin's older brother.

354 *Plas Edwards*: at Tywyn, Gwynedd on the west coast of Wales.

a Cimex mottled with red—the Zygena: *Cimex* is a genus of bedbug; *Zygena* is a genus of burnet moths with red-spotted forewings.

Pistol Rhyadwr: *Pistyll Rhaeadr* is a celebrated waterfall (the highest in England and Wales) in the north-east Welsh mountains.

355 [Table of contents]: page numbers in this table of contents have been changed to match those in this edition. Darwin's original numbering referred to the folios in his manuscript, as follows: 'From my birth to going to Cambridge p 1–25' / 'Cambridge Life p 25–47' / 'Voyage of the Beagle p. 47–57' / 'From my return home to my Marriage p. 58–73' / 'Religious belief p. 61–73' / 'From my marriage & residence in London to our settling at Down p 73–79' / 'Residence at Down p. 79' / 'An account how several books arose p. 81' / 'An estimation of my mental powers p. 111.'

A German editor: in a letter dated 20 Sept. 1875 (CUL: DAR 166: 94) Ernst von Hesse-Wartegg asked Darwin for accurate information to serve as the basis of an essay to appear in the Leipzig *Pionier*.

the mind of my grandfather written by himself: Darwin published a short biography of his paternal grandfather, Erasmus Darwin (1731–1802), in 1879.

356 *My mother*: Susannah Darwin (1765–1817).

357 *burial of a dragoon-soldier*: William Matthew had been a hussar, not a dragoon, and was buried a month after his mother's funeral. See R. Colp, 'Notes on Charles Darwin's *Autobiography*', *Journal of the History of Biology*, 18 (1985), 357–401, at 364.

357 *D*ʳ *Butlers great school in Shrewsbury*: founded in 1552, the school's fortunes were revived through the appointment in 1798 of the Revd Samuel Butler (1774–1839).

358 *Maer*: the house of Darwin's uncle, Josiah Wedgwood.

359 *a remarkable man*: the long passage on Robert Darwin was written at a later date and printed in an earlier chapter of the *Life and Letters*, as was the sketch of his son (Charles Darwin's brother) Erasmus. See F. Darwin (ed.), *Life and Letters of Charles Darwin* (London, 1887), i, 11–22.

24 stone: about 152 kilogrammes.

360 *M*ʳ *B, a small manufacturer in Shrewsbury*: Charles Bage (*c.*1752–1823) was in fact a substantial local manufacturer, building several large flax mills, including the world's first multi-storeyed iron-framed building. He also served a term as mayor of Shrewsbury. The largest loan to Bage is in Robert Waring Darwin's account books for 1823, when the two men knew one another well, having had financial dealings over several decades. The amount was for £1,000, not the £10,000 remembered by his son, who was about 13 years old at the time.

362 *A connection of my Father's*: Henry Parker, who had married Robert Darwin's eldest daughter Marianne in 1824.

The Earl of——: either the first or the second Earl of Powis, with whom Darwin had extensive financial dealings (on Robert Waring's death, the third earl owed him £40,000).

Lord Sherburn: a reference to William Petty (*formerly* Fitzmaurice), William, second earl of Shelburne, British prime minister 1782–3, and much involved throughout his political career in foreign affairs.

363 *Athenæum Club*: a London club for men of literature, science, and learning, founded in 1824.

364 *a young Doctor in Shrewsbury*: probably Henry Johnson, a recently trained physician who moved to the area sometime after 1828, and would have found Robert Waring Darwin's Leyden training old-fashioned.

366 *Major B*ʸ, *whose wife was insane*: Major Thomas Bayley, a family friend (see *Correspondence*, i (1985) and xvii (2008)) and his wife Sarah Bayley.

367 *in a Bank*: almost certainly a reference to the Eyton family, owners of a prominent bank in Shrewsbury over several generations.

my father's father, the author of the Botanic Garden &c: Erasmus Darwin (1731–1802).

368 *Wonders of the World*: Revd C. C. Clarke, pseudonym for Sir Richard Phillips, *The Hundred Wonders of the World, and of the Three Kingdoms of Nature* (London, 1818).

369 *Tutor of the College*: throughout most of Darwin's time at Christ's the college tutor was the Revd John Graham.

the principle of 'noscitur a socio': know a man by the company he keeps.

not found in Shropshire: hemipterous insects are characterized by sucking mouthparts and forewings thickened and leathery at the base; Zygaenidae are a family of moths generally with a metallic sheen and prominent spots; Cicindela are tiger beetles.

370 *'poco curante'*: an indifferent, apathetic person.

materia medica: the study of the properties of substances used in medicine.

371 *Wernerian geologist*: follower of Abraham Gottlob Werner, who attributed almost all geological phenomena to the action of water.

372 *Pontobdella muricata*: in his first paper (delivered March 1827), Darwin showed that the ova of *flustra*—a kind of sea-mat—were in fact larvae which moved with hair-like cilia; in the second, he showed that the black peppercorn-like bodies found in oyster shells were not seaweed spores, but the eggs of a marine leech.

373 *intelligent man*: John Edmonston, originally a slave in British Guiana, and moved to Scotland with his master in 1817, first briefly to Glasgow and then to Edinburgh.

Royal Soc. of Edinburgh: the Royal Society of Edinburgh, founded in 1783 for 'the cultivation of every branch of science, erudition, and taste'.

erratic boulder: a large fragment of rock differing in type and size from those native to the area in which it rests. The old man who pointed it out was probably Richard Cotton.

indurated on each side: for Darwin, the rocks at this celebrated site in Edinburgh showed clear signs of being formed by fire rather than water. A trap-dyke is an igneous mass injected into a fissure in the rocks; amygdaloidal rocks have agates and simple minerals scattered in them like nuts in a cake; and indurated strata have been hardened by heat.

374 *Mr Owens at Woodhouse, & at my Uncle Jos' at Maer*: William Mosytn Owen and Josiah Wedgwood II.

375 *'nec vultus tyranni &c'*: a reference to Horace, book 3, ode iii, lines 1–4, 'non vultus instantis tyranni'; the relevant passage can be translated from the Latin as 'the just man who holds fast to his resolve | is not shaken in firmness of mind by the passion | of citizens demanding what is wrong, | or the menace of the tyrant's frown'. See Horace, *The Complete Odes and Epodes*, trans. D. West (Oxford, 1997), 78.

376 *'credo quia incredibile'*: credo quia incredibilis, I believe because it is unbelievable (Lat.), as said by the early Church father Tertullian of the death of Christ.

phrenologists: phrenologists argued that mental faculties were located in specific organs of the brain, and that these could be judged from the shape of an individual's skull.

a private tutor: George Ash Butterton (1805–91).

Little Go: examination held in the second year of residence at Cambridge.

377 *Natural Theology*: William Paley's *Natural Theology* (London, 1802) aimed to demonstrate God's existence and attributes from the evidence of the natural world. Unlike several of Paley's other books, it was not a set text for examinations.

twelfth name on the list: Darwin ranked tenth on a pass-list of 178. The phrase οι πολλοι is Greek for 'the many'.

378 *Senior Wrangler*: highest scoring student on the final year Mathematical Tripos at Cambridge.

Fitzwilliam Gallery: founded by a bequest in 1816, the Fitzwilliam collection was at this time housed in the Perse Hall on Free School Lane in Cambridge.

379 *Panagaeus crux-major*: the crucifix ground beetle, so called because of the cross pattern on its back.

Carabidous: the Carabidae are a family of large carnivorous beetles, including *Licinus*.

380 *39 Articles*: the Thirty-Nine Articles were established in 1563 as the defining statements of Anglican doctrine in relation to the Reformation.

living of Hitcham: a benefice provided by the church at the parish of Hitcham in Suffolk.

381 *exserted*: stretched forth or out from a sheath, projecting beyond the surrounding parts.

tutor of Jesus College: Marmaduke Ramsay (1795–1831).

382 *Volute shell*: spiral shell of a gastropod of the genus *Voluta*.

383 *lateral & terminal moraines*: the piles of debris left by a retreating glacier at its sides and line of furthest extent.

my M.S. Journal: Darwin's manuscript diary, which formed the basis for his published *Journal of Researches*, is available as *Charles Darwin's Beagle Diary*, ed. R. D. Keynes (Cambridge, 1988).

385 *first Lieutenant*: John Clements Wickham (1798–1864).

386 *the Purser of the Adventure*: George Rowlett (d. 1834).

387 *Cirripedia*: a division of crustaceans including the barnacles and acornshells. In maturity they are always attached to other objects either directly or by means of a stalk. Crustaceans are articulated animals including crabs and lobsters, with a skin hardened by calcareous matter, and breathing with gills.

389 *triturated*: finely ground.

391 *"facile princeps botanicorum"*: easily the prince of botanists (Lat.). Humboldt's actual words were 'botanicorum facile princeps'.

Parallel roads of Glen Roy: the linear regularity of these terraces along the sides of Glen Roy in the Scottish Highlands had led eighteenth-century antiquarians to argue that they had been constructed as hunting roads.

392 *& have never since doubted*: six months after Darwin's death, Emma Darwin marked the passage from '& have never since doubted' to 'a damnable doctrine' in a fair copy of the *Recollections* made by her son Francis. As she wrote, 'I should dislike the passage in brackets to be published. It seems to me raw. Nothing can be said too severe upon the doctrine of everlasting punishment for disbelief—but few w$^{d.}$ call that "Christianity," (tho' the words are there.) There is the question of verbal inspiration comes in too. E.D.' See *The Autobiography of Charles Darwin*, ed. N. Barlow (London, 1958), 87, where this passage of the *Recollections* was reinstated.

395 *Tyler*: Edward Burnett Tylor (1832–1917).

396 *This conclusion . . . with many fluctuations become weaker*: these words are written on the opposite page of the manuscript in a different hand, followed by the annotation: 'Copied from a note in Father's hand in Frank's copy'.

hatred of a snake: Emma Darwin particularly objected to publication of this sentence at the time the manuscript was being prepared for publication by her son Francis in 1885. 'There is one sentence in the Autobiography which I very much wish to omit, no doubt partly because your father's opinion that *all* morality has grown up by evolution is painful to me; but also because where this sentence comes in, it gives one a sort of shock and would give an opening to say, however unjustly, that he considered all spiritual beliefs no higher than hereditary aversions or likings, such as the fear of monkeys toward snakes.

'I think the disrespectful aspect would disappear if the first part of the conjecture was left without the illustration of the instance of monkeys and snakes. I don't think you need to consult William about this omission, as it would not change the whole gist of the Autobiography. I should wish if possible to avoid giving pain to your father's religious friends who are deeply attached to him, and I picture to myself the way that sentence would strike them, even those so liberal as Ellen Tollett and Laura, much more Admiral Sulivan, Aunt Caroline, &c., and even the old servants' (H. Litchfield, *Emma Darwin* (Cambridge, 1904), ii, 360–1).

Agnostic: a term introduced by T. H. Huxley in 1869 to describe a person who is not a materialist, but who holds that the existence of anything beyond and behind material phenomena is unknown and (so far as can be judged) unknowable.

397 *M$^{rs.}$ Barlow*: perhaps Mrs Elizabeth Barlow, a close family friend with financial dealings with Robert Waring Darwin.

Nothing is more remarkable. . . my Redeemer liveth: Darwin notes that this paragraph was 'written in 1879—copied out Ap 22 1881'.

your mother: Emma Darwin, née Wedgwood.

398 *shortly after our marriage*: Emma Darwin worried that her husband was following his brother Erasmus into religious scepticism. As she wrote to him early in 1839: 'May not the habit in scientific pursuits of believing nothing till it is proved, influence your mind too much in other things which cannot be proved in the same way, & which if true are likely to be above our comprehension. I should say also that there is a danger in giving up revelation which does not exist on the other side, that is the fear of ingratitude in casting off what has been done for your benefit as well as for that of all the world & which ought to make you still more careful, perhaps even fearful lest you should not have taken all the pains you could to judge truly. . . . I do not wish for any answer to all this—it is a satisfaction to me to write it & when I talk to you about it I cannot say exactly what I wish to say, & I know you will have patience, with your own dear wife. Don't think that it is not my affair & that it does not much signify to me. Every thing that concerns you concerns me & I should be most unhappy if I thought we did not belong to each other forever.' Darwin wrote at the end of this letter, 'When I am dead, know that many times, I have kissed & cryed over this. C. D.' (*Correspondence*, ii (1986), 171–3).

399 *denudation*: the wearing away of the surface of the land by water.

mould: soft, loose topsoil.

our life in London: the remainder of this section was added in April 1881.

400 *'Craters of Elevation' & 'Lines of Elevation'*: the French geologist Léonce Elie de Beaumont suggested that mountain chains arose catastrophically along particular axes of elevation. The 'craters of elevation' theory had originated with the Prussian geologist Leopold von Buch, who argued that volcanic cones were like blisters, produced by an upward bulging of lava trapped below the surface.

401 *facile Princeps Botanicorum*: see note to p. 391.

404 *Lady Caroline (?) Bell*: actually Lady Catharine Bell.

405 *Effie*: Katherine Euphemia Wedgwood (1839–1931).

406 *the old Earl*: Philip Henry Stanhope (1781–1855).

407 *Goethes views on light*: in *Zur Farbenlehre* (1792), Johann Wolfgang von Goethe argued that colours arose from the mixing of light with darkness, combating Newton's views on the composite nature of white light.

409 *Concholepas*: in his *Beagle* notes, Darwin identified this as *Concholepas peruviana*, a gastropod mollusc (*Correspondence*, iv (1988), 389).

hydropathic treatment: cure by water; Darwin was scrubbed with a rough towel in cold water, drank tumblers of water, and wore a soaked compress.

hermaphrodites: individuals possessing the organs of both sexes.

a German writer: the zoologist August David Krohn (*Correspondence*, xiii (2002), 214 n. 9).

411 *Columbus & his egg*: a reference to the old story of how standing an egg on its end was deemed impossible, until Christopher Columbus showed it could be done if the egg was slightly broken.

412 *Journal of the Proceedings of the Linn: Soc. 1858 p. 45*: the reference to the Darwin and Wallace joint presentation is 'On the tendency of species to form varieties; and on the perpetuation of varieties and species by natural means of selection. . .', *Journal of the Proceedings of the Linnean Society (Zoology)*, 3 (1859), 45–62.

415 *dimorphic & trimorphic plants*: dimorphic species appear in two distinct forms; trimorphic plants have three.

416 *tending towards abortion*: dioecious species have the female and male sexual organs on different plants. In the cases described, two forms were emerging, one in which development of the female organs (pistils) were being arrested or 'aborted'; the other in which this was occurring in the male organs (stamens).

Cucurbitacean plant: the family of plants including gourds, melons, and cucumbers.

417 *Pangenesis*: for a brief outline of this theory, see p. 297 of this edition.

418 *My first child*: William Erasmus Darwin (1839–1914). Darwin's observations were published as 'A biographical sketch of an infant', *Mind*, 2 (1877), 285–94.

near Hartfield, where two species of Drosera abound: Hartfield, in Sussex, was the home of Sarah Elizabeth Wedgwood (1793–1880), Emma Darwin's sister. Drosera are the sundews.

nitrogenous: containing the element nitrogen.

419 *"nunc dimittis"*: 'let us now depart', from the evening service of the Anglican Church.

"The Effects of Cross & Self-fertilisation". . . old geological thoughts: this was added 1 May 1881.

Heterostyled flowers: the style is the middle portion of the perfect pistil, which rises like a column from the ovary. In heterostyled flowers, the styles are of different lengths relative to the stamens.

420 *'that every Whale has its Louse'*: 'Hat doch der Wallfisch seine Laus | Muss auch die Miene haben'; quoted in Huxley to Darwin, 3 Feb. 1880, in *Autobiography*, ed. Barlow, 210–11.

Frank's assistance: a reference to Francis Darwin (1848–1925).

424 *my gardener, an old Kentish man*: probably a reference to the gardener Mr Comfort.

A good deal . . . vestige of evidence: this paragraph is marked in the manuscript to be added in the middle of the preceeding sentence, at 'the progress of science'.

Bos: the genus of wild or domestic cattle and oxen.

425 *consanguineous*: related by blood, of the same family.

Hopedene: the home in Dorking, Surrey, of Emma Darwin's brother Hensleigh Wedgwood.

BIOGRAPHICAL INDEX

GENERAL INDEX